Cover portrait by Jorrit de Boer

The Lesson of Quantum Theory

The Lesson of Quantum Theory

Niels Bohr Centenary Symposium
October 3–7, 1985

edited by

Jorrit de Boer
Ludwig-Maximilians Universität, Munich

Erik Dal
Royal Danish Academy, Copenhagen

Ole Ulfbeck
The Niels Bohr Institute, Copenhagen

NORTH-HOLLAND
for the
ROYAL DANISH ACADEMY OF SCIENCES AND LETTERS

ISBN: 0 444 87012 1

Published by:

North-Holland Physics Publishing
a division of
Elsevier Science Publishers B.V.
P.O. Box 103
1000 AC Amsterdam
The Netherlands

Sole distributors for the USA and Canada:
Elsevier Science Publishing Company, Inc.
52 Vanderbilt Avenue
New York, N.Y. 10017
USA

Library of Congress Cataloging in Publication Data

The Lesson of Quantum Theory.

   Papers from the Niels Bohr Centenary Symposium,
Oct. 3–7, 1985.
   Bibliography: p.
   1. Bohr, Niels Henrik David, 1885–1962--Anniversaries, etc.
--Congresses. 2. Quantum theory--History--Congresses.
I. de Boer, Jorrit, 1930–   . II. Dal, Erik.
III. Ulfbeck, Ole. IV. Niels Bohr Centenary Symposium.
(1985: Copenhagen, Denmark) V. Kongelige Danske
Videnskabernes Selskab.
QC16.B63L47 1986   530.1'2   86–16468
ISBN 0 444 87012 1 (Elsevier)

Printed in The Netherlands

Preface

The concepts of quantal states and complementarity introduced by Niels Bohr in an intuitive though definitive manner more than a generation ago have come to form cornerstones in our description of nature. The revolution in scientific thought brought about by these ideas is parallelled only by the leap in human thinking that characterized Antiquity and Renaissance. The Centenary Symposium celebrating Niels Bohr pays tribute to his work which gave birth to contemporary physics and nourished related fields of knowledge.

The symposium was held under the auspices of the Royal Danish Academy of Sciences and Letters, the Danish Physical Society, the University of Copenhagen, Nordita and the Niels Bohr Institute. The contributions forming the proceedings are arranged in the sequence in which they were presented. Space limitations have forced the shortening of some of the discussions.

The book is divided into three parts. Part One consists of two communications to the Royal Danish Academy of Sciences and Letters in memory of Niels Bohr.

Part Two containing the main theme of the proceedings is divided into three sections. The title of the first, *The Lesson of Quantum Theory*, is an expression often used by Niels Bohr as time and again he endeavoured an ever deeper analysis of this lesson and its succinct formulation. For Bohr the lesson of quantum theory went beyond physics encompassing not only the natural sciences but all human conceptions. Thus the second section, *Unity of Knowledge,* conveys the progress in neighbouring disciplines in which Bohr took a particular interest. The last section, *Niels Bohr*, is dedicated to his personality, his activities, and his world of thought.

Part Three renders speeches and anecdotes presented during the evening gathering at Ny Carlsberg Glyptoteket on Niels Bohr's one hundredth birthday. These reminiscences convey the lively spirit brought forth by Bohr, in earnest and in jest.

The structure of the symposium grew out of advice from a wide circle of colleagues. The threads were gathered up by the organizing committee chaired by Sven Bjørnholm, who took care of the numerous organizational problems, assisted by an outstanding secretarial team. Jan Ambjørn, Thomas Døssing, Peter Orland and Jan Teuber helped in collecting and editing the discussions. Jerry Garrett coordinated a complete videotaping of the symposium.

Jorrit de Boer   Erik Dal   Ole Ulfbeck

Copenhagen, February 1986

Acknowledgements

The organizers gratefully acknowledge generous financial support by

Augustinus Fonden

Carlsbergfondet

Fabrikant Mads Clausens Fond

Direktør Ib Henriksens Fond

Reinholdt W. Jorck og Hustrus Fond

Lundbeckfonden

Vera og Carl Johan Michaelsens Legat

Scandinavian Airlines System

The committee also expresses its gratitude for the hospitality extended by Ny Carlsberg Glyptoteket on the evening of October 7, 1985.

Organizing Committee

Ben R. Mottelson   Christopher J. Pethick   Jens Lindhard

Nordita   Aarhus University

Sven Bjørnholm   Jørgen Kalckar   Jens Lyng Petersen
Ole Ulfbeck   Aage Winther

The Niels Bohr Institute

Contents

Part One: Niels Bohr

Communications delivered to the Royal Danish Academy of Sciences and Letters, October 3, 1985, edited by Erik Dal

Part Two: The Lesson of Quantum Theory

Proceedings of the Niels Bohr Centenary Symposium, October 4–7, 1985, edited by Jorrit de Boer and Ole Ulfbeck

The Lesson of Quantum Theory

Unity of Knowledge

Niels Bohr

Part Three: Jest and Earnest

Addresses presented at the evening gathering in Ny Carlsberg Glyptoteket, October 7, 1985, edited by Jorrit de Boer and Ole Ulfbeck

Part One

NIELS BOHR

Communications delivered to the
Royal Danish Academy of Sciences and Letters
October 3, 1985

edited by

Erik Dal

Royal Danish Academy, Copenhagen

Two communications in memory of Niels Bohr delivered at a session of the Royal Danish Academy of Sciences and Letters in the University Solemnity Hall on the evening of October 3, 1985

The Lesson of Quantum Theory, edited by J. de Boer, E. Dal and O. Ulfbeck
© Elsevier Science Publishers B.V., 1986

Niels Bohr and the Royal Danish Academy of Sciences and Letters

Bengt Strömgren

Nordita
Copenhagen, Denmark

Contents

1. Prize problem and dissertation

The Royal Danish Academy of Sciences and Letters in February, 1905, offered its gold medal for an investigation of jet vibration aiming at a determination of the surface tension of liquids. In answer to the prize problem two essays were submitted in November, 1906, and both were awarded the gold medal. One author was P.O. Pedersen, the later director of the Danish Technical University, —the other was Niels Bohr, then a student at the University of Copenhagen. The prize essays were judged by two members of the Academy, physics professors C. Christiansen and K. Prytz.

On February 24, 1907, Niels Bohr was notified of the award through a letter signed by the president of the Academy, Julius Thomsen, and its secretary, H.G. Zeuthen. However, already on January 26, Niels Bohr had received a letter of congratulation, revealing that he would be awarded the prize. This letter was from Harald Høffding.

Niels Bohr continued his work on the surface tension of liquids and published the results in 1909 in a paper in the Philosophical Transactions of the Royal Society of London, entitled "Determination of the Surface Tension of Water by the Method of Jet Vibration" [1]. On May 13, 1911, Niels Bohr publicly defended his dissertation "Studier over Metallernes Elektrontheori" for the doctor of philosophy degree. In his publications during the following years, Niels Bohr did not directly return to the subjects of his first papers. However, the intimate familiarity with problems of classical physics that he had gained through his early work became, as has been emphasized, of great importance as part of the background for the pioneering work that was to follow.

2. University professor and academician

During the centenary year of Niels Bohr's birth we have been reminded, in a multitude of ways, of his research contributions during the years 1911–1916. In 1916 Niels Bohr returned to Denmark after a two-year stay in Manchester to assume the duties of a newly created professorship in theoretical physics at the University of Copenhagen, and in 1917 he was elected a member of the Royal Danish Academy of Sciences and Letters.

All through the years of his membership Niels Bohr regularly attended the meetings of the Academy, except for the period of his involuntary stay in the United States during the last years of the Second World War. He rendered the Academy a great service by refereeing the very large number of physics papers, submitted to the Academy for publication. During the years 1917–1955 this amounted to a total of 90 papers published in the mathematical–physical publications of the Academy.

Relatively few of Niels Bohr's papers appeared in the publications of the Academy. However, one of his fundamental contributions "On the Quantum Theory of Line-Spectra I–III" was published by the Academy in the period 1918–1922, being reprinted in 1927 [2]. In 1933 Niels Bohr, in collaboration with Léon Rosenfeld, published the paper "Zur Frage der Messbarkeit der elektromagnetischen Feldgrössen" in the memoirs of the Academy [3], as was the case with works on the penetration of atomic particles through matter, the last of these in 1954, in collaboration with Jens Lindhard [4].

During the years from 1917 to 1955, Niels Bohr gave 29 lectures before the Academy or, in the terminology of the Academy, he made 29 communications. Doing this, he gave the members of the Academy a most valuable opportunity to follow the development of physics through epoch-making decades.

In order to give, on this occasion, an impression of Niels Bohr's unique research contribution, I shall attempt to describe investigations and results that constituted the background for some of these lectures.

In Volume I of "Niels Bohr, Collected Works", Léon Rosenfeld in his "Biographical Sketch" [5] has given a synopsis of the principal problems that dominated in Niels Bohr's work in successive periods: In 1913 the Quantum Postulates, during the years 1914–1925 Atoms and Radiation, from 1925 to 1935 Quantum Mechanics and Complementarity, from 1936 to 1943 Nuclear Physics, and finally from 1943 to 1962 Public Affairs and Epistemology.

3. Quantum postulates

Shortly after having become a member of the Academy, Niels Bohr—in 1917 and in 1918—gave two lectures, the first entitled "On the Quantum Theory of Line-Spectra", the second "On the Foundation of Quantum Theory". Through these lectures he submitted Part I of his above mentioned paper for publication by the Academy.

Part I appeared in April, 1918, while Part II followed in December, 1918 and Part III in November, 1922 [2].

Part I opens with a statement of the two fundamental assumptions of the quantum theory of line-spectra that were the basis of the breakthrough in 1913, the first being the quantum postulate of the stationary states of an atomic system, the second the quantum postulate on radiation absorbed or emitted during a transition between two stationary states, which states the frequency condition $E' - E'' = h\nu$, where h is Planck's constant.

In 1913 the correspondence principle introduced by Niels Bohr had played a very important role. Niels Bohr always emphasized that the quantum postulates meant a decisive break with classical physics, but he also stressed the consideration that a quantum theory, valid for description of atomic systems in the domain of atomic dimensions, necessarily gives results that converge toward those of classical theory in the limiting case of conditions under which dimensions are large compared to atomic dimensions. These considerations found expression in the correspondence principle.

In 1913 Niels Bohr had utilized the correspondence principle for the description of transitions in the hydrogen atom's stationary states with large neighbouring quantum numbers, corresponding to electronic orbits of relatively large radius with quantum emission of low frequency and correspondingly large wavelength. The correspondence principle led to an equation for the frequencies of the spectral lines of atomic hydrogen. The calculated values agreed with observation as closely as could be expected considering the precision with which the natural constants that entered, including Planck's constant, were known at the time. The correspondence principle had also led Niels Bohr to the correct interpretation of certain spectral lines, found in stellar spectra as well as in laboratory spectra, the so-called Pickering–Fowler lines, which for a time had seemed to present a problem for the theory. He showed that the lines in question are due to ionized helium and not to hydrogen, as had been assumed previously. Guided by the correspondence principle, Niels Bohr had computed the wavelengths of these ionized helium lines from the observed wavelengths of the hydrogen lines, taking into account the fact that the mass of the helium nucleus is four times as large as that of the hydrogen nucleus. The agreement with observation was perfect, with five-digit precision. In 1913 this result had made a great impression in the physics community. It is known from a letter written by George Hevesy to Niels Bohr (cf. Niels Bohr, Collected Works, Volume 2 [6]) that Einstein, when he heard about this result, exclaimed: "This is an enormous achievement! The theory of Bohr must be right then."

In 1913 the correspondence principle had been applied to electronic motion which, according to classical theory, was describable by simple periodic orbits. For elements other than hydrogen the situation is more complicated. In Part I of the paper submitted to the Academy during meetings in 1917 and 1918, Niels Bohr developed methods that made it possible to draw conclusions on the basis of the correspondence principle, even in more complicated cases. Here, the basis is the description of the electronic motion according to classical theory by Fourier analysis, and the correspondence principle then suggests that transition probabilities between two stationary states characterized by quantum numbers $n'_1, \ldots n'_s$ and

$n_1'', \ldots n_s''$, respectively, are directly related to intensities of radiations of frequencies $\omega_1(n_1' - n_1'') + \cdots + \omega_s(n_s' - n_s'')$ to be expected according to classical elec- trodynamics from the motions in these states. Niels Bohr emphasized that the estimate becomes more uncertain, the smaller the values of the n's are. These ideas are further developed in Part III of the paper in a discussion of the spectra of elements of higher atomic number. Part II contains a discussion of the hydrogen spectrum in which the fine structure of the lines, as well as the Stark and Zeeman effects, are considered.

During the years 1918 to 1924, the correspondence principle played an important role in a series of investigations, leading to the development of the new quantum mechanics. We shall return to this matter in section 5.

4. Atoms and radiation

In 1921 Niels Bohr gave a lecture at the Academy entitled "The Periodic System of the Elements Elucidated through Considerations of Atomic Structure", followed in 1922 by a lecture, "Atomic Theory and the Properties of the Elements". A short time later, on December 11, 1922 in Stockholm, Niels Bohr delivered his Nobel Lecture "On the Structure of Atoms". In these three lectures he gave an account of the great progress made and results obtained in atomic theory.

In the Nobel Lecture [7] the possibilities of elucidating the Periodic System of the elements are discussed in detail:

> "The ideas of the origin of spectra have furnished the basis for a theory of the structure of the atoms of the elements which has shown itself suitable for a general interpretation of the main features of the properties of the elements, as exhibited in the natural system."

Niels Bohr went on to say that the theory is primarily based on considerations of the manner in which the atom can be imagined to be built up by the capture and binding of electrons to the nucleus, one by one, and further that the optical spectra of elements provide us with evidence on the progress of the last steps in this building-up process.

While the optical spectra are of decisive importance for the understanding of the problems concerning the outermost electrons—the valence electrons—the results of X-ray spectroscopy, as emphasized by Niels Bohr in the Nobel Lecture, contribute significantly to the understanding of the inner parts of the electron configurations.

On the basis of the spectroscopic data, Niels Bohr was able to draw conclusions regarding the grouping of electrons and their changes as one proceeds through the Periodic System of the elements. The finding that a grouping of the outermost electrons repeats itself at regular intervals explains the periodicity in the properties of the elements. Thus, Christian Møller and Mogens Pihl express themselves in their contribution to the book "Niels Bohr", published in 1967 [8], and they continue:

> "A characteristic feature of Bohr's way of considering the problems is revealed here, namely, that intuitive understanding of the clues yielded by the empirical data was particularly important in guiding him to these results—results that were to pave the

way for a more profound insight into the regularities, gained only after Pauli's formulation of the so-called exclusion principle."

From the theory, Niels Bohr concluded that an element with atom number 72, not yet discovered at the time, would have to be chemically similar to zirconium, and therefore could be expected to occur in certain minerals containing zirconium. This led to the discovery of the new element by Hevesy and Coster, then working at Bohr's Institute. The element was called hafnium. This development was the subject of a lecture that Niels Bohr gave at the Academy in 1923, entitled "A New Element".

During the last years of the period 1914–1925 Niels Bohr collaborated with an international group of outstanding scientists trying to find new approaches to the quantum theoretical description of atomic states and processes.

A lecture at the Academy in February, 1925, entitled "On the Law of Conservation of Energy" had as its background a paper published in 1924 by Niels Bohr, H.A. Kramers–who was a close collaborator of Bohr during these years—and the American physicist J.C. Slater [9]. This work, "The Quantum Theories of Radiation", took as its point of departure the dualism between the wave theory and the theory of light quanta. The waves were interpreted in terms of a probability field, even if this meant that the formalism developed here—in Bohr's words:

"would not seem to allow a detailed description of atomic processes which presumes the law of conservation of energy, which occupies a central position in the classical description of Nature."

Shortly afterward, however, Niels Bohr reached the conclusion that these ideas could not be upheld, and an experiment by W. Bothe and H. Geiger did indeed indicate that the law of conservation of energy was valid for the individual atomic processes in question.

5. Quantum mechanics and complementarity

In 1925 a decisive breakthrough took place leading to further development of atomic theory and quantum theory, starting with a paper by Werner Heisenberg [10], followed by a long series of other contributions. During this period the Institute on Blegdamsvej was an important centre, and Heisenberg cooperated closely with Niels Bohr and Kramers. Regarding this whole development, and in particular with a view to the paper by Bohr, Kramers and Slater, Heisenberg has given the following evaluation:

"This investigation represents the actual climax in the crisis of quantum theory, and although it did not show the correct way out of the difficulties, the paper contributed more than any other from this period to the clarification of the situation in quantum theory."

A second investigation of decisive importance, carried out at Bohr's Institute during the period immediately preceding the development of the new quantum mechanics, was Kramers' paper on the dispersion of radiation [11].

The new quantum mechanics was the subject of two lectures by Niels Bohr at the Academy in 1926 and 1927, one entitled "Atomic Theory and Wave Mechanics", the other "The Quantum Postulate and the Recent Development of Atomic Theory", A theory had now been developed on the basis of which it was possible to account quantitatively for the atomic states and the atomic processes in the electron configurations surrounding the nuclei of the atoms, and agreement had been obtained between theory and a broad range of observations.

The dualism between the wave picture and the particle picture entered the theory as an essential feature, closely connected with fundamental observations. To Niels Bohr this dualism became the starting point for an interpretation based on the idea of complementarity. In the lecture by Hendrik Casimir, certain aspects of the fundamental contributions by Niels Bohr to this central area will be discussed.

6. Nuclear physics

During the years 1937–1941, Niels Bohr gave a number of lectures at the Academy concerning his work on nuclear reactions (see also the review by B.R. Mottelson in this volume). In these lectures he discussed the problem of the penetration of a neutron into an atomic nucleus. The neutron, being neutral, is not repelled by the positively charged nucleus as is the case with protons or α-particles. The penetrating neutron will quickly lose its energy, which in the process is distributed over the nucleons. In this way what Niels Bohr called a compound nucleus is formed. Next, one possibility is that the neutron binding-energy is emitted in the form of radiation. If that occurs the neutron has been captured by the nucleus, and a new atomic nucleus with the same charge and larger atomic weight has been formed. Or, it may happen that the excess energy is concentrated on one of the nucleons of the compound nucleus which is then emitted. The probabilities corresponding to the two possible processes depend on the properties of the compound nucleus, and these are determined by its charge and weight. Because of the analogy between conditions in a compound nucleus and a drop of liquid, respectively, the model is referred to as the droplet model. It turned out to be of great value for the explanation of a broad category of nuclear processes.

In 1938 Otto Hahn and Fritz Strassmann carried out an experiment in which uranium atoms were bombarded with neutrons. Then, in December, 1938, Lise Meitner and Otto Frisch reached the conclusion that what had occurred in this experiment was a splitting of an uranium nucleus into two about equally heavy atomic nuclei, i.e. a fission of the uranium nucleus took place. Niels Bohr, thereafter, showed that the droplet model could serve as a basis for the explanation of fission. The two known processes, emission of radiation or of a nucleon, respectively, would here be in competition with a third process, namely the fission of the compound nucleus. The probabilities with which the three processes occurred would depend on the atomic weight and charge of the compound nucleus.

Shortly afterwards, it was shown in a paper by Niels Bohr and John Wheeler [12] that neutrons were emitted in connection with the fission process, and that these emissions could cause a chain reaction in the case where the bombarded nuclei belonged to the uranium isotope ^{235}U. However, for the uranium isotope ^{238}U, which on earth has a much higher abundance than the isotope ^{235}U, such chain reactions would not take place. Herewith, the theoretical basis for the future development was essentially given, a development that was to lead to a peaceful application of the release of nuclear energy as well as to the nuclear bomb.

In a lecture at the Academy in November 1939, entitled "The Theoretical Explanation of Fission of Atomic Nuclei", Niels Bohr gave an account of the development. In the lecture, which made a deep impression on the audience, he explained the possibilities of a peaceful use of nuclear energy as well as the terrifying, destructive application. I clearly remember this evening in the Academy, in particular also that Niels Bohr emphasized:

> "The enormity of the technical difficulties connected with the large-scale isotope separation that would necessarily have to precede the production of an atomic bomb."

7. Public affairs and epistemology

Niels Bohr's last lecture at the Academy was given on the occasion of a festive meeting held in October, 1955, to celebrate his seventieth birthday. The lecture, entitled "Atoms and Human Knowledge", was published in the Yearbook of the Academy, 1955–56 [13].

Niels Bohr opened his discourse thus [14]:

> "In the history of science, this century's exploration of the world of atoms has hardly any parallel in so far as the progress of knowledge and the mastery of that nature, of which we ourselves are part, are concerned. However, with every increase of knowledge and abilities is connected a greater responsibility; and the fulfilment of the rich promise and the elimination of the new dangers of the atomic age confront our whole civilization with a serious challenge which can be met only by cooperation of all peoples, resting on a mutual understanding of the human fellowship. In this situation, it is important to realize that science, which knows no national boundaries and whose achievements are the common possession of mankind, has through the ages united men in their efforts to elucidate the foundations of knowledge."

A detailed review of the development of physics followed and concluded thus:

> "Far from containing any mysticism foreign to the spirit of science, the notion of complementarity points to the logical conditions for description and comprehension of experience in atomic physics."

The remaining part of the lecture dealt with questions concerning the general conditions for human knowledge. At the end of the lecture Niels Bohr returned to the theme of the introduction:

> "We have here reached problems which touch human fellowship and where the variety of means of expression originates from the impossibility of characterizing by any fixed distinction the role of the individual in the society. The fact that human cultures, developed under different conditions of living, exhibit such contrasts with respect to established traditions and social patterns allows one, in a certain sense, to call such

cultures complementary. However, we are here in no way dealing with definite mutually exclusive features, such as those we meet in the objective description of general problems of physics and psychology, but with differences in attitude which can be appreciated or ameliorated by extended intercourse between peoples. In our time, when increasing knowledge and ability more than ever link the fate of all people, international collaboration in science has far-reaching tasks which may be furthered, not least by an awareness of the general conditions for human knowledge."

8. Niels Bohr as president of the Royal Danish Academy of Sciences

In March 1939 Niels Bohr was elected president of the Academy, and he served it in this capacity until his death in 1962. Already in 1927, and later in 1934 and 1938, Niels Bohr had been approached with regard to the presidency of the Academy, but he had declined referring to his commitments to research.

The first meeting of the Academy at which Niels Bohr presided took place on October 20, 1939, shortly after the outbreak of the Second World War.

On November 13, 1942, the Academy celebrated its second centenary, and the meeting which took place on that day was, of course, limited in its scope because of the prevailing conditions in Denmark. However, three major Academy publications were presented on the occasion, namely the first volume of the History of the Academy, prepared by the archivist, Asger Lomholt [15], further volume 1 of a work on the topographical survey of Denmark, by N.E. Nörlund [16], and finally a book by Johannes Pedersen [17], the Chairman of the Carlsberg Foundation, on the history and activities of the Foundation. Following these presentations Niels Bohr spoke on the subject of the Academy's place in Danish society [18]:

> "The direct stimulus for the foundation of the Academy arose, as we know, from the need for solving certain specific problems which required a general scientific insight. Also, in the Academy's first century, few organisations existed in this country for the furtherance of science and its applications. In both respects our Academy had an important mission."

Niels Bohr continued:

> "About the middle of the last century this situation had largely changed, since many independent Danish institutions had gradually been established to deal with such tasks. In this development it was not the Academy itself, but many of its leading men who played the decisive part.
>
> The importance of the fact that there existed in our country a permanent centre, not so much for the carrying out of scientific research, but for its general appreciation, can scarcely be overestimated."

Regarding the role and the work of the Academy at the time of its second centenary, Niels Bohr expressed himself as follows:

> "The fact that representatives of widely differing branches of science come to our meetings and take part in all our discussions has given our Academy its special character. The unique opportunity which is given to humanists and scientists to tell each other what they consider important within their own fields, excludes the one-sidedness which may often occur in a larger body where a further division according to

the spheres of interest is necessary on practical grounds. Although not all branches of learning and research deserving the name of science are represented within the Academy, the contacts made here have definitely had and still have an importance for the harmonious development of scientific efforts in our country, in fact for the whole cultural life of Denmark."

Niels Bohr concluded his speech emphasizing the fact that the Academy had become an increasingly important link in international scientific cooperation.

Through his speech on the occasion of the second centenary of the Academy, in particular through his characterization of the role and the importance of the Academy, Niels Bohr did also clarify the general line which he as president, and in collaboration with the members, had followed and intended to follow during the years to come.

In the middle of the year 1943 Niels Bohr was forced by circumstances to leave Denmark, and he only returned in August, 1945. During his absence the chairmen of the two sections of the Academy—the section of science and the section of history and philosophy—presided over the meetings. The election of a president which, according to the rules of the Academy, should have taken place in the spring of 1944 was postponed, and at an extraordinary meeting in September, 1945 Niels Bohr was re-elected president. Re-elections again took place in 1949, 1954 and 1959, and so Niels Bohr remained president of the Academy until his death in 1962.

Through his efforts in careful planning and preparation of all the meetings over which he presided, Niels Bohr made a great contribution to the Academy. More than that, as president of the Academy he very effectively represented Danish science on important matters.

When UNESCO was established after the Second World War and a National Committee was set up in Denmark, Niels Bohr, on behalf of the Academy, took on the task of advisor to the committee, a task which at that time was of considerable importance.

Of particular importance, however, was an initiative which Niels Bohr took in 1951 in connection with the question of financial support for research in Denmark from the Danish government. In a critical situation in January, 1951, Niels Bohr, as president, proposed that the Academy would make a direct approach to the government. His proposal was unanimously adopted. There is no doubt that this initiative was of decisive importance to the further development which resulted in the establishment of a "National Science Foundation" in Denmark.

At a meeting of the section of science of the Academy in January, 1947, August Krogh, who as an outstanding scientist could speak with great authority, had proposed a radical change of the Academy to be brought about through a large increase in the number of members of the Academy, the aim being the election of many younger scientists. Through such a change the Academy should be put in a position to serve Danish society and to speak more effectively for science. The proposal, in this form, did not win approval by the Academy, and August Krogh then, in January, 1949, chose to relinquish his membership of the Academy.

In this matter the Academy followed Niels Bohr's line, and as a result there was during the following years a moderate increase in the number of members, corresponding to the increased number of people active in research in Denmark.

During the years after 1962 the Academy in its development continued to follow the lines in accordance with Niels Bohr's general ideas. The Royal Danish Academy of Sciences and Letters which Niels Bohr described in his speech before the Academy in 1942 has thus preserved its character and continued its work, and at the same time it has been able to strengthen its activities and position in Danish society.

Following Niels Bohr's death on November 18, 1962, the Academy held a meeting in his memory. Lectures were given by Christian Møller, Léon Rosenfeld and Johannes Pedersen.

Johannes Pedersen concluded with these words:

"We felt secure and confident having as our leader a man such as Niels Bohr, outstanding in intelligence as well as in character. His memory will live for ever in our hearts, his name will remain for all times an honour to our Academy."

References

[1] N. Bohr, Determination of the surface tension of water by the method of jet vibration, Philos. Trans R. Soc. London 209 (1909) 281–317.

[2] N. Bohr, On the quantum theory of line-spectra I–III, Dan. Vid. Selsk. Skrifter Naturv.-Mat. Afd. 8. Række IV-1 (1918–1922) 1–3.

[3] N. Bohr and L. Rosenfeld, Zur Frage der Messbarkeit der elektromagnetischen Feldgrössen, Mat.-Fys. Medd. Dan. Vidensk. Selsk. 12-8 (1933) 1–65.

[4] N. Bohr and J. Lindhard, Electron capture and loss by heavy ions penetrating through matter, Mat.-Fys. Medd. Dan. Vidensk. Selsk. 28-7 (1954).

[5] L. Rosenfeld, Biographical sketch, in: Niels Bohr, Collected Works, Vol. 1, Early Work (1905–1911), ed. J.R. Nielsen (North-Holland, Amsterdam, 1972).

[6] G. Hevesy to N. Bohr, in: Niels Bohr, Collected Works, Vol. 2, Work on Atomic Physics (1912–1917), ed. U. Hoyer (North-Holland, Amsterdam, 1981) p. 532.

[7] Niels Bohr, Om Atomernes Bygning, Les Prix Nobel en 1921–22 (P.A. Norstedt & Söner, 1923). English translation by Frank Hoyt in Nature 112 (1923) 29; also in: Niels Bohr, Collected Works, Vol. 4, ed. J.R. Nielsen (North-Holland, Amsterdam, 1977) p. 467–482.

[8] C. Møller and M. Pihl, in: Niels Bohr, His life and work as seen by his friends and colleagues, ed. S. Rozental (North-Holland, Amsterdam, 1967) p. 240.

[9] N. Bohr, H.A. Kramers and J.C. Slater, The quantum theories of radiation, Philos. Mag. 47 (1924) 785–802.

[10] W. Heisenberg, Z. Phys. 33 (1925) 879.

[11] H.A. Kramers, Nature 113 (1924) 673–674 and 114 (1924) 310–311.

[12] N. Bohr and J. Wheeler, Phys. Rev. 56 (1939) 426–450.

[13] Niels Bohr, Atoms and human knowledge, Dan. Vid. Selsk. Oversigt (1955–56) 112–124.

[14] English translation by J. Pedersen, in: Niels Bohr, Atomic Physics and Human Knowledge (Wiley, New York, 1958) p. 83.

[15] A. Lomholt, Det Kongelige Danske Videnskabernes Selskab 1742–1942, Samlinger til Selskabets Historie, Bind I (København, 1942).

[16] N.E. Nörlund, Danmarks Kortlægning. En historisk Fremstilling, Bind I: Tiden til Afslutningen af Videnskabernes Selskabs Opmaaling (København, 1942).

[17] J. Pedersen, The Carlsberg Foundation (Copenhagen, 1956).

[18] Quoted in J. Pedersen, Niels Bohr og Det Kongelige Danske Videnskabernes Selskab, in: Niels Bohr, Hans liv og virke fortalt af en kreds af venner og medarbejdere (J.H. Schultz Forlag, København, 1964) pp. 261, 262. English translation in: Niels Bohr, His life and work as seen by his friends and colleagues, ed. S. Rozental (North-Holland, Amsterdam, 1967), the chapter by J. Pedersen, Niels Bohr and the Royal Danish Academy of Sciences and Letters, p. 268–269.

The Lesson of Quantum Theory, edited by J. de Boer, E. Dal and O. Ulfbeck

Epistemological Considerations

H.B.G. Casimir

Heeze, The Netherlands

Contents

1. Introduction

It is a great privilege to have the opportunity to speak on this occasion at Videnskabernes Selskab, but the task before me is by no means an easy one. Shall I be able to do more than add a few platitudes to what Bohr has said himself and to what others have said about Bohr? Let me first of all explain my choice of subject.

I often had the impression that Bohr himself attached more importance to his ideas about the fundamental principles of the description of nature, to his contribution to the "philosophia naturalis" and to philosophy in general, than to his numerous more concrete triumphs. The confirmation of his notion of stationary states by the experiments of Franck and Hertz, the exact agreement with observations of his formula for the line-spectrum of ionized helium, the discovery of hafnium, an element with properties in agreement with the predictions of his theory of the Periodic System, must have given him immense satisfaction, but from the very beginning he was concerned with the question how his bold approach could be made to supplement, rather than to contradict the notions of classical theory.

I came for the first time to Copenhagen in the spring of 1929 and was introduced to Bohr by my teacher Ehrenfest with the words "... er kann schon etwas, aber braucht noch Prügel" (he has already some ability but still needs thrashing). How true this was—the second part that is. As a matter of fact, I had to a certain extent mastered the formalism of quantum mechanics, which by then had reached a fairly definitive form, but that was about all I knew. I soon discovered that Bohr's interest in those days centered on a further clarification and elaboration of his thoughts on complementarity. He was not particularly interested in the further development and refinement of mathematical techniques, although he encouraged me to do some work in that direction. Neither did solid state physics and other applications of quantum mechanics play an important role in his own work or in that of his institute. A few years after I had left—I stayed until the spring of 1931 with interruptions to pass my examinations at Leiden—nuclear physics began to dominate both the work at the institute and the yearly conferences, and Bohr himself made important contributions to the theory of nuclear reactions and the theory of fission, almost a second youth one might say, but even then Bohr returned time and time again to his epistemological considerations. Therefore, I want to make these considerations the main topic of my talk. But I shall not try to give a comprehensive account of Bohr's thinking: it would be both presumptuous and useless if I tried to reformulate and summarize the ideas Bohr himself was at such great pains to express as well as possible. My purpose is more modest: I shall say something about the discussions between Bohr and Einstein, but first of all I want to make some remarks about the methods Bohr used when dealing with these fundamental questions.

It is a striking fact that in pondering the most profound aspects of quantum mechanics Bohr did always consider simple cases and used only the simplest mathematics. Deceptively simple I should like to say, for there are many pitfalls to be avoided, and it took Bohr's grasp of classical physics to reduce the essence of complicated mathematics to easily grasped concepts. Such simplification is not only a matter of convenience: it meets a more profound requirement. Bohr always emphasized that the final result of a measurement must be something we can tell other people about and that we can describe in simple everyday language: the position of a pointer, a black spot on a film and so on.

2. Magnetic moment of a free electron

Although the most sophisticated application of Bohr's methods is to be found in his work with Rosenfeld on measurements of electromagnetic fields [1], in my opinion —and in the opinion of many others—the clearest summary of his thoughts can be found in his "Discussion with Einstein on epistemological problems in atomic physics", his contribution to the volume "Albert Einstein: Philosopher-Scientist" [2], and I shall presently look more closely at the examples discussed therein. However, my first contact with Bohr's way of thinking related to a slightly different theme. In his opening talk at the 1929 meeting Bohr had shown that it is impossible to determine the magnetic moment of a free electron by a "classical" Stern–Gerlach experiment, and one of my first assignments after I came to Copenhagen was to assist Bohr in writing a note on the subject. Bohr liked to have someone to talk to, who could say yes or no at the right moment and who would write down the sentences he pronounced while pacing the room. Stenography was not required, but one had to get used to his rather soft and somewhat slurred voice. Several drafts were produced—they are still in the Bohr archives, and here and there my handwriting appears—but the work was never considered fit for publication.

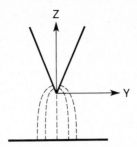

Fig. 1. The wedge-shaped polepiece for the Stern–Gerlach experiment has its edge along the x-axis.

However, Pauli used part of it in his report to the 1930 Solvay Conference (published in 1932) [3]. Figure 1 shows a usual arrangement for a Stern–Gerlach experiment. A wedge-shaped polepiece with its edge along the x-axis is facing a flat one. In the symmetry plane the field is strictly along the z-axis, but as soon as we get out of that plane there is a component H_y. Now we have

$$\text{div } H = 0,$$

which means in this case

$$\frac{\partial H_z}{\partial z} + \frac{\partial H_y}{\partial y} = 0.$$

The force on the magnetic moment of the electron is given by

$$F_s = \frac{eh}{4\pi mc} \cdot \frac{\partial H_z}{\partial z}$$

and this has to be large compared with the uncertainty of the Lorentz force, which is given by

$$\delta F_L = \frac{ev}{c} \delta H_y.$$

Since

$$|\delta H_y| = \left| \delta y \cdot \frac{\partial H_z}{\partial z} \right|$$

it follows from

$$F_s \gg \delta F_L$$

that

$$\delta y \ll \frac{h}{4\pi mv} = \frac{\lambda}{4\pi}.$$

But, if a wavepacket has to be much smaller than the wavelength it is entirely impossible to speak about a trajectory in the classical sense.

The whole trick is to have a clear picture of what a Stern–Gerlach experiment really is and to introduce an equation for the magnetic field at the right moment. Any student can follow the reasoning, but no one thought about it before Bohr.

3. Thought experiments

Let us now look at four thought experiments discussed in the paper I mentioned before.

3.1. First case

A screen with one hole or slit is illuminated by a plane electron-wave. If a is the radius of the hole (or half the width of the slit) and λ the wavelength of the electron, the wave emerging from the hole will show a spread given by $\delta\varphi \sim \lambda/a$ (fig. 2). Since $\lambda = h/mv$ we have $\delta p \simeq p\delta\varphi \sim h/a$. In principle one can measure the recoil of the slit. Then we know how much transverse momentum is imparted to the electron, but in order to carry out such a measurement the screen has to be mobile and the position of the hole becomes uncertain. Here I want to call attention to the following. Bohr uses here and elsewhere a simple formula for $\delta\varphi$. He does not exactly define this $\delta\varphi$. It might be a halfwidth or the width of the first diffraction peak. In textbooks one usually defines Δq as the root-mean-square deviation

$$(\Delta q)^2 = \int (q - \bar{q})^2 |\Psi|^2 \, dq$$

and similarly for Δp. Bohr does not use this definition and that is just as well for in the case of a slit with sharp edges this Δp becomes infinite. Recently, attention has been drawn to the fact that in many cases the root-mean-square deviations are useless and that a more refined analysis is called for. Bohr always steered clear of such difficulties.

3.2. Second case

A screen with two holes illuminated by a point source is considered. Qualitatively, we can easily see what a wave will do. Waves will emerge from the two holes and on

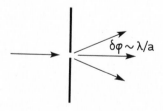

Fig. 2. Diffraction of electron wave by hole or slit.

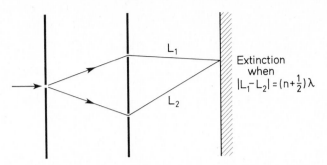

Fig. 3. Screen with two holes illuminated by a point source.

a distant observation screen they will strengthen or annihilate each other, depending on their phase difference. We may use a photographic plate or some kind of particle detector and we shall never find an electron at a node of the wave pattern. But as soon as we find an electron, somewhere between the nodes, we are tempted to ask: "Through which hole has it come? But if it came through one hole, how could the position of the other hole have exerted any influence?" Now it is not too difficult to show that any arrangement that makes it possible to tell through which hole the electron has come does away with an interference pattern. There exists no experimental arrangement that makes it possible to observe interference and thus to demonstrate the wave character of the electron and also to observe through which hole the electron has come (fig. 3). Can experiments of the type described really be carried out? With light, interference patterns obtained with two holes or two slits can easily be observed. With electrons this would be difficult because of the much shorter wavelength. But one can perform and has performed experiments that amount to almost the same: one can study diffraction round a very thin wire in the electron microscope. The unanswerable question is then: On which side has the electron passed the wire?

With the two-hole experiment we are so to say in the very middle of complementarity. I remember an evening at the Carlsberg mansion with Harald Høffding, Bohr's predecessor there. Bohr explained among other things the two-hole experiment. Of course the remark was made: "but the electron must be somewhere on its road from source to observation screen," to which Bohr replied: "what is in this case the meaning of the word *to be?*" And I remember the reaction of the philosopher Jørgen Jørgensen, protesting: "man kan sgu ikke reducere hele filosofien til en skærm med to huller" (one can, damn it, not reduce the whole of philosophy to a screen with two holes).

3.3. Third case

Einstein points out that energy may in principle be determined by weighing. Suppose we have a box full of lightquanta or of electrons, provided with a timer that opens a shutter and placed on a balance or suspended by a spring (fig. 4). The timer can set the time with any desired accuracy, and weighing, both before and after the

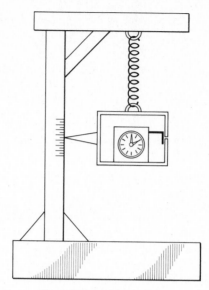

Fig. 4. Weighing apparatus.

electron or lightquantum have escaped, determines the energy, so we circumvent the relation $\delta E \cdot \delta t > h$. Bohr's answer is simple, but again it took Bohr to find it. How can we determine mass with a precision δm? Then $g\delta m$ must impart a measurable momentum to a balance in time t, available for the measurement. Therefore, the initial momentum of the balance must be determined better than $gt\delta m$. Hence, the place of the balance in the vertical direction is uncertain to the amount $\delta z \sim h/gt\delta m$. But general relativity tells us that this leads to an uncertainty of time, because of the red shift of the timer that activates the shutter. We have

$$\frac{\delta t}{t} = \frac{g\delta z}{c^2},$$

and it follows

$$\delta t \sim \frac{h}{\delta m \cdot c^2} = \frac{h}{\delta E}.$$

Again, extremely simple mathematics, but to see at once that weighing must involve uncertainty in place and that uncertainty in place involves uncertainty in time, and to extract from the imposing edifice of general relativity the one simple formula that settles the question, requires a penetrating understanding of the basic principles involved.

3.4. Fourth case

On page 229 of his "Discussion with Einstein" Bohr [4] refers to what Ehrenfest reported concerning some further objections of Einstein. I happened to be present at the Leiden colloquium when Einstein spoke about these objections. He again considered the box with shutter of the former example, but now he pointed out that after the electron or lightquantum had left, one still had the choice either to read the time immediately or to carry out a lengthy weighing procedure. The relation $\delta E \cdot \delta t > h$ is not violated, but the curious fact is that the particle whose energy or time of passage we want to determine is left untouched by the choice. Ehrenfest had given me the task of opening the discussion and I tried, to the best of my abilities, to explain the Copenhagen view on these matters. I still remember Einstein's reaction: "Ich weiss, widerspruchsfrei ist die Sache schon, aber sie enthält meines Erachtens doch eine gewisse Härte" (I know, the story is free from contradictions but in my opinion it contains all the same a certain unpalatability).

Einstein's famous paper with Rosen and Podolski [5] goes along similar lines. Two particles may interact temporarily and by measuring either the momentum or the place of the one, we can determine place or momentum of the other, without touching it. Therefore, Einstein concludes, this other particle must *have* a definite place and momentum. For this, there is no room in quantum mechanics, hence the quantum mechanical description is incomplete. In his discussion, Bohr slightly extends his usual mathematics. He points out that $P_1 + P_2$ commutes with $Q_1 - Q_2$ so we can know these two quantities simultaneously. Now we can measure Q_1 and find Q_2 or measure P_1 and find P_2. There is no way to assign a meaning to P_2 and Q_2, unless we specify the measuring equipment that has always to be included in any system considered.

Of course, in all these cases it is possible to go into more mathematical detail, and if one does, one is again struck by the power of Bohr's simple arguments.

4. Complementarity and completeness of description

Is there really a difference of opinion between Einstein and Bohr? Is it not a question of words? When Bohr says that quantum mechanics offers a complementary description is that not tantamount to saying that from a classical point of view the description is incomplete? And is that not exactly what Einstein is complaining about? Personally, I think that it is a legitimate use of language to say that the limited applicability of classical concepts to atomic and subatomic phenomena shows that the quantum mechanical description is incomplete, but that does not mean that there is no serious difference between Bohr and Einstein. Bohr argues that the quantum mechanical description is as complete as it can possibly be and he is ready to accept the limitations of our pictures of reality and of our language, ready also to renounce strict causality such as we find in classical mechanics and electrodynamics. He is willing to accept these limitations, because he regards them as essential features of nature and of our human existence. New forces may be revealed, surprising new phenomena may be brought to light, but Bohr considered it impossible that they would ever transgress the limitations imposed by quantum mechanics. Einstein on the other hand, though admitting that quantum mechanics is a powerful discipline that provides a valid description of many phenomena, was convinced that one should search for something beyond, for a theory that would to a certain extent re-establish the notions of classical physics.

A simple analogy may be in order. Thermodynamics is a powerful discipline, providing a satisfactory description of many phenomena, but we know now that behind thermodynamics there are innumerable atoms and molecules at work, and physicists have successfully looked for phenomena where the atomic structure reveals itself. Shouldn't we in a similar way look for something behind the statistical laws of quantum mechanics? Bohr's answer would certainly be an emphatic NO!

What will the future be? So far, the followers of the Copenhagen School can point to greater success. Accepting the limitations of the quantum mechanical description as inviolable laws of nature they have enormously enriched our understanding of nature and our ability to create new phenomena and new devices, whereas Einstein and his followers have made little headway. I am convinced that also in centuries to come atoms and molecules will be studied by means of the same Schrödinger equations we use today, just as we can use Newtonian mechanics to calculate the orbits of planets and satellites, with only minor and in most cases negligible relativistic corrections. But, since Einstein, we talk in a different way about gravitation, although Newton's formula for the attraction remains an excellent approximation. Will one in centuries to come talk about quantum mechanics in the same way as we talk today? Who am I to make a prediction?

The battle of wits between Bohr and Einstein did never lead to personal antagonism, to insinuations or intrigues, things alas not unknown in the history of science. Bohr has on many occasions expressed his admiration for Einstein and his indebtedness to his critical objections. And Einstein, though instinctively averse the Copenhagen School, frankly admitted its consistency and its importance and did not grudge it its successes. We, physicists, should be grateful for this, but that is only part of my gratitude for having known the great physicist and even more for having known the great human being that was Niels Bohr.

References

[1] N. Bohr and L. Rosenfeld, Zur Frage der Messbarkeit der elektromagnetischen Feldgrössen, in: Mat. -Fys. Medd. Dan. Vidensk Selsk. 12-8 (1933) 1–65; English translation in: Selected Papers of Léon Rosenfeld (Reidel, Dordrecht, 1979) p. 357–400.
[2] N. Bohr, Discussion with Einstein on epistemological problems in atomic physics, in: Albert Einstein, Philosopher-Scientist, Vol. VII of The Library of Living Philosophers (Evanston, IL, 1949).
[3] W. Pauli, in: Le Magnétisme, Rapports et Discussions du 6 ème Conseil de Physique, Bruxelles, 20–25 octobre 1930 (Gauthier-Villars, Paris, 1932) p. 175–238.
[4] N. Bohr, Discussion with Einstein (see ref. [2]).
[5] A. Einstein, B. Podolsky and N. Rosen, Phys. Rev. 47 (1935) 777.

Part Two

THE LESSON OF QUANTUM THEORY

Proceedings of the Niels Bohr Centenary Symposium
October 4–7, 1985

edited by

Jorrit de Boer

Ludwig-Maximilians Universität, Munich

and

Ole Ulfbeck

The Niels Bohr Institute, Copenhagen

The Lesson of Quantum Theory

Thirteen communications delivered on October 4 and 5, 1985, in the lecture hall of the Panum Institute, University of Copenhagen

Unity of Knowledge

Four communications delivered on October 6, 1985, in the lecture hall of the Panum Institute, University of Copenhagen

Niels Bohr

Four communications delivered on Niels Bohr's one hundredth birthday, October 7, 1985, in the University Solemnity Hall

The Lesson of Quantum Theory, edited by J. de Boer, E. Dal and O. Ulfbeck

Measurement in Quantum Theory and the Problem of Complex Systems

Philip W. Anderson

Princeton University
Princeton, New Jersey, USA

Contents

1. An approach to the Bohr–Einstein debate

Every physicist who has studied the quantum theory is familiar with the great debate on the uncertainty principle between Bohr and Einstein which began at the 1927 Solvay Conference but which then simmered on for many years. Perhaps the commonest view is that Bohr simply won hands down; in Einstein's somewhat mistranslated phrase, God *does* "play dice with the world" and the outcome of a quantum measurement is truly uncertain. In a way, this view corresponds to the truth—at least to the pragmatic truth that now, after 60 years, there is no doubt in any quantum physicist's mind that the whole quantum theory, including the original quantum prescriptions of measurement theory, are correct in describing the results of experiment. Why, then, re-open the question?

It may be slightly provocative, but surely in a spirit Bohr himself would have enjoyed, to suggest that many of those who have thought beyond the textbook level have come to the conclusion that the really correct answer to which one was right is —neither! It is not true that quantum mechanics is a purely probabilistic theory, any more so than statistical mechanics, for instance. That it is, seems to have been Bohr's point of view, in so far as we can follow his statements about complementarity as a general philosophical principle. In fact, generations of philosophers have been misled, probably against Bohr's intent, to give a deep meaning to this "uncertainty". On the other hand, so far as I can see, one cannot avoid the probabilistic results of measurements on atomic scale systems, as Einstein wished. Einstein's reasons are clear enough: he believed firmly that general relativity was the appropriate model for future theories, with its fundamental emphasis on the geometry of space-time, and it disturbed him deeply to have that geometry indeterminate and fluctuating—in particular, how can an indeterminate geometry properly maintain causality? The difficulties of quantizing general relativity vindicate Einstein's misgivings to some extent. Whether or not Einstein would have felt comfortable with the present liberties being taken with the structure of space-time is hard to know; one would have thought he would have felt a proprietary interest in four dimensions. It is true that we are beginning to return to his geometrical view and that at some ultimate scale everything may yet change. But really that is a bit of a diversion.

I want to talk here about simple non-relativistic quantum theory, although I believe everything I say is equally valid in relativistic field theory at any scale short of the Planck mass, e.g. during most of the events which took place during the "big bang" including the symmetry-breaking phase transitions.

What I consider to be the correct approach to the Einstein–Bohr debate began very much with one of my own personal idols, Fritz London. Interestingly enough, London trained originally as a philosopher, and one sees, in following his career, a certain philosophical theme: how can we deal with the quantum theory at the level of macroscopic objects? He is most justly famous for his ideas on the two macroscopic quantum phenomena which most severely test quantum mechanics in this regime, superfluidity and superconductivity; but I refer here to a seminal paper in the theory of measurement by London and Bauer (1939), who I believe were the first to take the point of view I advocate here, that the central problem of measurement theory is not the quantum mechanics of atoms, which is simple and easy, but the fact that macroscopic everyday objects are very difficult indeed for the quantum theory to deal with properly. To mis-quote another famous Dane: "The fault, dear Horatio, is not in our atoms but in ourselves."

The sticking point is twofold. The first problem is that the quantum theory is in principle *not* probabilistic but deterministic. The equation for the time variation is a simple deterministic one:

$$i\frac{\partial \psi}{\partial t} = H\Psi,\tag{1}$$

where H is a given—in principle—function of a maximal set of arguments of Ψ at the present time—or in relativistic theory, at a space-like surface. Of course, there is a mathematical equivalence between eq. (1) and a sum over classical trajectories, which resembles a stochastic process, but every attempt to replace quantum mechanics with a truly stochastic theory ends in failure. Thus in principle, given an initial $\Psi(t=0)$, we know precisely what Ψ is. What is more, if quantum mechanics is a complete theory, $\Psi(t)$ should be a complete specification of the physics including measurements, though not necessarily vice versa.

What London points out to be the second prong of the problem is that the Ψ involved must be taken to be the wave function describing the measuring apparatus and the experimenter as well as the experiment. It is philosophically repugnant to suppose that the prescriptions of quantum mechanics are different depending on whether we include the apparatus and the observer in the system or not. I follow Everett (1957) in his important, if perhaps somewhat incomplete paper, on "relative states" sometimes called "many worlds". Everett gives the following statement of the question: "The wave function must be taken to be the basic physical entity... interpretation comes only *after* an investigation of the logical structure of the theory", hence also only after an investigation of the quantum mechanics of large or rigid or sentient objects. A clear necessity is that Ψ must tell us what it may know about itself, not vice versa—there is no necessity that Ψ be a measurable thing in general, only that it describes what we actually observe.

The large objects which are involved in a conventional measurement behave in essentially non-quantum ways in at least three respects, each of which, at one time or another, has been conjectured to be the source of the anomalies of measurements.

(1) They must be or appear irreversible, so that at least the result of the measurement is recorded after it has taken place, not predicted before; measurement is of the essence of an irreversible process in actual experimental fact.

(2) They must be rigid; in order to make the appropriate kinds of measurements (we will see that "rigidity" can be defined in a much generalized way) they must have the properties of pointers.

(3) They must be sentient, to communicate it. Now (3) involves us in enormous difficulties with the different levels of computational complexity and with unknown aspects of the brain; fortunately, as London and others have pointed out, measurements can be made and recorded by purely automatic machinery, and one hardly sees physics depending on the final process of peeking at the instrument. I believe both (1) and (2), on the other hand, play an important role.

2. Canonical Stern–Gerlach experiment

It is important to think a little about the actual measurement process at this time. Let us start for instance with the canonical Stern–Gerlach experiment, which has been described as using the position variables of the atomic beam as an instrument to measure the spin of the atoms (fig. 1). But we must realize that just separating the two beams is not a measurement per se; a measurement must in some sense, as everyone agrees, destroy the possibility of interference between the two beams.

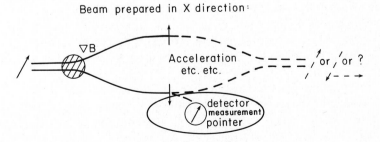

Fig. 1. Stern–Gerlach experiment.

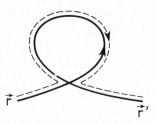

Fig. 2. A self-intersecting Feynman path for an electron to propagate from r to r'. Propagation along solid and dashed paths can interfere.

A measurement, according to the old viewpoint which agrees with experiment, definitely sets the relevant quantum number equal to that appropriate to the one beam alone, but interference with the other modifies the values of the quantum number and destroys the property of being in the one eigenstate, so if such an interference experiment remains possible, no true measurement in the canonical sense has taken place. This will be an important observation to the remainder of the chapter. In fact, in a Stern–Gerlach apparatus we very clearly can, even after accelerating the beams and passing them through all kinds of lenses etc., return them to coincidence and observe interference effects. This kind of experiment has been successfully carried out to demonstrate the effects of potentials on neutrons, for instance. (Macroscopic quantum coherence is *not* relevant here. That is enforced by generalized rigidity and minimization of free energy and is irrelevant to measurement theory.) The existence of an interference effect after subjecting beams of particles to all kinds of vicissitudes is the basis of the phenomena of "weak localization" in metals, where we demonstrate interference between electrons travelling in a state and its time—reversed, i.e. precisely backscattered, version. Strong localization, in fact, occurs when this backscattering prevents irreversible loss of coherence entirely (fig. 2). A most striking experiment was carried out first by Sharvin and Sharvin (1981) at the suggestion of Altshuler et al. (1981). In this experiment the interfering pairs of states are carried around an array of holes in the metal, or along an array of wires in a magnetic field, leading to Bohm–Aharanov interference phenomena at half-flux quanta periods between paths for these so-called "cooperons" which take opposite direction around the holes. An excellent review which describes these phenomena is by Lee and Ramakrishnan (1985), from which fig. 2 is borrowed, showing schematically the identical paths of the particle and the backscattered hole which interfere in such a way as to affect the response function which gives the resistivity due to elastic scattering. This response function is closely related to the electron–hole pair propagator. In fig. 2 we see two possible routes for this pair around a hole containing magnetic flux, which can interfere constructively or destructively depending on the flux. Figure 3 shows an example of the delicate microlithographic methods employed to fabricate a network containing many such tiny holes, and an example of such a network (from Bishop et al. 1985); fig. 4 shows some examples of the resulting interference pattern. The amplitude of this pattern measures the degree of coherence and vanishes if the path around the hole is larger

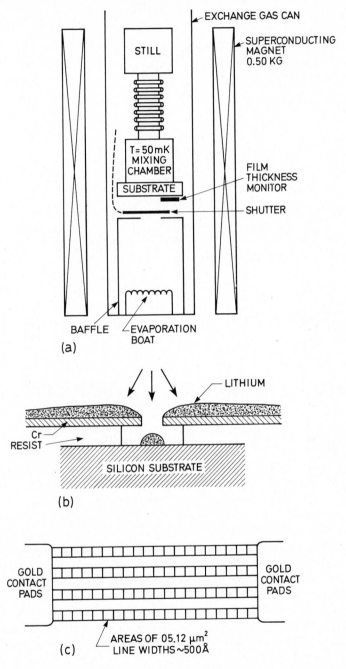

Fig. 3. (a) Apparatus shown schematically; (b) the overhanging photoresist and (c) the grids for our normal metal flux quantisation experiment.

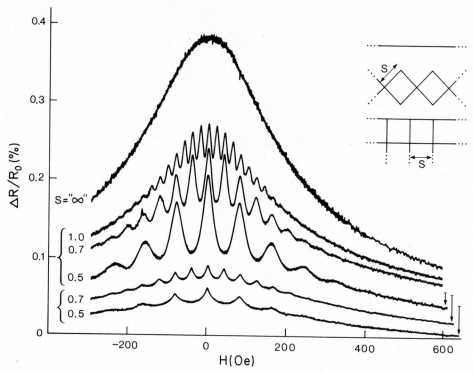

Fig. 4. Relative change of resistance showing interference between particle and backscattered hole as a function of magnetic flux enclosed by the path.

than the "Thouless length" $\sqrt{l_e l_i}$ [l_e and l_i are the elastic and inelastic (coherence-destroying) mean free paths]. This is a particularly clean system for studying the effects of inelastic scattering of the particles on the interference phenomenon. As suggested initially by Thouless (1977), in these experiments it is possible to measure quantitatively the rate of loss of coherence. Fukuyama and Abrahams (1983), Altshuler, Aronov and Khmelnitski (1982) and Fukuyama (1984) have been studying—with some controversy among themselves, but in essential agreement with experiment—the problem of how to quantify the definition of an inelastic collision event which destroys coherence between the electron and its backscattered partner, as opposed to carrying away a given amount of energy or momentum. This is most significant in that, to my knowledge, it is the first example of measurement and prediction of coherence - destroying inelasticity rates, in the true sense. A second interest is that the Russian group has emphasized that the relevent scattering can be thought of as coming from the Nyquist noise, generated by the density fluctuations of the electrons themselves. Because it deals with a completely isolated many-particle system, this work gives the feeling of having finally gotten down to the actual nuts and bolts of the phenomenon of irreversibility itself.

Once inelastic scattering has taken place, we would have to recohere the additional excitations in order to reconstruct the beam, and then each of our

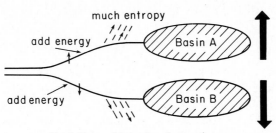

Fig. 5. Two stable basins of attraction.

excitations itself scatters inelastically, etc. Perhaps, in principle, a perfect reversible quantum computer (Bennett 1973) could keep track of every scattering, but in effect the two halves of the beam are by now following different, untraceable paths through wave function space and we must treat the outcome in terms of probabilities and not probability amplitudes.

So far, however, no "measurement" has taken place. For instance, no "particle" has been registered as having gone through a channel. In a general measurement process, we add energy—e.g. accelerate the beams, or cause them to impinge on metastable grains of silver salt, or some other similar scheme. This allows us to enhance the rate of dissipation—i.e. production of entropy—so that the trajectory of each piece of the wave function in Hilbert space can satisfy Liouville's theorem by spreading out in irrelevant variables, but contracting strongly in certain relevant ones which have been coupled to the variables to be measured (see fig. 5). The relevant variables, which are normally the coordinates of some macroscopic, rigid object (a "pointer"), become trapped in a "basin of attraction" which is different for the different values of the original quantum number. By virtue of registering that fact in his neurons, which are themselves switching devices with multiply stable basins of attraction, the observer too becomes entrapped in beam A or beam B, since in essence the Hilbert spaces of the macroscopic objects involved are essentially disparate and no physical operator acting on a few coordinates at a time, expressing the only possible physical processes internal to the system, can connect states in A to states in B. (Of course, an external force can move a rigid pointer by the application of an overall force coherently to every atom of the pointer simultaneously, but this merely redefines "A" and "B".)

3. Phase variable of superfluid helium. Broken symmetry

These properties of macroscopic systems are much more clearly shown if we look at an example. One example of measurement, and of quasi - degeneracy and of rigidity properties, which has always intrigued me, is that of the phase variable of superfluid helium. This is particularly attractive because this is a broken symmetry variable whose nature is not confused by the presence of many other objects with the same kind of broken symmetry. At this point let me insert a brief primer on broken symmetry in condensed matter physics. [see Anderson (1984) for more details.] This takes the form of five points.

(1) The initial Hamiltonian of our system has a large group G (e.g. local homogeneity and isotropy for a liquid).

(2) The lowest-energy state, or mean field fixed point, has a lower symmetry H (as a crystal, with discrete rotation and translation symmetries).

(3) Hence there is an "order parameter" describing the state, which has "phase angles" free to move in a space isomorphic with the factor group G/H.

(4) Interactions enforce generalized rigidity of these phase angles: there is a (free) energy $\propto (\nabla\phi)^2$. This allows dissipationless "action at a distance" via rigidity, supercurrents, etc.

(5) Hence all atoms are correlated, in highly non-quantum but very familiar behavior—*we* are broken symmetry objects with quasi-degeneracy and rigidity, after all, so we have no trouble in dealing with these concepts except when we try to make them compatible with quantum theory.

Now, in order to explore the relationship of these peculiar—if you are a quantum person—or ordinary—if you are macroscopic—properties to measurement problems, let us do the following sequence of "Gedanken" experiments. [With recent advances in technique perhaps these need not all remain "Gedanken" experiments forever (see Avenel and Veroquaux 1985).]

As a first stage, cool down isothermally a bucket of liquid helium from above T_λ to near $T = 0$, starting, if one likes, from a microcanonical ensemble so that the resulting state has a fixed number N of particles. The bucket will subside into its lowest state, which can be written as

$$\Psi = \int d\phi\ e^{iN\phi}\ \Psi(\phi),$$

where Ψ is a quasi-classical coherent state in which the order parameter is taken to be

$$\langle\Psi\rangle = |\Psi_0|\ e^{i\phi}.$$

The phase ϕ will be uniform everywhere because of the "generalized rigidity" term in the free energy

$$\tfrac{1}{2}\rho_s v_s^2,$$

where

$$v_s = \frac{\hbar}{m}\nabla\phi.$$

I would assert that at this point, already, the different components of the wave function have ceased to interfere and that our bucket is already in the "many worlds" representation; within very fine, calculable limits enforced by the N-ϕ uncertainty principle, ϕ has become a classical variable, and while no experiment can determine what the actual overall value of ϕ is—since it is gauge-

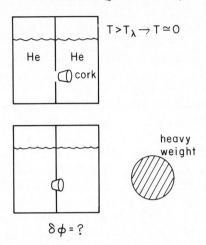

Fig. 6. Change in the condensate phase due to gravitational potential.

dependent—any future experiment will be interpretable as though ϕ was fixed. For instance, as a second stage of our experiment, let us suppose our original bucket was nearly divided into two, and at a certain time we close off the septum between the two halves of the bucket, and thereafter subject the two halves to different gravitational histories—e.g. by putting a large object like a locomotive near one half and not near the other half. If the difference in gravitational potential is ΔV, we expect to observe a change in phase of $\Delta\phi = \hbar^{-1}\int\Delta V \, dt$ (see fig. 6).

Each half of the bucket, we assert, is now in a state in which it has a wide variation of N, ΔN, and in which there is coherence between the wave functions belonging to different values of iN:

$$\langle N_{1+1} \mid \Psi^* \mid N_1 \rangle = \Psi_0 \, e^{i\phi_1}$$

and similarly for the other half of the bucket. When we reunite them, a measurable interference current will initially flow depending on their relative gravitational histories. This interference current, of course, violates Wigner's "superselection rule" but that is not our problem. What is relevant to measurement theory is that when the two are reunited the experimentalist does not see any interference between different values of the overall phase. This is not quite exactly the case; the phase in each bucket does diffuse slightly, because compressibility forces make the states of different N slightly inequivalent. The time scale, for that, however, is on the order of 10^8 years.

A third experiment illustrates this point. I would suppose that if the experimenter now cools down two entirely different, non-communicating buckets, each of which initially had a fixed N and no preferred phase, that he, upon opening an orifice between the two, would see initially with equal probability any *fixed* value of the phase difference, and thereafter no experiment he tried could recover the components of the wave-function which started out with different relative phases. He would not see *zero* interference current, which would be the result if he was to average over all of his many worlds.

That is, we have done a macroscopic Stern–Gerlach experiment, dividing the states of a macroscopic object into beams with different quantum numbers.

The above experiments essentially represent the equivalent of a "reference system" for phase. It is rather like the development in the early universe of the first gravitational inhomogeneities—prior to that time, "location" would not have had a measurable meaning.

In another sense this experiment resembles the original big bang. During the earliest moments, present theory suggests that one or several symmetry-breaking phase transitions take place leaving us with a present set of particle interaction symmetries which are much reduced, as well as with values of certain "Higgs" fields which have, in principle, arbitrary phase angles—otherwise, of course, they would have broken no symmetry. Coincidences which, to me, suggest that the whole story is not yet in, leave us with no experimental handles on that phase-angle and no corresponding rigidities; nonetheless we can be quite sure that physics averaged over all possible values of that phase-angle is not what we live with, even though in the initial big bang—corresponding to our helium bucket above T_λ—the symmetry-breaking field averaged to identically zero and its phase was meaningless, so that below T_λ we must be in a linear superposition of all possible quasi-degenerate worlds with different values of that phase-angle. (This is of course, quite independent of the existence of monopoles and the singularities of the symmetry-breaking fields, which are disturbances of the relative, not absolute, values of the symmetry-breaking field.)

4. Conclusions

It is symptomatic of the depth of Bohr's thinking that his simple resolution of this deep problem still stands up as the practical way to deal with these hard questions. What I have tried to show is that nonetheless they remain a fascinating subject for subtle experiment and intriguing theory.

Let me close with a quotation from Niels Bohr remarking on the early days of quantum mechanics: "Although the spectroscopic successes of the quantum theory were more spectacular, the explanations of the macroscopic properties of matter were more satisfying and more fundamental". This has always been my view.

References

Altshuler B.L., A.G. Aronov and B.Z. Spivak, 1981, JETP Lett. **33**, 94.
Altshuler, B.L., A.G. Aronov and D.E. Khmelnitski, 1982, J. Phys. **C15**, 7367.
Anderson, P.W., 1984, Basic Notions in Condensed Matter Physics (Benjamin–Cummings, New York).
Avenel, O., and E. Veroquaux, 1985, Phys. Rev. Lett. (to be published).
Bennett Jr, C.H., 1973, IBM J. Res. Dev. **17**, 525.
Bishop, D.J., J.C. Licini and G.J. Dolan, 1985, Appl. Phys. Lett. **46**, 1000.

Everett III, H., 1957, Rev. Mod. Phys. **29**, 454–462.
Fukuyama, H., 1984, J. Phys. Soc. Jpn. **53**, 3299.
Fukuyama, H., and E. Abrahams, 1983, Phys. Rev. **B27**, 5976.
Lee, P.A., and T.V. Ramakrishnan, 1985, Rev. Mod. Phys. **57**, 287.
London, F., and E. Bauer, 1939, in: Actualités Scientifiques et Industrielles: Exposé de Physique Géneral, no. 775. ed. P. Langevin (Hermann, Paris) reprinted in: J.A. Wheeler and W.H. Zuerek, 1982, Quantum Theory of Measurement (Princeton Univ. Press, Princeton, NJ).
Sharvin, D.Yu., and Yu.V. Sharvin, 1981, Pis'ma v Zh. Eksp. Teor. Fiz. **34**, 285 [JETP Lett. **34**, 272].
Thouless, D.J., 1977, Phys. Rev. Lett. **39**, 1167.

Discussion, session chairman W. Kohn

Thirring: Don't you think that some axioms of the orthodox interpretations of quantum mechanics should be modified, namely that the measuring device has to be described in terms of classical variables only. These variables should certainly not be limited to the p's and q's alone because, as you pointed out, the phase of liquid He-II can be used for measurements.

Anderson: I turn your question around to a statement and say: I agree with it.

Weisskopf: You have connected your remarks with Everett's theory of many worlds, and I am not quite clear how you interpret this. For example: In your experiment with the two buckets one could say that there was an equal probability for any phase difference. But the one observed is the one that happens to be realized in this case. I personally cannot see why one has to assume that all others are also present, but your reference to Everett makes me think that you made this assumption.

Anderson: I am afraid I do mean that. I agree with John Wheeler who once said that that is much too much philosophical baggage to carry around, but I can't see how to avoid carrying that baggage.

Weisskopf: Why are you forced to carry it along in this case?

Anderson: Let's say I carry it along in the philosophical, and therefore not very serious, part of my mind.

Ambegaokar: While I thoroughly agree that, for the problems treated by most physicists, many paradoxes are avoided by treating subsystem *and* environment as a single quantum-mechanical system, is it not a rather large extrapolation to apply quantum mechanics to the universe as a whole? Is there some *evidence* that this is a justified or prudent assumption?

Anderson: No, there is no evidence. So far there is, in fact, a disturbing absence of evidence in that people have tried to predict a number of singularities in the hypothetical symmetry-breaking field left around: for instance, transition monopoles and things like that, which have not been seen. It is hard to know whether the phenomenon of inflation is to be considered as a confirmation or a further puzzle and, as will be discussed later, there is also the strange question of the cosmological constant. Note, incidentally, that all of this takes place only in the *relative* phases and says nothing about all the other universes.

Casimir: I am afraid I don't quite understand the remarks by Thirring and Anderson about a measurement not always having to be reduced to classical, macroscopic observations. I accept that one can use the phase in liquid helium for measurements. But how do you determine that phase? How do you read that instrument? You have got to arrive at something classical. As an intermediate step, you can certainly use things which are not classical, not macroscopic. But I like to maintain that you can always, in the last instance, reduce the measurements to classical phenomena.

Anderson: You are, of course, right. I guess I would make two caveats: One is at what point the measurement actually takes place; and the other is that someday we may make a computer entirely out of Josephson junctions, which is itself capable of making the relevant observations.

Kohn: You have made extensive reference to macroscopic quantum phenomena. In the last two or three years we have seen this remarkable development of the electron-tunnelling microscope where we see quantum phenomena, namely tunnelling, literally on a one-atom scale. This has been an exciting experimental development. A lot of work has been done since its invention and all kinds of applications have emerged. But I wonder what your thoughts are on the future relevance of that direction of experimental work for measurement theory?

Anderson: I haven't really thought about it. The only thing I have thought about a little bit is that it would be amusing to use some of the electronics in the Josephson mode. I don't think that anyone has done it yet. There could be some things that you might be able to do with adjustable junctions of that sort, in terms of turning Josephson currents on and off.

Ginzburg: If I understand correctly you have in mind to measure a difference (or gradient) of the phases between two buckets with He-II. But this means that you measure the velocity or the mass flux of helium. These, however, are macroscopic quantities, measured in a macroscopic way. So I do not understand what is the difference here with the usual interpretation of quantum-mechanical measurements using macroscopic devices.

Anderson: see answer to Casimir above.

Peierls: At the very beginning you stated that quantum mechanics is deterministic because the wave function satisfies a causal equation. Does this not imply that one regards the wave function as a physical object? This leads to all kinds of difficulties.

Anderson: I guess my answer to that is: If it's not a physical object—what else have we got? Quite seriously, I am saying that it contains all the physics but some parts of it are hard or impossible to observe because we are in it, and if you like, we don't observe alternative versions of ourselves.

The Lesson of Quantum Theory, edited by J. de Boer, E. Dal and O. Ulfbeck
© Elsevier Science Publishers B.V., 1986

Quantum Mechanics at the Macroscopic Level

Anthony J. Leggett

University of Illinois at Urbana-Champaign
Urbana, Illinois, USA

Contents

1. The principle of complementarity

It is a great honor and privilege to have been invited to speak at this symposium celebrating the centenary of the birth of Niels Bohr. The topic of my talk is essentially the foundations and interpretation of quantum mechanics, a subject on which Bohr probably thought longer and harder than anyone else in history, and as far as possible I shall therefore try to motivate what I have to say by direct reference to his own ideas. Nevertheless. I hope to persuade you that the foundations of quantum mechanics is not, as is so often thought, a topic of interest only to historians and philosophers, where ancient and hallowed ground is continually trodden and retrodden, but rather that in the last fifteen years or so it has become an *experimental* subject in a way which in Bohr's day might well have been difficult to imagine.

Throughout his life, Niels Bohr maintained a consistent point of view on the interpretation of quantum mechanics, which can be summarized in four principal theses:

(1) Microscopic entities (such as electrons and atoms) are not even to be thought of as possessing properties in the absence of specification of the macroscopic experimental arrangement.

(2) Macroscopic experimental arrangements, and the results of experiments, are to be expressed in classical, realistic terms.

A.J. Leggett

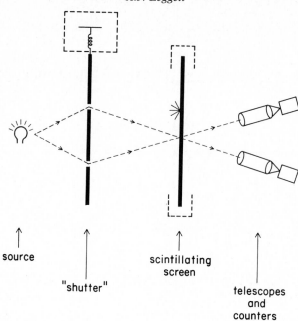

Fig. 1. The Young's slits experiment. Broken lines indicate parts of the apparatus that can be removed.

(3) There exists what Bohr repeatedly refers to as an "unanalyzable link" between the microsystem and the macroscopic measuring apparatus.

(4) The principle of complementarity: different experimental arrangements exclude one another, and the measurement of one property may therefore be "complementary" to the measurement of another. (Bohr regarded this principle as a universal feature of nature, and attempted to apply it not only to physics but to biology, psychology, etc.)

As an illustration of Bohr's point of view, let us consider the Young's slits (thought-) experiment so familiar from elementary quantum mechanics textbooks (see fig. 1). The apparatus consists of: (1) a source of electrons (or photons); (2) a "shutter", that is a screen containing two apertures which can be opened or closed at will, with the option of either bolting the shutter down or suspending it by a weak spring; (3) a removable scintillating screen; and (4) a pair of telescopes (or the equivalent device for electrons), one focussed on each of the two slits, and appropriate counters attached to each. It is evident that we cannot simultaneously arrange for the electrons (or photons) to enter the telescopes *and* for them to hit the scintillating screen. The basic "paradox" is, of course, that if we decide to employ the telescopes, i.e. (in the geometry of the figure) to remove the screen, then any one electron or photon will be registered in one counter and not in the other, i.e. it will appear to have definitely passed through one slit and not through the other; whereas, if we use the scintillating screen (so that the particles never reach the telescopes) then the accumulated statistical distribution of the particles on the screen will show diffraction effects which indicate *interference* between the two possible paths.

A number of features of this experiment are worth noting. First, it involves at least three different types of macroscopic apparatus, with quite different functions: the source, the shutter (which until further notice I shall regard as firmly bolted down) and the detection apparatus (scintillating screen *or* counters). Secondly, as pointed out by Wheeler (1978) it is, in principle at least, possible to delay the choice of which experimental arrangement to use (screen or telescope) for any single given electron until after that electron has certainly passed the shutter. Thirdly, it is possible to do a "negative-result" experiment, for example by placing a telescope and counter opposite one slit only; if the counter does *not* click, then one can infer that the microsystem passed through the other slit, without having caused (or so one would think!) any physical interaction between the microsystem and the detection apparatus.

Niels Bohr's interpretation of this situation is, of course, that the two different experimental arrangements are *complementary*: in any given experiment one is faced with the choice of measuring the path (i.e. which slit the particle passed through) or measuring the interference pattern. Contrary to what is often inferred from Heisenberg's discussion (1930) of his "γ-ray microscope" thought-experiment, there is in Bohr's view no question of an *interaction* between the system and the apparatus: the quantum-mechanical "state" of the system *is* nothing but a link between the microsystem and the macroapparatus (and hence between the various pieces of the latter), not an intrinsic property of the microsystem itself. As the philosopher Paul Feyerabend (1962) has put it, the conceptual mistake made by someone who attributed the dependence of the microsystem's behavior on the experimental apparatus used to an interaction with that apparatus would, in Bohr's view, be similar to the mistake of attributing the difference in the properties of a system observed in a moving frame of reference to an interaction with the frame. This point of view seems entirely self-consistent internally, and has been put on an even more complete logical foundation by the philosopher Hans Reichenbach (1944) in his book "The Philosophic Foundations of Quantum Mechanics". However, it is essential to remember that in the account given so far the notion of "measurement", and of "macroscopic measuring apparatus", has been taken for granted as logically unproblematical. I return to this point below.

2. Verification of thought-experiments

What has changed in this field since Niels Bohr's death? First, there have been dramatic verifications of some of the more counter-intuitive predictions of quantum mechanics at the atomic level—and hence, one might perhaps reasonably conclude, of the correctness of his interpretation at this level at least—in a number of experiments which in his days were merely "thought-experiments", but have now been actually carried out in the laboratory. Two classes of experiments seem to me particularly spectacular. The first is a set of experiments using the neutron interferometer, and in particular those reported by the Vienna group (Summhammer et al. 1983). The experimental setup is shown schematically in fig. 2a. The source is weak enough that only one neutron at most is normally present in the apparatus at

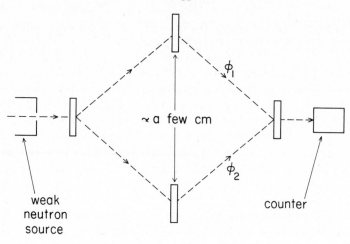

Fig. 2a. Schematic diagram of the neutron interferometer.

any given time. The de Broglie wave representing the neutron is split by the first silicon crystal (shown in the figure as a rectangle) into two packets which, as remarked by Greenberger (1982), are roughly of the size and shape of a small postage stamp. Each of these two wave packets then travels freely until it reaches one of the two intermediate silicon crystals, which are a distance of a few centimeters apart. Thus, at this point the separation of the two packets is much larger than their size and they are quite distinct. The intermediate crystals deflect the wave packets in such a way that their trajectories meet again at the final silicon crystal, where the total neutron flux is measured in a way not shown explicitly in the

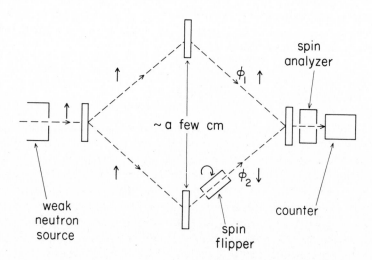

Fig. 2b. Spin-flip experiment in the neutron interferometer.

figure. * By adding suitable selection devices it is possible to measure not only the total flux, but also the flux of neutrons with spin polarization along a particular axis.

The simplest type of experiment one can carry out using this apparatus does not involve the neutron spin and is the exact analog of the classical Young's slits experiment. We simply block off beam 2 (by inserting a suitable absorber in its path) and measure the flux in beam 1, and vice versa. We then allow *both* beams to propagate freely to the detection apparatus and measure the total flux, which turns out not to be a simple sum of the measured fluxes in beams 1 and 2 separately, and, moreover, to show the quantum-mechanically expected diffraction pattern (see footnote). This already shows that the properties of the two beams combined are not the sum of those of the individual beams, even though there was only one neutron in the apparatus at any given time and the spatial separation of the two components of the wave packet was large compared to their individual extent.

The experiment just described is, of course, conceptually identical to some which had already been done with photons and even with electrons. However, by exploiting the fact that a neutron has a nonzero spin and magnetic moment one can do a good deal more. Suppose that, as shown in fig. 2b, we insert a spin analyzer in front of the detector, and moreover introduce in one beam, but not in the other, a device which rotates the neutron spin by π; for the moment, we assume that this device is a static magnetic guide which exchanges no energy with the neutron, so there is no question of any "measurement" having taken place at this point. Suppose now that the neutrons entering the apparatus are polarized along the positive z-axis. Using the notation $|\uparrow\rangle$ for such a spin state, and $|\downarrow\rangle$ for the orthogonal state of negative z-polarization, we expect that the wave function of the neutrons entering the detector will have the general form

$$\psi = \phi_1(r)|\uparrow\rangle + \phi_2(r)|\downarrow\rangle, \tag{1}$$

i.e. the neutrons "in" beam 1 will be unflipped and have spin up, whereas those "in" beam 2 will have had their spins flipped by the magnetic guide and therefore have spin down. If we now select for detection only those neutrons with spin up, this is equivalent to selecting only those in beam 1 and we should expect no diffraction pattern: similarly, if we select only those with spin down. This is confirmed experimentally. Moreover, if we measure the *total* flux (without spin selection) then we expect that the two beams, being associated with different spin states, will now add incoherently, so that again we expect no diffraction pattern: again, experiment confirms this prediction. However, suppose that we select only neutrons polarized along the x-axis (the corresponding spin state is denoted $|\rightarrow\rangle$ and the orthogonal state $|\leftarrow\rangle$). Using the fact that the state $|\rightarrow\rangle$ can be written

---

* In practice, rather than observing the flux as a function of position in the vertical plane, one observes the details of the interference between the two beams by inserting a wedge of material in the path of one beam, which shifts the relative phase of the two wave packets by a variable and controllable amount. Moreover, one actually observes the intensity transmitted in various directions, in general not forward as in the figure. These differences are of no importance in the present context and I shall ignore them in the ensuing discussion.

as a linear combination of $|\uparrow\rangle$ and $|\downarrow\rangle$, we see that the wave function (1) becomes on this basis:

$$\psi = 2^{-1/2}\{(\phi_1(r) + \phi_2(r))| \rightarrow \rangle + (\phi_1(r) - \phi_2(r))| \leftarrow \rangle\} \tag{2}$$

and we would expect the flux of selected neutrons to be proportional to $|\phi_1(r) + \phi_2(r)|^2$, that is, to show a diffraction pattern. Once more, this prediction is verified by the experimental results.

One might argue that this experiment too, is conceptually no different from one that can be carried out with photons (with the spin variable replaced by polarization). However, the final twist is the most amusing of all, and to the best of my knowledge has no (practicable) analog using either photons or electrons. Suppose we replace the static magnetic guide by a radiofrequency cavity, which is so tuned that a neutron entering with spin up will have its spin rotated by exactly π (as in the guide). With regard to the measurement of up-spin, down-spin or total neutron flux the situation is unchanged. The interesting question is: If we measure the flux of neutrons selected to be in the spin state $| \rightarrow \rangle$, do we or do we not expect, according to the standard formalism of quantum mechanics, to see a diffraction pattern? This question (or, more precisely, the significance of the answer for the interpretation of the formalism) has provoked a lively discussion in the recent literature [see e.g. Dewdney et al. (1984), Rauch (1983)]. At first sight, it is tempting to argue that the RF field in the cavity has transferred a finite amount of energy to the neutron, which could in principle be measured, and that we have therefore in effect measured at this point which of the paths the neutron followed (if the cavity is found to have lost energy, we know that the neutron has come via beam 2, whereas if it has not, the neutron is in beam 1). Now the standard quantum measurement axioms, at least as presented in most textbooks, tells us that measurement "collapses" the wave packet and that, if as here the result of the measurement is not inspected, the various amplitudes cannot interfere coherently (i.e. the state of the system is a mixture, not a superposition, of eigenstates of the measured quantity). This would seem to imply that, when the spin state $| \rightarrow \rangle$ is selected, the amplitudes $\phi_1(r)$ and $\phi_2(r)$ would add incoherently and we should get no diffraction pattern. Thus, prima facie we would conclude either that no diffraction pattern will occur or that, if it does indeed occur, we have refuted one of the standard measurement axioms.

I believe this argument is fallacious [see also the remarks of Rauch (1983)]. In the first place, the first conclusion is definitely incorrect: the diffraction pattern certainly is seen experimentally. * Secondly, the conclusion that the quantum measurement axioms are at fault is invalid, because by flipping the spin in the way described we have not in fact made a measurement or even provided the possibility of doing so. The crucial point is that the electromagnetic field in the RF field in a *coherent state*, and therefore its final state, after flipping the neutron spin, is virtually identical to its initial one. A correct application of quantum mechanics to the whole setup (cavity plus neutron) then predicts a diffraction pattern. That the

---

* Because the two interfering states have different energies, the interference is time-dependent, so that stroboscopic detection is required.

cavity has lost some energy is irrelevant, because the initial uncertainty in the energy was much larger than the amount lost, and there is therefore no way of inspecting the system to find out whether or not the loss has occurred. Note, however, that in principle (though perhaps not in practice) it would be possible to modify the experiment so that such inspection is possible. For example, if the radiation in the cavity is not in a coherent state but rather corresponds to a precise number of photons (hence a precise energy), there seems no objection in principle to determining whether or not the energy loss has occurred. In this situation, however, the final state of the cavity is orthogonal to its initial state, and application of the formalism of quantum mechanics to the combination of neutron plus cavity then predicts quite unambiguously that no diffraction pattern should be seen, whatever the spin of the neutrons selected for detection. While this particular experiment has not been done, it is conceptually similar to the experiment with a static guide in which the total neutron flux is measured: there, we could have (but did not choose to) measure the z-component of spin of the incident neutrons directly and thereby infer their path, and sure enough no interference occurred. Thus, this series of experiments, taken as a whole, provides rather spectacular evidence that we must indeed make a choice between observing the path of the neutron and observing the interference between the possible paths.

3. Aspect's experiment

The second class of experiments which is specially significant for the foundations of quantum mechanics is the series of photon polarization–correlation experiments which was initiated by Freedman and Clauser (1972) and culminated in the experiment by Aspect and co-workers (1982b). The experimental setup is shown schematically (for the Aspect experiment) in fig. 3. The source consists of a set of atoms which are pumped into an excited state from which they decay by a two-photon cascade process: the lifetime of the intermediate state is on the order of a nanosecond. According to quantum mechanics, the polarizations of the emitted photons will be correlated. In the experiment we are interested only in that subclass of processes in which the two photons are emitted approximately in the $+z$ and $-z$

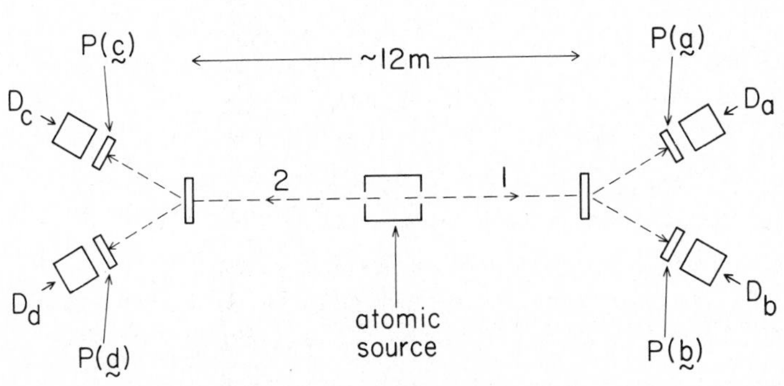

Fig. 3. The photon polarization–correlation experiment.

directions. Photon 1 (emitted in the $+z$ direction) travels a certain distance (a few meters) and enters the commutator C_1 (represented in the figure by a vertical rectangle), which can switch it into one of two distinct beams. If it is switched into the upper beam it encounters a polarizer $P_{\hat{a}}$ "set for polarization direction $\hat{a}$", i.e. so oriented that the photon will pass if its polarization is along the direction $\hat{a}$ (in the plane perpendicular to its wave vector) and will be rejected if the polarization is in the direction orthogonal to $\hat{a}$ in this plane. If it passes, the photon will be detected in the photomultiplier detector $D_{\hat{a}}$: otherwise, the detector will not register. Were we being less self-conscious, we would of course describe the whole operation as a "measurement of the photon polarization along $\hat{a}$", or more specifically a measurement of whether or not the photon is polarized along the $\hat{a}$ direction. In an exactly similar way, if the photon is deflected into the lower beam, it enters an apparatus which will, in a similarly abbreviated description, measure the polarization along $\hat{b}$. Exactly similar arrangements are made for photon 2, which is switched by commutator C_2 so that its polarization is measured along either the $\hat{c}$ or the $\hat{d}$ direction. Note that for a single given photon we can measure the polarization either along $\hat{a}$ or along $\hat{b}$, but not both (similarly for $\hat{c}$ and $\hat{d}$). In the Aspect experiment the commutators C_1 and C_2 are activated in what is (one hopes) a random manner, and the dimensions of the apparatus, and the lifetime of the intermediate state, are such that the events of photon 1 passing polarizer $P_{\hat{a}}$ or $P_{\hat{b}}$ and entering detector $D_{\hat{a}}$ or $D_{\hat{b}}$, and photon 2 doing the same with $P_{\hat{c}}$ or $P_{\hat{d}}$, etc., have spacelike separation (as do the events of passing the commutators). Thus, given the postulates of special relativity theory, there is no opportunity for transmission of information to photon 2 (or the associated apparatus) about which measurement we have made or intend to make on photon 1, nor vice versa.

To discuss this experiment it is convenient to introduce *operationally defined* quantities $A, B, C, D = \pm 1$ as follows: suppose that we "measure the polarization of photon 1 along direction $\hat{a}$", i.e. photon 1 is switched into the upper beam. Then we assign a value $+1$ to A if the photomultiplier $D_{\hat{a}}$ clicks, and -1 if it does not *. If the photon is switched into the lower beam, so that it is the polarization along $\hat{b}$ which is measured, then for the moment A is undefined. Similarly, if the photon indeed is switched into the lower beam, we define B to have the value $+1$ if the photomultiplier $D_{\hat{b}}$ clicks, otherwise -1; if, however, it is the polarization along $\hat{a}$ which is measured, then B is undefined. Similar definitions are given, in the obvious way, for the variables C and D. *For any given pair of photons* one clearly measures, according to the above definitions, one and only one of the four pairs of quantities (A, C), (A, D), (B, C) and (B, D), thus one and only one of the four products AC, AD, BC and BD, which can each obviously take only the values ± 1.

To analyze the significance of this experiment, I follow a line of reasoning similar (but not identical) to that given by Stapp (1985) [see also d'Espagnat (1979)]. [The

---

* In the setup as described, there is clearly no way of knowing in practice, in any individual event (i.e. for any individual pair of emitted photons) that a photon has been rejected by the polarizer. This lacuna could in fact be filled by ancillary apparatus: cf. the earlier experiment of Aspect and coworkers (1982a). However, it is of little importance in the present context, since with perfect detector efficiencies a very slight generalization of the postulates of locality and induction allow the relevant statistics to be inferred from a measurement of detector counting rates *without* polarizer P_a in place.

original and classic paper on this subject is of course the famous paper of Bell (1964).] Let us make three assumptions which at first sight seem completely natural and obvious, namely the assumptions of *locality, counterfactual definiteness* and *induction*. For example, suppose that on a particular pair of photons we measured A and C, i.e. photon 1 was directed into $P_{\hat{a}}$ and photon 2 into $P_{\hat{c}}$. Then we assume that had we decided to measure A and D rather than A and C, that is, to direct photon 2 into $P_{\hat{d}}$ rather than $P_{\hat{c}}$, then we would still have obtained the result $A = +1$ (locality). Moreover, we assume that had we decided to measure B rather than A, and either C or D, then the result we *would have* got for B is definite, though of course unknown, (counterfactual definiteness) and moreover is independent of whether it was C or D which was simultaneously measured (locality). Similar assumptions are made for the quantities C and D. Note that the assumption of counterfactual definiteness is weaker than that of hidden variables or even of local objectivity at the microscopic level (it says nothing about any "properties" of the microscopic entities as such, only about the macroscopic events which we "would have" observed). Finally we assume that the statistical properties of the subset of events in which we actually measure a given pair of the quantities A, B, C, D are representative of the statistical properties of the whole ensemble of events. *

With the assumptions of counterfactual definiteness and locality, we can assign to any definite event (associated with emission of a particular pair of photons) definite values of all four of the quantities A, B, C, D. For example, suppose that for this event we measure A and C. Then the values of A and C are, trivially, the values we actually get, i.e. are definite and known, while B and D are now defined and have the values which we "would have" got, i.e. are definite but unknown. In the words of Einstein et al. (1935) we can regard all four of A, B, C, D as "elements of reality". Since by their definitions A, B, C and D can each take only the values ± 1, it is obvious (if necessary by exhaustion of the 16 possibilities!) that we have for each individual event the relation

$$AC + AD + BC - BD \leqslant 2, \tag{3}$$

and hence that for the ensemble of events as a whole we must have

$$\langle AC \rangle_{\text{ens}} + \langle AD \rangle_{\text{ens}} + \langle BC \rangle_{\text{ens}} - \langle BD \rangle_{\text{ens}} \leqslant 2, \tag{4}$$

which is the generalization by Clauser et al. (1969) of the celebrated inequality originally proved by Bell (1964). Finally, by the assumption of induction the expectation value of e.g. $\langle AC \rangle$ for the ensemble may be replaced by the average of the values obtained in that subset of events in which A and C were actually measured, i.e. by the "experimental" value. Thus we have a clear prediction, which follows from our three general assumptions above, about the results of the experimental measurements. The punch-line is, of course, that the quantum-mechanical predictions for the quantities $\langle AC \rangle$ etc., *violate* the inequality, and that the

---

* The rather clumsy language of this formulation is used to emphasize the point that the argument does *not* require the ascription of definite properties to the photons themselves.

experiments find agreement with quantum mechanics. Because of various technical problems associated with imperfectly efficient detectors etc. [see the experimental papers, and the review paper by Clauser and Shimony (1978)], a few subsidiary assumptions are necessary before we can claim that the experiments conclusively refute all theories having the three properties assumed above. Most workers in the area [though not all—cf. e.g. Marshall (1983)] would regard these subsidiary assumptions as so plausible that the loopholes left by them are of little interest, and if one takes this point of view one can say that any theory which makes simultaneously the assumptions of locality, counterfactual definiteness and induction is conclusively refuted by the experiments.

The class of theories so excluded certainly contains, as a subclass, not only all theories in which the microscopic entities (photons) possess properties which are independent of the experimental arrangement, but also all theories which ascribe to them properties which depend only on the *local* experimental arrangement: if we wish to talk about "properties" of the individual photons at all, then the experiment shows that these properties must be a function of the *global* experimental arrangement, including events (such as the switching of the "distant" photon into one beam or the other) which, within the assumptions of special relativity, could not have physically influenced the microsystem in question. It is therefore certainly extremely difficult, if not impossible, to interpret the outcome in terms of an *interaction* between the system and the apparatus—at least within the framework of very basic "common-sense" assumptions which most of us would be very loth to give up. Thus, the polarization-correlation experiments, and in particular the most recent experiment of Aspect and co-workers (1982b), may be regarded as a spectacular confirmation of the correctness of Bohr's point of view at the microscopic level. Whether or not these experiments tell us anything about the nature of reality at a *macroscopic* level is a question to which I shall return below.

4. Describing the measuring apparatus in plain language

Now let us turn to the main topic of this talk and explore the consequences of following Niels Bohr's views in a different direction. Bohr repeatedly stresses in his writings the distinction between the quantum system under investigation and the measuring apparatus, which must be described classically. Often, but not always (cf. below), he seems to imply that the distinction between "quantum" and "classical" coincides with the distinction between "microscopic" (or "atomic") and "macroscopic".

Bohr repeatedly stresses that

> "the description of the experimental arrangement and the recording of observations must be given in plain language, suitably refined by the usual physical terminology" [this particular quotation is from his essay on "Quantum Physics and Philosophy" (Bohr 1958)].

To the question "Why must it?" his answer is:

> "This is a simple logical demand, since by the word 'experiment' we can only mean a procedure regarding which we are able to communicate to others what we have done and what we have learned" [ibid.].

Now, when someone says that something "must" be done, he clearly believes that it *can* be done. The question now arises: Is it possible, within the framework of quantum mechanics, to assign a classical description to the measuring apparatus? This question, which has of course been a subject of endless debate for the last fifty years, poses itself for the following reason. If we consider a typical apparatus—say for example a Geiger counter—it is nothing but a complicated assembly of atoms and molecules put together in a certain way and therefore, one would think, it must obviously be *possible*, even if it is not necessary, to describe it within the framework of quantum mechanics. In particular, its macroscopic state (e.g. of being triggered or not) must in principle be describable in quantum-mechanical language. In practice, a realistic description would almost certainly require the use of a density matrix, but as Wigner (1963) showed conclusively, this fact in no way affects the point of the paradox I am in process of developing, so for simplicity of notation *only** let us assign to it a pure state. Thus, for example, let us consider the Young's slits experiment as above, with, for simplicity, a single telescope and counter set up opposite slit 2. Suppose the probability amplitude (wave function) of the (microscopic) particle is written in the general form $\psi = a\psi_1 + b\psi_2$, where ψ_1 (ψ_2) represents a wave propagating through slit 1 (slit 2) only. Consider first the case $a = 1$, $b = 0$, so that $\psi = \psi_1$: then the particle certainly propagates through slit 1 and misses the counter. The final wave function of the latter (cf. above) is then some wave function Ψ_1 which corresponds to the state of not being triggered. Conversely, if $b = 1$ and $a = 0$ ($\psi = \psi_2$) then the particle certainly propagates through slit 2 and enters the counter, which is thereby triggered. The final state of the counter is therefore described by some wave function Ψ_2, which is not only orthogonal to Ψ_1, but corresponds to macroscopically different properties. Thus, symbolically, we can write:

$$\psi_1 \rightarrow \Psi_1, \qquad \psi_2 \rightarrow \Psi_2.$$

So far, there is no particular difficulty: the quantum fluctuations of physical quantities in a macroscopic system in either of the states Ψ_1, Ψ_2 are extremely small relative to their thermodynamic average values in these states, so there is no inconsistency in describing each of the states in ordinary classical thermodynamic terms (cf. also footnote).

The problem arises when the initial state of the microsystem is a linear combination of the states ψ_1 and ψ_2 (as we know it must be in the Young's slits experiment if we are ever to produce a diffraction pattern): $\psi = a\psi_1 + b\psi_2$. In that case, by the linearity of the quantum formalism, it follows directly from the above equations that we have

$$a\psi_1 + b\psi_2 \rightarrow a\Psi_1 + b\Psi_2,$$

that is, the correct quantum-mechanical description of the final state of the counter

* This point needs some emphasis, since those physicists who deny the existence of a "quantum measurement paradox" frequently seem to labor under the illusion that those who assert it are unaware of the necessity of a density-matrix description (cf. below).

is a superposition of *macroscopically distinct* states. Such a state does not appear, prima facie, to correspond to *any* classical description (the "Schrödinger's Cat" paradox). It is all very well to say "we *must* describe the apparatus by classical mechanics, because otherwise we cannot communicate with one another", but in the light of the above considerations that begins to sound rather as if I should say: "My old car *must* be able to do 45 mph on the interstate freeway". If, after inspecting it sceptically, you say: "I doubt if it can; why are you so sure it must be able to?" it is hardly an adequate answer for me to say: "because that is what the law requires"!

It seems to me that Niels Bohr never gave a quite explicit response on this point. However, in his controversy with Einstein he did comment on a rather different point, namely that in the two-slit experiment it would in principle be possible to let the shutter be suspended freely and, by measuring its recoil, determine which slit the photon passed through. His comment is:

> "It is not relevant that experiments involving an accurate control of the momentum or energy transfer from atomic particles to heavy bodies like diaphragms and shutters would be very difficult to perform, if practicable at all. It is only decisive that, in contrast to the proper measuring instruments, these bodies together with the particles would in such a case constitute the system to which the quantum-mechanical formalism has to be applied."

From this it is possible to infer his probable response to the problem posed above. As the historian of quantum mechanics, Max Jammer, puts it in his book "Philosophy of Quantum Mechanics" (1974):

> "This statement, and similar passages, suggest that Bohr, recognizing the insufficiency of the phenomenalistic position, regarded the measuring instrument as being describable both classically and quantum mechanically. By concluding that the macrophysical object has objective existence and intrinsic properties in one set of circumstances (e.g. when used for the purpose of measuring) and has properties relative to the observer in another set of circumstances, or, in other words, by extending complementarity on a new level to macrophysics, Bohr avoided committing himself either to idealism or to realism. Summarizing, we may say that for Bohr the very issue between realism and positivism (or between realism and idealism) was a matter subject to complementarity."

Were he alive today, and had he read the voluminous literature of the last two decades on the quantum measurement paradox, I suspect Bohr might have made his argument explicit as follows: No doubt there are or could be circumstances (e.g. if the voltage across the counter were such that it could be ionized but not triggered) in which it is not excluded a priori that we could see interference between macroscopically different states of the counter. However, if we wish the counter to work *as a measuring apparatus*, then we must introduce a considerable degree of irreversibility, simply to stabilize the result of the measurement. Then we can argue (step 1) that there is now no possibility of exhibiting any interference between the macroscopically distinct states, and (step 2) that there is "hence" no inconsistency in assigning a definite macrostate to the counter, that is, in giving a classical description. Now I want to stress that in that (small) subset of the physics community which worries about the quantum measurement paradox, step 1 is not usually regarded as controversial or in need of further emphasis: Those who feel that there is a problem do *not* feel it because they believe that in practice, or perhaps even in principle, one can demonstrate interference between the different final states of the counter (and by the same token will not be reassured by further demonstrations that this is impossible). What worries them is step 2. It worries them because they feel that in the final analysis, physics cannot forever refuse to give an account of *how* it is that we obtain definite results whenever we do a particular measurement. This problem should not be confused with the question of why we get the particular result we do: the present issue is how it is possible for us to get a definite result at all. For those who believe there is a problem, this is perhaps *the* basic question in modern physics, and it has of course been subjected to endless debate in the literature: see, for example, part 4 of d'Espagnat's book "Conceptual Foundations of Quantum Mechanics" (1976).

However that may be, it is in any case interesting to raise the question: Is it possible, in practice, to prepare a macroscopic system (not necessarily something which is itself suitable to be a measuring apparatus) which can be in one of two or more *macroscopically distinct* states, and then present ourselves with the choice of either measuring which of the two states it is in, or of observing the interference between the two possibilities? Further, could we then demonstrate that in the second case it *could not have been* in a definite macrostate? If the answer to these questions is yes, and if the experiment is done and comes out in favor of the quantum-mechanical predictions, then this would be a spectacular confirmation of Bohr's viewpoint as interpreted by Jammer, i.e. of the extension of the idea of complementarity to the macroscopic scale. The whole reason that I am here giving this talk is that I believe that the answer to the above two questions is probably yes.

Before embarking on the discussion of this, let me digress for a moment to correct what seems to be a quite widespread misconception: the Aspect experiment, fundamental as it is for the interpretation of the quantum formalism at a microscopic level, sheds no direct light on the above questions. The main reason is that there is no question, in this experiment, of observing any interference between macroscopically distinct states of the apparatus. Perhaps the easiest way to see this is to note that all the experimental results are completely compatible with a theory in which each of the macroscopic counters is always in a definite macrostate, but that state produces a nonlocal effect on the distant photon. Needless to say any such theory would be incompatible with the universality of the quantum formalism and would involve some obvious difficulties, e.g. connected with the Lorentz-noninvariance of the time order of spacelike-separated events, but it is not clear that they would be any worse in principle than in the usual interpretation. In any case my point here is not the plausibility or otherwise of such theories, but merely that they are obviously not excluded by the experimental results, which therefore cannot be used as evidence for the superposition of different macrostates. To put it another way, the Aspect experiment forces us to sacrifice *either* counterfactual definiteness or locality (or of course induction), not necessarily both. The standard extrapolation of quantum mechanics to the macrolevel in effect sacrifices counterfactual definiteness, a choice which is exactly in the spirit of Bohr's thinking: My point is that at the macrolevel, even if not necessarily at the microlevel, the sacrifice of locality is not only not logically excluded but may arguably be no more unattractive.

5. *Interference between macrostates*

Returning now to my main theme, let me imagine that I had been able to put the above idea, of observing interference between macrostates and checking the complementarity principle at the macroscopic level, to Niels Bohr. I suspect he would have commented that the experiment, while certainly interesting in principle, would be in practice impossible. This, indeed, is the dogma which has developed in five decades of the literature on the quantum measurement problem. Probably, Bohr would have given two reasons. First, as he often emphasized, quantum effects are important only when the action S is on the order of the quantum of action $\hbar$. If E is the energy of the system in question and $\tau_{cl} \sim \omega_{cl}^{-1}$ is the order of magnitude of its classical period of motion, then in a typical experiment we have $S \gtrsim E\tau_{cl}$, and so the condition $S \lesssim \hbar$ implies $E \lesssim \hbar\omega_{cl}$; since for a macroscopic system the relevant classical frequencies are certainly not greater than, say, 10^{16} s^{-1} (and are often much less), this in turn implies that the characteristic energy scale for the macroscopic motion can be no greater than a few eV, i.e. on the order of the ionization energy of a single atom. It seems at first sight totally out of the question that any macroscopic variable should be associated with so tiny an energy.

What has changed the situation dramatically here is an effect which Bohr did not quite live to see, namely the effect predicted theoretically by Josephson (1962) and bearing his name. The really exciting thing about the Josephson effect is that, through it, the motion of a *macroscopic* variable (such as the current or trapped flux in a bulk superconducting ring, see below) can be controlled by a *microscopic* energy, on the order of the thermal energy of an atom at room temperature or even smaller. Although the Josephson effect itself is *not* an example of superposition of macrostates, it lays the basis for an attempt to build such a superposition (other systems, such as the charge density waves in a one-dimensional metal, or the photon field in a ring laser, may also be candidates, but for technical reasons I have myself tended to concentrate on Josephson devices and will therefore take my concrete examples from this area).

The second objection to the proposal to observe interference at the macrolevel, which has become a standard theme of the literature on the quantum theory of measurement, is that macroscopic systems, by their nature, always interact so strongly and so dissipatively with their environments that the quantum phase coherence between the different macrostates is inevitably washed out. The only way to combat this objection is to try to formulate as realistic a model as possible of (for instance) a specific type of superconducting device, incorporating from the beginning its interactions with an open environment, and to actually carry out a fully quantum-mechanical calculation of its behavior. [For example, in the case of a superconducting device the open environment includes the normal (unpaired) electrons, the radiation field, the phonons, the nuclear spins and so on.] There has been a considerable amount of work along these lines over the last five or six years [for a partial review, see e.g. Leggett (1986)], and the provisional conclusion is that while the conditions to see the effects of phase coherence at the macrolevel are indeed extremely stringent, it is not obvious that with modern state-of-the-art levels of cryogenics, microfabrication and noise control they could not be met in a purpose-built system. There are two principal reasons why the standard arguments to be found in the literature on quantum measurement theory fail: One is that even for a system interacting very strongly with its environment, much of the effect of the interaction may be adiabatic in nature and merely renormalize the system parameters without destroying the phase coherence (a point which seems to have been missed in much of the literature). The second is that the residual "dissipative" part of the interaction, which certainly *is* fatal to phase coherence, can be extremely small in a system such as a superconducting device because of the extremely low entropy (e.g. in a 1 cm niobium ring at 10 mK, if we assume that the standard Gibbs formulae apply, the electronic entropy of the whole macroscopic body is actually less than Boltzmann's constant!).

6. Macroscopic quantum tunneling and coherence

The system which I believe to be conceptually the simplest candidate for these experiments is an RF SQUID ring, that is, a bulk superconducting ring interrupted by a single Josephson junction (see fig. 4). An external magnetic flux can be imposed through the ring: we treat this provisionally as a c-number. It will induce screening currents which in turn produce their own magnetic fields, and hence the *total* flux Φ trapped through the ring, or equivalently the circulating current, is a dynamical variable. If we believe that quantum mechanics does indeed apply at the macrolevel, then the variable Φ must be treated as a quantum-mechanical operator; in fact, were we to treat this degree of freedom as totally decoupled from the rest, we could write down a wave function $\psi(\Phi)$ for it which should satisfy Schrödinger's equation, with the junction capacitance playing the role of mass. (However, in reality it is essential, as stressed above, to give a description which, while fully quantum-mechanical, adequately takes into account the effect of the environment.) By varying the parameters of the system, in particular the external flux, we can produce different forms of "potential energy" associated with this variable, and look for various specifically quantum-mechanical effects. Two kinds of behavior are of particular interest (see fig. 4): the first, which has become known in the literature as "macroscopic quantum tunnelling" (MQT) * is the macroscopic analog of the decay of a heavy nucleus by emission of an α-particle: the system starts in a metastable potential well and tunnels out through the potential barrier into what is effectively a continuum. What distinguishes this phenomenon from α-particle decay and other "microscopic" tunneling effects is that the classically accessible regions which are separated by the barrier are, by most reasonable criteria, macroscopically distinct. The second phenomenon of interest, which is generally known as "macroscopic quantum coherence" (MQC) is the macroscopic analog of the inversion resonance of the ammonia molecule; the system tunnels coherently backward and forward between two degenerate potential wells. Again, the classically accessible regions between which tunnelling takes place are macroscopically distinct.

---

* In restrospect "macroscopic quantum decay" might have been a better name.

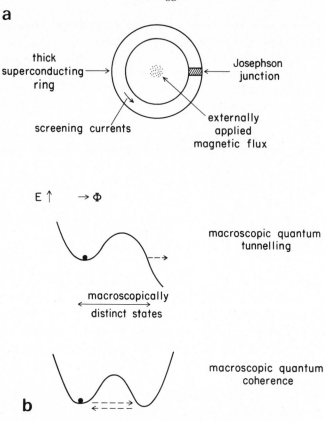

Fig. 4. (a) SQUID ring; (b) macroscopic quantum tunnelling and coherence.

There are now at least six experiments in the literature on MQT, the most recent and sophisticated being those of the IBM, SUNY Stony Brook and Berkeley groups (Washburn et al. 1985, Schwartz et al. 1985, Devoret et al. 1985). * These experiments confirm that tunneling of a macrovariable does occur, and that its dependence on the parameters, including these which describe its dissipative interactions with its environment, agree with the theoretical predictions at least qualitatively. In addition, the Berkeley group have recently carried out a very pretty experiment (Martinis et al. 1985) in which the tunneling out of discrete excited states can be observed by exciting the system into these states by microwave radiation. Taken together, these experiments offer strong circumstantial evidence that, provided the interaction with the environment is adequately taken into account, the behavior of a macroscopic variable is compatible with the predictions of quantum mechanics extrapolated to the macroscopic scale. However, these experiments do not directly test the principle of complementarity at this level: for this we need an experiment of

---

* Most of these experiments were actually performed not on the SQUID ring described above but on a closely related system, a single junction biased by a fixed external current. The theory is however very similar.

the MQC or a similar type. Although the MQC experiment has in fact been attempted by Bol and co-workers (Bol et al. 1983) they were unable to reach the parameter regime where spectacular (i.e. oscillatory) results are predicted by the theory, and indeed none were seen. * Let us assume that in the future it will be possible to reach this regime, and examine the significance of the experiment.

The system is a reasonably (geometrically) macroscopic RF SQUID ring, in an external magnetic flux which is adjusted so that the ring has available to it two degenerate states, corresponding to a current I_0 circulating in either the clockwise or the anti-clockwise sense. The magnitude of the current is typically on the order of a few microamperes, so that by any reasonable criterion the two states are macroscopically distinct. To discuss the significance of the experiment, we introduce a variable $P(t)$ which is defined to be $+1$ if a measurement is made at time t and a current of magnitude I_0 observed with a clockwise sense, and -1 if the measurement is made and an anti-clockwise current of magnitude I_0 observed. If no measurement is made at time t, then the variable $P(t)$ is for the moment undefined.

As applied to this system, quantum mechanics makes two important predictions: (a) If the current is measured, then, except in exponentially rare cases (corresponding to finding the system "under the barrier"), the value is always found ** to be $\pm I_0$, i.e. any measurement will yield the result $P(t) = \pm 1$. (b) For an ideal (noninteracting, nondissipative) system the experimentally observed *correlation* between the values of $P(t)$ at different times (with no measurement made in the meantime) is predicted to be given by the simple formula

$$\langle P(t_i)\, P(t_j) \rangle = \cos\big[\Delta(t_j - t_i)\big], \tag{5}$$

where Δ is the characteristic resonance frequency of the system ($\hbar\Delta$ is the energy splitting between the even- and odd-parity eigenstates). Note carefully that the result (5) is a direct consequence of the existence, at times *intermediate* between t_i and t_j, of a superposition of eigenstates of P. For example, if $P(0)$ is $+1$ and we measure P after half a cycle, the prediction of eq. (5) is that P is then found with certainty to be -1. This results *only* if we take the state of the system at, say, a quarter cycle to be a *linear superposition* of $P = \pm 1$, even though we know that an observation at this time would certainly have given one of these two values or the other: had we taken the state to be a classical mixture of those eigenstates, then at half-cycle the expectation value of P would have been 0, not -1.

What this argument shows is that one cannot modify quantum mechanics piecemeal and expect to preserve the results. To that extent it is parallel to the argument given by Furry (1936) about nonlocal correlations shortly after the original paper by Einstein et al. (1935); he showed that if one replaced the pure

* Various experiments on an RF SQUID ring coupled to a tank circuit have also been interpreted as evidence for MQC-type effects (Prance et al. 1983, and earlier references cited therein; cf. also Dmitrenko et al., 1984). These experiments are so indirect, and involve so many unknowns, that I believe it would be rash to draw any firm conclusions from them in the present context.
** Strictly speaking we need to generalize the definition of $P(t)$ slightly to allow for the zero-point quantum fluctuations in each well, i.e. for small deviations of the current magnitude from I_0 (cf. Leggett and Garg 1985).

state of two separated particles by a mixture, but otherwise preserved the axioms of quantum mechanics, one would get results which violate the original quantum predictions and (as we know) also contradict experiment. Clearly, the significance of the MQC experiment would be enhanced if one could generate the analog, in this context, of the much stronger result embodied in the famous theorem of Bell (1964), i.e. prove that any experiment which agreed with quantum mechanics must ipso facto be inconsistent with a few "common-sense" assumptions. I shall now attempt to do just that, referring to a published paper by myself and Garg (Leggett and Garg 1985) for the details of the argument.

7. Macrorealism

Let me define a general class of theories about the world, which I shall call "macrorealistic", by the following postulates:

(1) Macrorealism: a macroscopic body which has available to it two or more *macroscopically distinct* states must at all times * "be" in a definite one of these states, whether or not it is observed.

(2) Noninvasive measurability at the macrolevel: by a sufficiently careful measurement (in particular, by an "ideal negative result" measurement in which we throw away the positive results) we can determine which of the above macrostates the system is in without affecting its subsequent dynamics (at least as regards these states).

(3) Induction: results obtained on the subset of an ensemble on which a given property is actually measured are representative of the properties of the ensemble as a whole.

The postulate (1) allows us to infer that $P(t)$ exists for all times t, whether or not measured, and (nearly) always takes one of the two values ± 1. A trivial adaptation of the inequality (4) then allows us to conclude that, for an ensemble of runs *in which the system is undisturbed* (e.g. by measurement) we have the inequality [where $P_i \equiv P(t_i)$]

$$\langle P_1 P_2 \rangle_{\text{ens}} + \langle P_2 P_3 \rangle_{\text{ens}} + \langle P_3 P_4 \rangle_{\text{ens}} - \langle P_1 P_4 \rangle_{\text{ens}} \leqslant 2, \tag{6}$$

and postulates (2) and (3) then allow us to apply this prediction also to the experimentally measured correlations. Finally, by taking (for instance) $t_4 - t_3 = t_3 - t_2 = t_2 - t_1 = \pi/4\Delta$, we demonstrate that the quantum-mechanical prediction (5), violates the above inequality.

The quantum-mechanical prediction (5) holds of course only for the unrealistic case of a system totally decoupled from its environment, and one might wonder whether a realistic degree of dissipative coupling would not modify the correlations enough that they satisfy the "macrorealistic" inequality (6). This question has motivated some fairly detailed calculations (Chakravarty and Leggett 1984, Leggett et al. 1986) and the upshot is that for a degree of dissipation, which may not be

---

* More accurately: at "nearly all" times; see the cited reference.

unrealistically low from an experimental point of view, the quantum-mechanical predictions continue to violate the inequality. Thus the experiment can in principle *discriminate unambiguously between macrorealism and complementarity at the macro-level.*

Several experimental groups are now actively studying the feasibility of this experiment, and the probability is that it will at least be attempted within the next two or three years. If it works, and confirms quantum mechanics, it will be a spectacular confirmation of the validity of extending Niels Bohr's thinking to the macroscopic level.

Now, the reaction of 99.9% of the physics community to this experiment may well be a bored shrug. We all know quantum mechanics is right, it will be said: so if the experiment comes out in favor of quantum mechanics, it only tells us what we already know, whereas, if it comes out against quantum mechanics, it is obviously a bad experiment. But is it? In the last five minutes of my talk I am going to try to throw an intellectual hand-grenade into the discussion. I start with an awful confession: If you were to watch me by day, you would see me sitting at my desk solving Schrödinger's equation and calculating Green's functions and cross-sections exactly like my colleagues. But occasionally at night, when the full moon is bright, I do what in the physics community is the intellectual equivalent of turning into a werewolf: I question whether quantum mechanics is the complete and ultimate truth about the physical universe. In particular, I question whether the superposition principle really can be extrapolated to the macroscopic level in the way required to generate the quantum measurement paradox. Worse, I am inclined to believe that at *some* point between the atom and the human brain it not only may but *must* break down. I am inclined to believe this for a simple but overwhelming reason: try as I may (and I have tried for many years) I simply cannot convince myself that any of the solutions proffered to the quantum measurement paradox is philosophically satisfactory, and to pretend otherwise, even in this place and on this occasion, would be intellectually dishonest.

But, you will say impatiently, is it not *obvious* that if atoms and molecules obey the superposition principle, and if Geiger counters, cats and even ultimately our brains are composed of atoms and molecules, then these macroscopic objects must themselves obey the principle? Indeed it is obvious—just as it was obvious, until 1957, that the laws of nature had to be the same in a right-handed system of coordinates as in a left-handed one. Just as in that case, the proper question to ask is: *Where is the evidence*? After all, whatever our theoretical prejudices, physics is supposed to be an experimental subject! Well, until quite recently there simply was no evidence on this point [despite a widespread misconception to the contrary, the so-called macroscopic quantum phenomena such as the Josephson effect itself or circulation quantization in liquid helium are quite irrelevant in this context—see e.g. Leggett (1980)]. As we have seen, recent experiments on MQT and on energy level quantization do provide strong circumstantial evidence that, as regards these phenomena, at least, quantum mechanics is still working at the level of superconducting devices. However, they do not directly test the principle of superposition of macroscopically distinct states, so that it is not a foregone conclusion that the MQC experiment, which is different in substantial ways, will come out in its favor.

8. The ultimate lesson of quantum mechanics

How might the principle break down? One logical possibility is that there is some effect (not of course accounted for in Schrödinger's equation) which is a strong function of the *number* of particles which behave differently in the two "branches" of the superposition. * Then, for experiments at the atomic level, where this number is 1 or 2, it could be totally negligible, whereas for macroscopic superpositions of the "Schrödinger's Cat"-type, where the number is $\sim 10^{23}$, it could totally dominate the behavior. This proposal involves various technical difficulties and I suspect it is in any case much too conservative. In fact, on nights when the full moon is *very* bright, I suspect that what may be needed is nothing less than a radical revision of the reductionist prejudice which has served us so well for 200 years, that is, the prejudice that *all* the properties of macroscopic systems can be explained, in principle, in terms of their constituent atoms and molecules. In fact, could it be that Bohr was right in his insistence that the macroscopic instruments "define the very conditions" under which the atomic phenomena appear, but wrong to conclude that the way in which this happens is itself unanalyzable and beyond the laws of physics? Could the ultimate "lesson of quantum mechanics" be that we eventually need to go *beyond* quantum mechanics? It would not be the first time that such a shift of viewpoint has happened in physics: for pretty well 100 years the orthodoxy was instantaneous action-at-a-distance—an axiomatic and unanalyzable concept—and the minority who felt the concept to be metaphysically objectionable were no doubt told to get back to their calculations and forget about such "philosophical" questions. In that case, as we all know, the doubters were eventually proved right and their doubts led to qualitatively new physics. Should the corresponding outcome some day occur with quantum mechanics, it would of course in no way contradict Bohr's viewpoint at the atomic level, nor would it negate his repeated insistence that the quantum theory has blocked off forever a return to our old classical common-sense picture of the world. Rather, any such theory of the future would almost certainly be a logical extension of his line of thinking, taking us even further away from classical notions in a way which at present we can hardly imagine.

Obviously, these are wild and vague speculations: but the experimental and theoretical program I have described in the main body of this talk is perhaps at least a small step towards exploring them. Of course, if you ask me to bet on the possibility that a well-conducted MQC experiment will come out against quantum mechanics, then when sober (and particularly after contemplating the results of the tunnelling and other experiments) I would probably not take odds of less than 100 to 1; I suspect that the solution of the measurement paradox, when it comes, will

---

* In view of some of the comments made in the discussion of this lecture, it is worth emphasizing that this is a quite different variable from anything connected with the geometrical *scale* of the phenomenon. Thus arguments of the type "We know (do we in fact?) that quantum mechanics works from the Planck length up to the atomic scale (25 orders of magnitude): what is so special about the extra ten orders of magnitude needed to get up to the human scale, that it should suddenly break down?" seem to me quite irrelevant. To the best of my knowledge neither cosmology nor particle physics offers any evidence against (or for!) the hypothesis discussed here.

come at a much deeper and more subtle level. But, after all, a journey of 10 000 miles starts with a single step. Were he alive today, Niels Bohr would surely have offered us odds of much more than 100 to 1 in favor of quantum mechanics continuing to hold at this and indeed at any level: but, as a scientist whose own life's work was based on the overthrow of what were then common-sense ideas, I believe he would at least have encouraged us to ask the questions and to do the experiments, and it is in this spirit that I dedicate this talk to his memory.

References

Aspect, A., P. Grangier and G. Roger, 1982a, Phys. Rev. Lett. **49**, 91.

Aspect, A., J. Dalibard and G. Roger, 1982b, Phys. Rev. Lett. **49**, 1804.

Bell, J.S., 1964, Physics **1**, 195.

Bohr, N., 1958a, Quantum physics and philosophy, in: Essays 1958–62 on Atomic Physics and Human Knowledge (Interscience, New York) p. 1.

Bohr, N., 1958b, Discussion with Einstein, ibid. p. 50.

Bol, D., R. van Weelderen and R. de Bruyn Oubober, 1983, Physica **122B**, 1.

Chakravarty, S., and A.J. Leggett, 1984, Phys. Rev. Lett. **52**, 5.

Clauser, J.F., and A. Shimony, 1978, Rep. Prog. Phys. **41**, 1881.

Clauser, J.F., M.A. Horne, A. Shimony and R.A. Holt, 1969, Phys. Rev. Lett. **23**, 880.

d'Espagnat, B., 1976, Conceptual Foundations of Quantum Mechanics, 2nd Ed. (Benjamin, Reading, MA) Part 4.

d'Espagnat, B., 1979, Sci. Am. November 1979, 158.

Devoret, M.H., J.M. Martinis and J. Clarke, 1985, Phys. Rev. Lett. **55**, 1908.

Dewdney, C., A. Garuccio, A. Kyprianidis and J.P. Vigier, 1984, Phys. Lett. **104A**, 325.

Dmitrenko, I.M., G.M. Tsoi and V.I. Shnyrkov, 1984, Fiz. Nizk. Temp. **10**, 211 [Sov. J. Low Temp. Phys. **10**, 111].

Einstein, A., B. Podolsky and N. Rosen, 1935, Phys. Rev. **47**, 777.

Feyerabend, P.K., 1962, in: Frontiers of Science and Philosophy, ed. R.G. Colodny (Univ. of Pittsburgh Press, Pittsburgh) p. 219.

Freedman, S.J., and J.F. Clauser, 1972, Phys. Rev. Lett. **28**, 938.

Furry, W.H., 1936, Phys. Rev. **49**, 393.

Greenberger, D.M., 1983, Rev. Mod. Phys. **55**, 875.

Heisenberg, W., 1930, The Physical Principles of the Quantum Theory (University of Chicago Press, Chicago).

Jammer, M., 1974, The Philosophy of Quantum Mechanics (Wiley–Interscience, New York) p. 207.

Josephson, B.D., 1962, Phys. Lett. **1**, 251.

Leggett, A.J., 1980, Prog. Theor. Phys. Suppl. **69**, 80.

Leggett, A.J., 1986, in: Directions in Condensed Matter Physics, ed. G. Grinstein and G. Mazenko (World Scientific, Singapore).

Leggett, A.J., and Anupam Garg, 1985, Phys. Rev. Lett. **54**, 587.

Leggett, A.J., S. Chakravarty, A.T. Dorsey, M.P.A. Fisher, Anupam Garg and W. Zwerger, 1986, submitted to Rev. Mod. Phys.

Marshall, T.W., 1983, Phys. Lett. **98A**, 5.

Martinis, J.M., M.H. Devoret and J. Clarke, 1985, Phys. Rev. Lett. **55**, 1543.

Prance, R.J., J.E. Mutton, H. Prance, T.D. Clark, A. Widom and G. Megaloudis, 1983, Helv. Phys. Acta **56**, 789.

Rauch, H., 1983, in: Proc. Int. Symp. on the Foundations of Quantum Mechanics in the Light of New Technology, eds S. Kamefuchi H. Ezawa, Y. Murayama, M. Namiki, S. Nomura, Y. Ohnuki and T. Yajima (Japanese Physical Society, Tokyo).

Reichenbach, H., 1944, The Philosophic Foundations of Quantum Mechanics (University of California Press, Berkeley, Los Angeles).

Schwartz, D.B., B. Sen, C.N. Archie and J.E. Lukens, 1985, Phys. Rev. Lett. **55**, 1547.

A.J. Leggett

Stapp, H., 1985, Am. J. Phys. **53**, 306.
Summhammer, J., G. Badurek, H. Rauch, U. Kischko and A. Zeilinger, 1983, Phys. Rev. **A27**, 2523.
Washburn, S., R.A. Webb, R.F. Voss and S.M. Faris, 1985, Phys. Rev. Lett. **54**, 2712.
Wheeler, J.A., 1978, in: Mathematical Foundations of Quantum Theory, ed. R. Marlow (Academic Press, New York) p. 9.
Wigner, E.P., 1963, Am. J. Phys. **31**. 6.

Discussion, session chairman W. Kohn

Peierls: The experiments discussed are interesting and I hope will be done, but I would bet heavily on their confirming quantum mechanics. This is because I do not see that on the way from the atomic to the macroscopic there can be any point which makes a qualitative distinction. I do sympathize with Leggett about the difficulty of understanding how the observer can be described in terms of quantum mechanics. I believe he cannot be so described—not because he is macroscopic, but because he is alive. I believe that it is not certain that biology is a branch of physics in the sense in which chemistry is a branch of physics. This is a revolutionary thought, too difficult to be spelt out in the discussion.

Leggett: If you asked me today I would certainly bet myself that the solution would not be at the level of "inert" physical devices such as SQUIDs, but would more likely be found at the level of complexity and organization which must be necessary for biology, let alone psychology. But one has to start somewhere!

Thirring: I would like to strengthen what Rudolf Peierls has said. If one accepts the present view then the fundamental length scale where the real action takes place is the Planck length or something a little bigger, say 10^{-28} cm to be generous. In these units even an atom is an immense structure, 10^{20} times bigger, and we are another 10 powers of ten bigger. So why should the boundary between microscopic and macroscopic be just between 10^{20} and 10^{30} times the real microscopic length and the laws of nature change drastically in this region?

Leggett: I feel that this is a difficult theme. I think it is a difficulty which makes it very obvious how deeply ingrained are our reductionist prejudices, even when we try consciously to override them. I think it is not obvious that the behaviour of large collections of atoms must obviously be determined by the behavior of their individual constituents. This indeed goes against the scientific thinking of the last 200 years. But I personally think that there is a good chance that we may have to revise our ideas in this direction.

Bleuler: Underlining Walter Thirring's remark I would like to emphasize that a neutron (usually described by Schrödinger's equation) is, in fact, a most complicated bound system (valence quarks, seaquarks, gluons) having certain similarities with macroscopic systems. The same holds, in principle, for electrons (surrounded by the photons) and also for the quarks (surrounded by the gluons).

Leggett: I too would like to believe that fundamental laws of physics don't show dependence on the scale of things. But this is in the end a matter for experiment.

There is in any case a fundamental difference between the neutron and the macroscopic system in that the latter corresponds to states of large baryon number B rather than B about equal to one.

Jens Bang: I would like to recall Bohr's stressing the *concepts* and the *language*, and I would like to do it in connection with Schrödinger's cat. As we remember, it was in a superposition of being alive or dead. One must ask: Is there a meaning to this? Are concepts like life and death really describable as quantum states? Maybe they are hardly describable as classical physical states!

Leggett: Of course a proper description of the state of the cat in Schrödinger's thought-experiment, if it can be given at all, must be given in the language of density matrices rather than wave-functions. However, as I tried to emphasize, Wigner has shown, in my opinion conclusively, that doing this in no way blunts the force of the paradox. If it is the question of whether concepts like "life" and "death" can ever be adequately described, even in the language of density matrices, this doesn't seem to me a real problem since we can always replace the cat by a macroscopic but inert object such as a counter.

Weisskopf: I would personally put a bet even higher than $100:1$, perhaps $10\,000:1$, that quantum mechanics is the only relevant theory in the macroscopic world. But still I have my doubts about those remarks that tell us that the atom is actually much nearer to man than the quark world or whatever happens at the Planck length.

When we go from atoms and molecules to us human beings, complexity, organization and disorder enters, concepts that are rarely found at the lower level. We do not yet understand too well complexity and organization and disorder, and therefore it is possible that some new unexpected fundamental principle may enter. The probability is not zero because of our limited and inappropriate knowledge of what happens in complex organizations, in particular in the brain.

Anderson: In connection with complexity we should really also discuss the problem of the purely quantum-mechanical computer. This was a deliberate omission in my talk because I am not knowledgeable enough and am also sceptical. That is really at the heart of this basically philosophical problem: Can one make a quantum computer that has essentially no dissipation? I think I can imagine a scenario that solves all of the measurement paradox difficulties, if I can say that dissipationless switching processes are taking place throughout the whole complex system. If one could actually make a device that could work as a computer, i.e. totally reversible and totally quantum-mechanical—that would disturb me very much and I am sceptical about it.

Leggett: I would agree with that statement, but perhaps not with the interpretation you put on it. In particular, I would strongly disagree that your scenario, even if attainable, would in any way solve the quantum measurement paradox.

The Lesson of Quantum Theory, edited by J. de Boer, E. Dal and O. Ulfbeck
© Elsevier Science Publishers B.V., 1986

Strange Quantum Numbers in Condensed Matter and Field Theory

J. Robert Schrieffer

University of California,
Santa Barbara, California, USA

Contents

Aside

It is a great honor and pleasure for me to join in celebrating the hundredth anniversary of the birth of Niels Bohr. I had the good fortune of spending the spring of 1958 in the Summer House on Blegdamsvej. There I met many exciting physicists and was moved by the tradition of openness and enthusiasm for science which pervaded the Institute. On several occasions Niels Bohr invited me to discuss with him the theory of superconductivity. We were joined by Aage Bohr and Léon Rosenfeld. The discussions focussed on Bohr's ideas concerning superconductivity, some of which trace back to the late 1920s. We also discussed the theory just proposed by Bardeen, Cooper and myself. These discussions made a very deep impression on me and remain one of the high points of my life.

In addition to benefiting from the science and hospitality at Blegdamsvej, I learned of Weisskopf's theorem, and soon provided another proof of its validity by marrying a lovely Danish girl who continues to provide ties to Denmark. It is in this sense of extended family that I join with you and the Bohr family in this joyful celebration.

1. Introduction

From the early days of quantum theory, a fundamental principle of measurement theory has been the indivisibility of an elementary particle. For example, consider the traditional Stern–Gerlach experiment illustrated in fig. 1, in which a particle of spin $s = \frac{1}{2}$ is prepared in an eigenstate of σ_x. The particle passes through a magnetic field gradient along z, such that the incoming wave splits into two separated beams, $+$ and $-$, which pass through slits in a shutter and strike a screen which detects the beams.

If the wavefunction $\psi(r, t)$ describes a classical field, one would expect one half a particle to pass through each slit, each time a particle is projected toward the shutter. On the contrary, quantum theory and experiment show that either:

(1) a full particle is observed to pass through the $+$ slit with charge $-e$, and no particle passes through the $-$ slit, or

(2) a full particle passes the $-$ slit and not through the $+$ slit.

The probability for observations (1) and (2) are both $\frac{1}{2}$. Thus, an elementary particle cannot be split into two half particles.

A related question is the following: can stable particles (excitations) of sharp fractional charge exist in systems composed solely of particles of integer charge? Several clarifications are in order. Firstly, by sharp we mean that every time the charge of the particle is accurately measured, the same fractional result will occur, rather than a distribution of values, the average of which is fractional. Secondly, while vacuum polarization effects renormalize all charge, these effects cancel out in the ratio of the dressed fractional charge and the dressed primitive charge from which the system is constructed.

The first step in this story occurred in 1976 when R. Jackiw and C. Rebbi [1] discovered that if a spinless Dirac field is coupled to a nonlinear scalar background field which supports kink-like solutions in one space dimension, the negative energy sea in the vicinity of the kink is depleted by one half a fermion. Also, a zero-energy fermion state (zero mode) was found to be bound to the kink, which, if filled, leads to a total excess of one half a fermion associated with the kink.

Independently, W.P. Su, A.J. Heeger and the author [2], studying quasi one-dimensional conductors, discovered that soliton excitations can occur whose charge

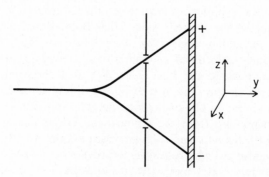

Fig. 1. Stern–Gerlach experiment.

per spin is $\pm e/2$, where $-e$ is the electronic charge in the medium. Since two spin directions are equally populated in these systems, the soliton carries a total charge $Q = -e$, but a spin $s = 0$. This surprising result appears to violate Kramer's theorem relating the electron number and the allowed spin value, i.e. one electron $(-e)$ must correspond to spin $\frac{1}{2}$. Fundamentally, these peculiar effects arise from the same physical effect: vacuum charge flows at domain walls in quantized amounts in systems having discretely degenerate vacua. The sharp quantization of charge is a consequence of this discrete symmetry breaking. In essence, the wave-functions of the negative-energy (occupied) fermion states are distorted by the Bose field without change of the fermion occupation numbers [3].

A third example of fractional charge arises in the fractional quantum Hall effect. In this case, electrons confined to two space dimensions are subject to a strong magnetic field perpendicular to the plane. While semiclassical theory predicts the transverse conductivity σ_{xy} to be a linear function of the electron density ν, experiments by von Klitzing et al. [4] showed that σ_{xy} exhibits plateaus at integer values of ν, corresponding to filled Landau levels. Steps at fractional ν were observed by Tsui et al. [5]. To account for the latter result, Laughlin [6] proposed a fluid-state theory possessing fractionally charged quasi particles. Recently, an alternative theory based on a Wigner-crystal approach has been advanced by Kivelson et al. [7] and this also exhibits fractionally charged excitations. In these theories, as in those mentioned above, only vacuum flow of current is involved in the fractional charge. However, the charge quantization comes about from a somewhat different mechanism in the Hall effect, namely local energy stabilization near the core of the excitation rather than discretely degenerate vacua extending over large regions of space (nonlocal energy stabilization).

Below, we briefly discuss these examples of "charge splitting without violating quantum mechanics".

2. A classical example

Consider the infinite line with the integers marked off as in fig. 2(top). We place particles, each of charge q, on the odd sites starting at $-\infty$, leaving the even sites vacant. Having filled a given site, say number 1, we make an error and place the next particle at 2 instead of 3. We continue placing charges on every other site, i.e. the even sites to $+\infty$. Note that we have made two domains: $-\infty$ to 1 with odd sites occupied, even sites empty (termed the A phase); and 2 to $+\infty$ with the reverse occupancy (termed the B phase). A domain wall separates the two phases and is located at the midpoint $x = \frac{3}{2}$ between the two adjacent occupied sites.

Suppose we move the particle initially located on site 2 to site 3, as shown in fig. 2(bottom). Notice that the midpoint between the two adjacent occupied sites has moved to $x' = \frac{3}{2} + 2$, that is, the domain wall moves *two units* even though we moved the particle only *one unit*! Let us now determine the effective charge Q associated with the domain wall, i.e. the charge which will be observed in long-wavelength experiments. We do this by equating the charge in the electric dipole moment Δp of the system calculated in two ways, namely through the particle motion

J.R. Schrieffer

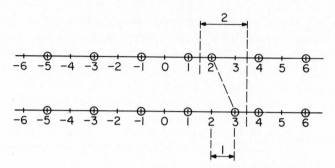

Fig. 2. A classical example of a fractionally charged excitation, illustrating the difference between the particle and domain-wall interpretation of configuration changes.

$q \cdot (+1)$, and through the domain wall motion $Q \cdot (+2)$. Since Δp must be independent of how we describe the system we have $q = 2Q$ or $Q = q/2$. Thus from integer charges q, we have discovered an "excitation" whose charge Q is fractional, $Q/q = \frac{1}{2}$. While this example appears to be trivial, it illustrates one mechanism for the occurrence of fractional charge, namely domain-wall or more generally topological solitons separating degenerate ground states which have different negative-energy fermion wavefunctions. The charge conjugate wall with $Q/q = -\frac{1}{2}$ is given by leaving two vacant spaces rather than no spaces between particles at the wall.

Clearly, we could generalize the model to charge $Q/q = \frac{1}{3}$, by placing charges on every third site, with one rather than two spaces between particles at the wall, etc. [8]. Below we discuss fractionalization in quantum systems.

3. Solitons and chain conductors

The simplest example of charge fractionalization in a quantum system occurs in the linear polymer [9] trans-polyacetylene $(CH)_x$, illustrated in fig. 3. While each carbon has four valence electrons, three of these are dynamically mute, being involved in strong bonds with its three neighboring atoms (two C's and one H). The remaining electron is in a p_z orbital and is free to wander along the chain to form a one-dimensional metal. In band language, this p_z band is half filled, because of two spin states, and should lead to large electrical conductivity. In fact, precisely the reverse is true, as first discussed by R.E. Peierls [10]. He showed that any one-dimensional Fermi gas, when coupled to lattice distortions or phonons, leads to a spontaneous symmetry-breaking. In this case the coupling breaks the combined symmetry operation of translation by one unit along x, and π rotation about x. The broken-symmetry state has a periodicity of 2 units along x and corresponds to a modulation of carbon–carbon bond lengths, long, short, long, short,.... . In practice these bond-length changes are very small, less than 5% of the unstretched bond; however, they have a dramatic effect on the electron spectrum, opening up a gap of $2\Delta \simeq 1.4$ eV at the Fermi surface, converting the undistorted metal to a large-gap

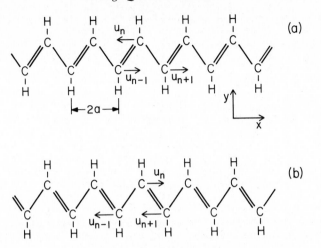

Fig. 3. Trans-polyacetylene $(CH)_x$ in its two degenerate ground states.

semiconductor. The fact that the gap opens at precisely the Fermi surface is no accident since the system energy would increase were it elsewhere in momentum space. Chemists term the Peierls instability dimerization or bond alternation.

To obtain a more quantitative understanding of these effects consider the model Hamiltonian [2] describing the coupled electron–phonon system in trans $(CH)_x$:

$$H = \sum_{n,s} t_n \left(C^+_{n+1,s} C_{ns} + \text{h.c.} \right) + \sum_n \left\{ \frac{\Pi_n^2}{2M} + \frac{K}{2} \left(u_{n+1} - u_n \right)^2 \right\}, \tag{1}$$

where C^+_{ns} creates a p_z electron of spin s on site n. u_n is the displacement of the nth (CH) group along the x-axis, with the momentum Π_n conjugate to u_n. The electronic hopping matrix element t_n is modulated by the phonons and is well approximated by

$$t_n = t_0 + \alpha \left(u_n - u_{n+1} \right), \tag{2}$$

where t_0 generates the bare-electron band structure and α is the electron–phonon coupling constant. In field theoretic terms, H describes a Fermi field linearly coupled (α) to a Bose field on a lattice. The finite lattice spacing provides a natural cutoff for the theory.

Despite the simplicity of H, its ground state and excitations are known only approximately at present. Fortunately, typical electron frequencies are large compared to phonon frequencies so that the electrons can be integrated out within the adiabatic approximation (one-loop level) to obtain an effective potential V_{eff} for the phonon field. It is convenient to introduce the staggered Bose field

$$\phi_n \equiv (-1)^n u_n, \tag{3}$$

to remove the rapid spacial oscillations of u_n which occur near the minimum-energy

J.R. Schrieffer

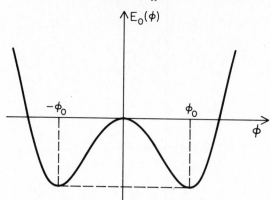

Fig. 4. The effective potential V_{eff} as a function of the amplitude of the staggered displacement (Bose field) ϕ, for ϕ spacially uniform. The peak at $\phi = 0$ exhibits the Peierls instability with degenerate vacua at the broken-symmetry values $\pm \phi_0$.

configuration. For the special case $\phi_n = \phi$, i.e. perfect alternating bond length with amplitude ϕ, V_{eff} is illustrated in fig. 4. The negative curvature of V_{eff} at $\phi = 0$ reflects the Peierls instability, while minima at $\pm \phi_0$ represent the two-fold degenerate broken-symmetry vacua. The physical origin of this dynamical symmetry breaking is clear. Nonzero $\langle \phi \rangle$ leads to a gap 2Δ (or mass $m \equiv \Delta$) in the electron spectrum, where $\Delta = 4\alpha \langle \phi \rangle$, as shown in fig. 5. The energies of the occupied states are lowered by the existence of the gap, leading to the decrease of V_{eff} for small ϕ. For large ϕ, the lattice-strain effects represented by the positive K term in H, dominate the interaction terms. Thus, within the harmonic fluctuation approximation about each ground state, the system appears to be a conventional semiconductor.

However, a closer look reveals the system to be highly unconventional, in that in addition to phonons, the stable excitations are not electrons and holes as in conventional semiconductors but topological solitons with peculiar charge–spin relations. To understand how this comes about, consider the displacement pattern shown in fig. 6 in which a domain wall is located at site 1 (fig. 6a). If the displacements were large, a pair of p_z electrons ($\uparrow \downarrow$) would be localized on each short bond and none would be localized on long bonds, just as in the classical model. When the electron pair localized on the bond joining 1 and 2 moves one unit

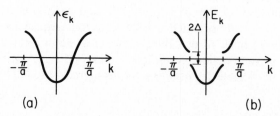

Fig. 5. The π electronic structure: (a) for $\langle \phi \rangle = 0$, and (b) for $\langle \phi \rangle = \pm \phi_0$. The gap 2Δ arises from and drives the spontaneous symmetry-breaking by negative energy states, being lowered when $\langle \phi \rangle \neq 0$.

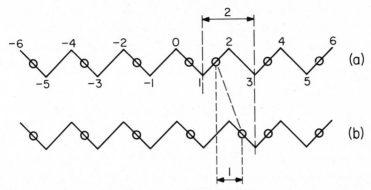

Fig. 6. The motion of a kink in trans-$(CH)_x$ is illustrated in analogy with the classical example shown in fig. 2. In practice, the kink width is on the order of fourteen lattice spacings rather than two spacings as shown, reflecting electrons being in scattering states rather than in the localized states used for this schematic illustration.

to the bond joining 2 and 3 as in fig. 6b, the domain wall moves from site 1 to site 3, i.e. two units. Therefore, we expect the effective charge of the kink to be $-e/2$ per spin direction, or a total charge of $-e$.

Another way of understanding this result is illustrated in fig. 7 in which N (even) (CH) groups form a ring. Initially the system is in the A phase, shown in panel (a) with ϕ plotted radially. The spectrum has $N/2$ levels in the positive and $N/2$ levels in the negative energy regions. In panel (b), ϕ is distorted to form a soliton S and an antisoliton $\bar{S}$. If S and $\bar{S}$ are widely separated, the electronic spectrum exhibits two zero-energy states, each split symmetrically from the positive and negative energy continua. Furthermore, since the spectral sum rule for the local density of

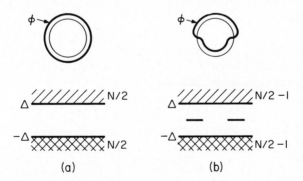

Fig. 7. A trans-$(CH)_x$ chain having N (CH) groups in a ring configuration. In (a) the system is in the A vacuum and in (b) the system is distorted to have a soliton S and a widely spaced antisoliton $\bar{S}$. While the positive and negative continua each have $N/2$ states in (a), two states near zero-energy appear in (b), one localized near S and the other near $\bar{S}$. Since the total number of states is conserved, the positive and negative energy continua are each depleted by one state when S and $\bar{S}$ are created. The essential point is that the state depletion (and hence charge depletion per spin) of the negative-energy states is $\frac{1}{2}$ from the vicinity of S and $\frac{1}{2}$ from $\bar{S}$. This is the origin of the fractional charge of these objects.

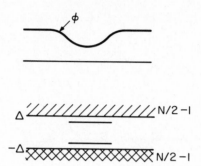

Fig. 8. A polaron P or bag state leading to two split off states in the gap. Note that P can be interpreted as a bound S$\bar{\text{S}}$ pair (see fig. 7).

states $\rho_{nn}(E)$ requires

$$\int_{-\infty}^{\infty} \rho_{nn}(E) \, dE = 1, \tag{4}$$

we see that spectral weight for the zero-mode wavefunction ψ_0 for S (or $\bar{\text{S}}$) is stolen from the immediate vicinity of S (or $\bar{\text{S}}$). This depletion occurs symmetrically from the $+$ and $-$ energy continua since H is invariant under charge conjugation. Thus, the negative sea is depleted by a $\frac{1}{2}$ state per spin near S and the same holds true near $\bar{\text{S}}$. Clearly, this occurs by ϕ_n acting as a scattering potential which distorts or phase shifts the continuum states near S and $\bar{\text{S}}$.

In addition to soliton or kink states, trans-$(CH)_x$ supports polaron, P, or bag states [11,12]. In fig. 8, ϕ_n is shown having a dip which splits off two states, at $\pm\epsilon$, from the continuum. Only when the total number of electrons in $\pm\epsilon$ is either 1 or 3 is the bag stable. Three electrons in $\pm\epsilon$ corresponds to an electron polaron which has a charge $-e$ (since vacuum depletion leads to a charge $+2e$ when both $\pm\epsilon$ are empty) while 1 electron in $\pm\epsilon$ corresponds to a hole polaron, of charge $+e$. The charge and spin relations are summarized in table 1.

The doubly charged polarons P^{2+} and P^{2-} decay rapidly to $S^+\bar{\text{S}}^+$ or $S^-\bar{\text{S}}^-$. Thus, P is essentially a strongly bound S$\bar{\text{S}}$ pair.

Table 1
Charge (Q) and spin (s) relations in a conventional semiconductor.

Object	Symbol	Q	s
Electron	e^-	$-e$	$\frac{1}{2}$
Hole	e^+	$+e$	$\frac{1}{2}$
$(CH)_x$ soliton	S^-	$-e$	0
	S^0	0	$\frac{1}{2}$
	S^+	$+e$	0
Polaron	P^-	$-e$	$\frac{1}{2}$
	P^+	$+e$	$\frac{1}{2}$

Detailed calculations [2] predict that for trans-$(CH)_x$ the soliton width 2ξ is approximately 14 lattice sites or about 18 Å, and the soliton rest energy [13] is $E_S \cong (2/\pi)\Delta$, while the soliton mass M_s is approximately six electron mass units. As discussed below, many experiments confirm these general results [14], although electron–electron Coulomb interactions, quantum fluctuations of ϕ, etc., lead to some quantitative changes [9,15,16].

4. A relativistic model

Closely related to the $(CH)_x$ model is the relativistic model studied by Jackiw and Rebbi [1]. They considered a neutral scalar ϕ^4 field weakly coupled to a spinless Dirac field in one space dimension,

$$H = \int dx \left\{ \psi^+(\alpha p + \beta g\phi)\psi + \tfrac{1}{2}\dot{\phi}^2 + \frac{1}{2}\left(\frac{d\phi}{dx}\right)^2 + V(\phi) \right\}, \tag{5}$$

where

$$V(\phi) = \frac{-a\phi^2}{2} + \frac{b\phi^4}{4}. \tag{6}$$

As shown in fig. 9, a kink occurs when ϕ moves from the $-\phi_0$ to the ϕ_0 vacuum. Treating ϕ as a background field, Jackiw and Rebbi found a zero mode ψ_0 which if empty leads to a fermion number $n_f = -\tfrac{1}{2}$ associated with the kink, while $n_f = +\tfrac{1}{2}$ if ψ_0 is occupied. Again, this is due to the flow of vacuum charge as negative-energy states distort at fixed occupation in response to the distortion of $\phi(x)$.

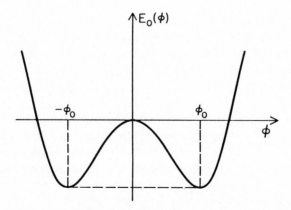

Fig. 9. The potential $V(\phi)$ for a ϕ^4 relativistic model. For $\langle\phi\rangle = \pm \phi_0$, a mass appears in the fermion spectrum. When a kink is introduced a zero mode occurs and the kink is fractionally charged. Precisely the same physical origin of fractional charge occurs here as in the $(CH)_x$ case.

5. Sharpness of fractional charge

The question has been raised if fractional charge is a sharp quantum observable or if only the expected value of Q is fractional. This has been answered by direct calculation [17] as follows. When one defines the charge of an extended object, it is the long-wavelength limit Q_f of the charge form-factor which is relevant. Thus, if $\rho(x)$ is the charge-density operator, we have

$$Q_f = \int f(x)\, \rho(x)\, \mathrm{d}x, \tag{7}$$

where for convenience we choose a Gaussian sampling function

$$f(x) = \mathrm{e}^{-x^2/L^2} \tag{8}$$

and let $L \to \infty$ for an infinite system. One can prove that the mean square fluctuations of Q_f about its mean value $\langle S|Q_f|S\rangle \equiv Q_s$ vanish exponentially as $L \to \infty$,

$$\langle S|(Q_f - Q_s)^2|S\rangle - \langle 0|Q_s^2|0\rangle \to e^{-L/\xi}, \tag{9}$$

where ξ is the half-width of the soliton. Note that the fluctuations of the vacuum are subtracted since these are unrelated to soliton fluctuations. In any event the vacuum fluctuations also vanish as the sampling size $L \to \infty$. Thus, for smooth sampling of the charge, a sharp fractional charge occurs.

6. The experimental situation

While fractional charge of isolated objects (as opposed to bound quarks) has not been observed in particle physics, strong evidence exists that soliton excitations, as discussed above, do exist in quasi one-dimensional conductors. Heeger et al. observed the rapid transport of spin without charge in NMR and ESR experiments. For example, electron spin resonance experiments on prestine trans-$(CH)_x$ show that objects with spin $\frac{1}{2}$ are moving rapidly, near the speed of sound, rather than being localized on a given (CH) group. The solitons' motion is reflected as a motional narrowing of the spin resonance line width. At liquid helium temperatures, the width grows since the soliton presumably comes to rest, however, a residual motional narrowing occurs. This residual width is a direct measurement of the delocalization of the zero mode wavefunction ψ_0 over the size $2\xi \sim 14$ Å of the soliton.

In addition to showing S^0 and $\bar{S}^0$ have spin $\frac{1}{2}$, Heeger et al. showed that $S^\pm$ and $\bar{S}^\pm$ have spin 0, in agreement with table 1. Finally, polarons $P^\pm$ have been observed to have spin $\frac{1}{2}$. The charged soliton and the polarons are created by doping the material with donor or acceptor impurities such as Na or Cl, etc., with one soliton being created per impurity atom.

Another elegant method of creating solitons in trans-$(CH)_x$ is by photoproduction of $S\bar{S}$ pairs. A photon of energy $\geq 2\Delta$ is absorbed creating a bare electron–hole pair across the gap. These excitations are highly unstable and decay in one phonon period $\sim 10^{-13}$ s into an $S\bar{S}$ pair, some fraction of which separates near the speed of sound. Experiments [14] suggest that only charged solitons $S^{\pm}\bar{S}^{\mp}$ are formed in this way. Optical absorption of these photoproduced $S^{\pm}$ and $\bar{S}^{\mp}$ shows transitions between the top of the negative energy sea and the zero mode (called the gap state) or from this state to the positive energy sea. There is also evidence of direct $S\bar{S}$ production below the 2Δ threshold. In addition, the shape-oscillation mode of the soliton (in essence ξ oscillation) has been observed.

Another intriguing experiment by Dalton et al. [18] is electron nuclear double resonance (ENDOR) on prestine trans-$(CH)_x$, which probes the spacial distribution of spin in a neutral soliton. Their results are consistent with the soliton spin per site oscillating with a period of two lattice sites, as predicted by theory [19]. The reverse (down) spin between up-spin sites is due to Coulomb exchange effects. Remarkably, as the soliton moves, the up-spin sites remain up-spin and vice versa so that this distinction between up and down sites (or down and up sites for $\bar{S}^0$ with spin up) is preserved, precisely as predicted by theory.

There are many other experiments supporting the soliton model of $(CH)_x$, including transport properties, luminescence and photoconductivity, etc. [14] Finally, if the degeneracy of the minima of V_{eff} at $\pm\phi_0$ is split, one would predict a confinement potential $V_c(x) = -c|x|$ binding the $S\bar{S}$ pair. This is observed, for example, in cis-$(CH)_x$ where the A and B phases are not degenerate by symmetry, the chemical structure being square wave rather than zig zag as in the trans material.

7. Quantum Hall effect

A second mechanism leading to quantization of non-integer charge occurs in the so-called fractional quantum Hall effect. While the Hall effect historically was observed in three-dimensional materials, such as the semiconductors Ge and Si, recent interest focussed on systems such as MOSFET and heterojunction semiconductor devices, in which a two-dimensional (x, y) electron gas exists on a surface or at an interface. A very strong magnetic field $B_0 \simeq 10^5$ Oe is applied along z, as sketched in fig. 10. If an electric current j_x is made to flow along x, an electric (Hall) field E_y proportional to j_x is observed.

7.1. One-electron theory

The classical theory of the Hall effect is extremely simple. Each electron is acted on by the Lorentz force

$$F = -e\left(E + \frac{v \times B}{c}\right),$$
(10)

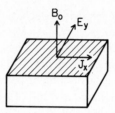

Fig. 10. Configuration for the two-dimensional Hall effect.

which vanishes in steady state. Thus, $v_x = cE_y/B_0$ and the current is given by

$$j_x = -nev_x = \left(\frac{-nec}{B_0}\right)E_y \equiv \sigma_{xy}E_y, \tag{11}$$

where σ_{xy} is the Hall conductivity. For fixed B_0, σ_{xy} is linear in the two-dimensional electron density n.

von Klitzing, Dorda and Pepper [4] discovered that for devices having extremely long electron relaxation times τ, so that $\omega_c\tau \gg 1$, with $\omega_c = eB/mc$ being the cyclotron frequency, σ_{xy} is not a linear function of n but has plateaus at "integer" values of n, corresponding to the complete filling of a Landau level, i.e. $n = \nu B_0/\varphi_0$ where $\nu = 1, 2, 3\ldots$ and $\varphi_0 = hc/e$ is Dirac's flux quantum. Since Planck's constant appears, at least a semiclassical theory is required. If one neglects electron–electron interactions, the system is described by electrons independently filling the Landau levels $E_l = (l + \frac{1}{2})\hbar\omega_c$, $l = 0, 1, 2\ldots$. Using the classical expression for σ_{xy} with n given by $\nu B_0/\phi_0$ for general ν, one finds

$$\sigma_{xy} = -\nu\frac{e^2}{h}. \tag{12}$$

Therefore σ_{xy} is predicted to directly measure e^2/h. At present the relative error in determining e^2/h by this method is on the order of 10^{-7}, comparable with the Josephson-effect measurement of this constant.

What produces the plateaus at $\nu = 1, 2, 3\ldots$? It is thought that for ν near an integer, the extra electrons (or holes) about integer occupation are localized in bound states, presumably due to crystal defects. Only when all of the localized trap states are full, do the extra electrons add to the current. Remarkably, sum rules show that the depletion of the continuum states to form bound states does not reduce the Hall conductivity for $\nu = $ integer [20].

Following the discovery of the integer effect, Tsui, Störmer and Gossard [5] discovered plateaus at fractional $\nu = p/q$, where q is an odd integer and p is integer. These experiments require even greater care than integer ν since the effects are quite subtle. While the gaps between one-electron Landau levels give a natural basis for accounting for the integer effect, it is clear that electron–electron interactions must be involved in the fractional ν effect.

7.2. Liquid-phase theory

Soon after the discovery of Tsui et al., Laughlin [6] proposed a trial wavefunction for the interacting electron gas in the form of a Jastrow-type state, i.e. a product of pair correlation factors f_{ij},

$$\Psi_0 = \left(\prod_{ij} f_{ij}\right) \exp\left(-\sum_k \frac{|Z_k|^2}{4l_0^2}\right), \tag{13}$$

where $Z_k = x_k + iy_k$ is the complex coordinate of electron k, and l_0 is the magnetic length $\pi l_0^2 \equiv \varphi_0/B_0$. For $\hbar\omega_c \gg e^2/\epsilon l_0$ ($\epsilon =$ the dielectric constant), only the lowest Landau level $l = 0$ need be included in the many-body basis states. It is seen that Ψ_0 must be a polynomial in Z_j. Laughlin chose $f_{ij} = (Z_i - Z_j)^m$, where $m = 1, 3, 5 \ldots$, so that Ψ_0 is properly antisymmetric. For m non integer, Ψ_0 cannot describe the system since higher Landau levels are admixed. The density is $\nu = 1/m$ for the Laughlin state.

To accommodate extra electrons or holes near $\nu = 1/m$, Laughlin proposed including a factor $\prod_{j'}(Z_{j'} - Z_0)$ for a quasi-particle centered at Z_0. This factor acts as a raising operator for the angular momentum of each particle about Z_0. One finds that the charge of the quasi-particle (hole in this case) is $Q = \nu e = e/m = e$, $\frac{1}{3}e$, $\frac{1}{5}e$, $\ldots$, while a conjugate factor acting on Ψ_0 produces quasi-particles of negative fractional charge. Note that the quantization of charge arising in this case is not due to the discrete degeneracy of the ground state, but rather from the local quantization implied by the restriction of remaining in the lowest Landau level imposed by large B_0. The raising of all angular momenta about Z_0 by one unit depletes the region surrounding Z_0 by precisely ν electrons. Therefore, since no charge accumulation occurs away from the vicinity of Z_0, we see $Q = e/m$. However, Laughlin has argued that his state describes an incompressible fluid so that the charge νe swept out from the origin cannot be screened by polarization effects away from Z_0. Spin is not considered here since the spin Zeeman energy is also assumed large compared to $e^2/\epsilon l_0$. Another way of phrasing the issue is that the factor $\prod_{j'}(Z_{j'} - Z_0)$ represents a singular gauge transformation which arises from a conceptual point magnetic vortex tube threading the plane at Z_0. Only if the flux ϕ of the line is a multiple of ϕ_0 does the state remain in the lowest Landau level and has low energy.

Finally, we note that higher-order plateaus, $\nu = p/q$ where $p = 2, \ldots$ have been interpreted by Haldane [21] and by Halperin [22] as arising from Laughlin condensation of quasi-particles from a $\nu = 1/m$ state to form a new state which in turn has quasi-particles which condense, etc. This forms a hierarchy of p/q charged particles.

7.3. Quasi-particle statistics

To work with fractionally charged quasi-particles, one must know their statistic. Instinctively, since electrons gain a phase of π on interchange, it seems reasonable that two quasi-particles i and j of charge νe would gain a phase $\nu\pi$ on interchange in two dimensions, i.e.

$$\Psi(i, j) = e^{i\pi\nu} \, \Psi(j, i). \tag{14}$$

That this is true in two dimensions was proved by Arovas, Wilczek and the author [23] using the extended adiabatic theorem [24,25]. Namely, as one adiabatically interchanges 1 and 2, the wavefunction picks up a phase $\gamma(t)$ given by

$$i\dot{\gamma} = \langle \Psi(t) | \dot{\Psi}(t) \rangle. \tag{15}$$

By using Laughlin's state for two quasi-particles i and j one finds the result given in eq. (14).

The free energy of a non-interacting gas of such quasi-particles is peculiar [26].

7.4. Crystalline-phase theory

While the liquid-state theory of Laughlin along with the Haldane–Halperin hierarchy give an explanation of much of the data on the fractional quantum Hall effect, a difficulty occurs. One can prove that the correct state at low electron density, $\nu \ll 1$ is a Wigner crystal in which particles are located on a triangular lattice in two dimensions. An important question is at what density ν_c do the crystal and liquid phases have equal energy? Starting with the Wigner crystal state, Maki and Zotos [27] calculated the vibration spectrum and found the crystal is differentially stable except in the range $0.45 < \nu < 0.55$. Since the Laughlin description of the liquid only holds near $\nu = \frac{1}{3}, \frac{1}{5}, \ldots$, the differential instability of the liquid has not been calculated. Presumably, the lower boundary is between $\nu_c = 0.2$ and 0.45 although this question remains open at present.

Recently, Kivelson, Kallin, Arovas and the author (KKAS) [7] proposed an alternative theory of the fractional Hall effect in terms of a collective ring-exchange mechanism starting from the crystalline phase. Using a path-integral formulation, they find important processes are those in which a ring having L electrons collectively tunnels to a new configuration, such that $Z_j \to Z_{j+1}$. Because flux is enclosed by the tunneling current, each ring enters with a phase given by ϕ/ϕ_0, where ϕ is the flux enclosed by the area A of the ring. For $\nu = p/q$ with p and q integer, these phases are such that exchange energy of large loops on average add in phase. However, for other values of ν, the crystal and magnetic lattices are incommensurate and the exchange energy of large rings is random in sign and cancels. Thus, the ground state energy $E_0(\nu)$ is found to exhibit cusps at $\nu_i = p/q$ with the cusp-like part of E_0 varying as $|\delta\nu| \ln |\delta\nu|$ near each cusp, where $\delta\nu = \nu - \nu_i$. The strength of the cusp is largest for $p = 1$ and q small, i.e. $\nu = \frac{1}{3}$, $\frac{1}{5}, \ldots$.

To determine $E_0(\nu)$ one writes

$$E_0(\nu) = - \lim_{\beta \to \infty} \frac{\partial}{\partial \beta} \ln Z, \quad Z \equiv \mathrm{Tr} \, e^{-\beta H}, \tag{16}$$

where the trace is over many-body states constructed solely from the lowest Landau level. The effective Hamiltonian is given by

$$H = P_0 \left(\sum_{i \neq j} V_{ij} \right) P_0, \tag{17}$$

where P_0 projects onto the lowest Landau level. Roughly, V_{ij} is given by

$$V_{ij} = \frac{e^2}{\epsilon \left[|Z_i - Z_j|^2 + \lambda^2 \right]^{1/2}}, \tag{18}$$

where λ is the extent of the one-electron orbitals in the z direction.

The so-called coherent states $|R\rangle$ form a convenient one-electron basis set for taking the trace in eq. (16). These states have wavefunctions

$$\langle r | R \rangle = \frac{1}{\sqrt{2\pi l_0^2}} \exp \left\{ -\frac{1}{4l_0^2} |r - R|^2 + \frac{i}{2l_0^2} (r \times R) \cdot \hat{z} \right\}, \tag{19}$$

and satisfy the completeness relation

$$\int \frac{d^2 R}{2\pi} |R\rangle \langle R| = P_0. \tag{20}$$

By writing $e^{-\beta H} = (e^{-\epsilon H})^M$ with $\epsilon = \beta/M$ and inserting between each factor of $e^{-\epsilon H}$ the product of projectors P_0 from eq. (20) for all particles, one finds, as $M \to \infty$, a path-integral representation of Z:

$$Z = \mathcal{N} \sum_{\mathcal{P}} \text{sign } \mathcal{P} \int \prod_{j=1}^{N} \mathcal{D} R_j(\tau) \, e^{-S[R(\tau)]}, \tag{21}$$

with the boundary conditions $R_j(0) = R_{\mathcal{P}(j)}(\beta)$. Here $\mathcal{P}$ permutes the particle coordinates. The action S is given by

$$S[R] = \tfrac{1}{2} \int_0^\beta d\tau \left\{ -i \sum_{j=1}^{N} (\dot{R}_j \times R_j) \cdot \hat{z} + \sum_{i \neq j} V_c(R_i - R_j) \right\}, \tag{22}$$

where V_c is the effective Coulomb interaction eq. (18) in the coherent state basis.

To evaluate Z we consider the saddle point (semiclassical) approximation in which one continues $X_j(\tau)$ and $Y_j(\tau)$ to complex values, where $R_j = (X_j, Y_j)$. One finds those paths $\{R_{jc}(\tau)\}$ which minimize the classical action S. These extremal paths satisfy

$$i\frac{dX_j}{d\tau} = \frac{\partial V_j}{\partial Y_j}, \qquad i\frac{dY_j}{d\tau} = -\frac{\partial V_j}{\partial X_j}. \tag{23}$$

J.R. Schrieffer

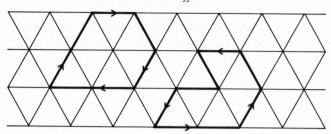

Fig. 11. Collective ring-exchange processes in which a set of L electrons simultaneously rotate so that each moving electron finally occupies the initial position of its moving nearest neighbor. Contributions from different rings add in phase only for the Landau-level filling factor $\nu = p/q$, where p and q are integers.

Then, Z is approximately given by the sum over external classical paths c:

$$Z = \sum_c D[R^c] \exp\{-S[R^c]\},\tag{24}$$

where D is the fluctuation determinant arising from fluctuations quadratic in δR^c about each path. As is well known, tunneling processes are often well treated by such an approximation.

Using the above procedure, KKAS find an important set of paths are those in which a ring of L electrons tunnels from an initial configuration R^0 to a final configuration R^f with $R^f_j = R^0_{j+1}$ or R^0_{j-1}, corresponding to a clockwise or counterclockwise rotation of the ring, as illustrated in fig. 11. In this path, all tunneling electrons move simultaneously, staying out of each other's way as well as away from non-moving electrons. This path is in contrast to a sequence of conventional pairwise exchange processes in which electrons must surmount high Coulomb barriers produced by non-moving electrons. Thus, while ring exchange via pairwise exchange processes is very weak for interesting values of ν, the above collective ring exchange processes is considerably stronger, e.g. by a factor on the order of 10^4–10^5 even for a ring containing three electrons, for $\nu = \frac{1}{3}$.

For a large number of particles L in a ring of area A, one finds the contribution to Z in time $d\tau$ is

$$Z_{LA} = \tau_0^{-1}\, d\tau \exp[-\alpha(\nu)L + ihN_A + O(\ln L)],\tag{25}$$

where $\alpha(\nu)$ is the tunneling exponent per particle, $h \equiv \pi(\nu^{-1} - 1)$ and $N_A = \nu A/\pi l_0^2$ is the number of Wigner-lattice unit cells contained in area A. The ν^{-1} term in h gives the flux contribution to the phase while the factor -1 accounts for the Pauli principle factor for even and odd L loops, since L and N_A are both even or both odd for a triangular lattice. Calculations show $\alpha\left(\frac{1}{3}\right) \simeq 0.81$.

While the contribution of large L rings is essential to obtain cusps, it might appear that large L is exponentially suppressed due to the tunneling factor $e^{-\alpha L}$. However, path counting shows that the number of closed paths of length L varies as $e^{\kappa L}$ so that aside from the phase question, the sum over rings would become large for $\alpha(\nu) < \kappa$, i.e. when ν increases to the point where the tunneling is sufficiently probable that "path entropy" dominates "path energy". Thus, we expect nonanalytic structure in $E_0(\nu)$ when rings add in phase on the average and in addition ν is larger than a critical density ν_c, such that $\alpha(\nu_c) = \kappa$.

The ring summation is conveniently carried out by mapping the problem onto the discrete Gaussian spin model [28]:

$$H_{\mathrm{DG}} = \alpha \sum_{(\lambda,\gamma)} (S_\lambda - S_\gamma)^2 + ih \sum_\lambda s_\lambda, \tag{26}$$

in an imaginary magnetic field h. As illustrated in fig. 12, the spins S_λ live on the dual lattice, where S_λ is defined as the number of clockwise minus the number of counterclockwise collective exchange-rings encircling the dual lattice point λ during the time interval τ_0 centered at time τ. It is helpful to think of the rings as domain boundaries separating regions having different values of S_λ. For $\nu \ll 1$, α is large and the ring density is low. In this case rings rarely overlap. However for larger ν, α decreases (tunneling increases) and the ring density increases. Since two rings cannot share a common edge during one time slice τ_0, a ring repulsion must be included. The $(S_\lambda - S_\gamma)^2$ factor leads to such a repulsion since by crossing two separated boundaries one would obtain a factor of $1^2 + 1^2 = 2$ in the energy while crossing a double boundary leads to $2^2 = 4$. While a power higher than 2 in eq. (26) might be preferable, the results are likely to be insensitive to this change.

From studies of the discrete Gaussian model [28] and related models [29], it is believed that $E_0(\nu)$ has cusps of the form $|\delta h| \ln(2\pi/|\delta h|)$ or $|\delta\nu| \ln|\delta\nu|$ at all rational $h/2\pi$ for $\alpha(\nu) < \alpha_c[h(\nu)]$. Estimates of α lead to cusps at densities $\nu = \frac{1}{3}$, $\frac{1}{5}$, $\frac{2}{5}$, $\frac{2}{7}$, $\frac{1}{4}$, $\frac{3}{7}$ and $\frac{4}{9}$ although some phases may be unstable with respect to competing phases. By charge-conjugation symmetry, cusps are also expected at the conjugate densities $1 - \nu$. We note that the crystal lattice approach produces the higher-order plateaus, e.g. $\nu = \frac{2}{5}, \frac{2}{7}, \frac{3}{7}, \frac{4}{9}$, without constructing a Haldane–Halperin hierarchy.

Finally, we turn to quasi-particle excitations in the collective ring-exchange scheme. As discussed above, the apparent reason for the plateaus in $\sigma_{xy}(\nu)$ at

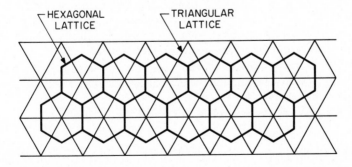

Fig. 12. The dual lattice for the pseudo-spins on to which the cooperative ring-exchange problem is mapped. The walls between domains having different spin orientation are the ring-exchange paths of the original problem.

fractional ν_i is the stability of the condensate for these particular densities. That is, there exists an energy gap for adding or subtracting charge about each fractional ν_i such that $\alpha(\nu) < \alpha_c[h(\nu)]$. To understand how such a gap comes about consider a uniform compression of the density, about a cusp at ν_i. The elastic modulus

$$K = \frac{\partial E_0}{\partial \nu}\bigg|_{\nu_i} \to \infty \tag{27}$$

so that the crystal is incompressible at the cusp. However, suppose we thread a point flux tube φ through the plane, say at a point Z_0 on the dual lattice. As in Laughlin's approach, this singular gauge transformation "blows a hole" in the electron density near Z_0 with charge being transported to the sample boundary. Direct calculation shows that the charge transport to the boundary is νe for $\varphi = \varphi_0$, the Dirac flux quantum. Furthermore, the quasi-particle charge is localized within the magnetic unit cell of size $\sim l_0$ at Z_0. For the opposite sign of φ, the electron density accumulates near Z_0 to form a quasi-particle of charge $-\nu e$.

A question remains: why choose $\varphi = \pm\varphi_0$? It is readily seen that if φ is not an integer multiple of φ_0, cooperative exchange rings encircling Z_0 will have phases which do not add coherently near the cusp densities ν_i. Therefore there will be a discontinuous energy increase as φ deviates from $\pm\varphi_0$. It is this phenomenon which leads to quantization of the fractional charge in this scheme. An estimate [7] of the quasi-particle energy is $\Delta \equiv E_{qp}(\nu) \sim 0.5\, \nu^2 e^2/\epsilon l_0$.

As in other descriptions, the plateaus of $\sigma_{xy}(\nu)$ presumably arise from added density $\delta\nu$, occupying localized defect states inside the gap $-\Delta < E < \Delta$.

8. Conclusion

While quantum mechanics tells us that accurate observations of an isolated elementary particle will always produce integer charge, excitations in systems or fields of integer charge can carry sharp fractional charge Q. The fractional part of Q generally arises from flow of charge in the vacuum, without producing added excitations. Thus, it is the negative-energy states which are deformed, not the occupation numbers which are changed, when fractional charge is formed. The sharp quantization of Q can arise from either a discretely degenerate broken symmetry vacuum, as in one-dimensional conductors or in field theoretic models, or by local energy constraints coupled with incompressible flow of vacuum charge to the boundaries, as in the fractional quantum Hall effect.

While direct observation of an isolated fractional charge in a medium remains for the future, strong evidence exists for the effects discussed above in quasi one-dimensional conductors such as $(CH)_x$.

I hope these ideas would have pleased Niels Bohr.

Acknowledgements

The author would like to acknowledge the hospitality of the Niels Bohr Institute on the occasion of the Niels Bohr Centenary and the support of the National Science Foundation Grant no. DMR82-16285.

References

[1] R. Jackiw and C. Rebbi, Phys. Rev. D13 (1976) 3398.

[2] W.P. Su, J.R. Schrieffer and A.J. Heeger, Phys. Rev. Lett. 42 (1979) 1698; Phys. Rev. B22 (1980) 2099.

[3] R. Jackiw and J.R. Schrieffer, Nucl. Phys. B190 (1981) 253.

[4] K. von Klitzing, G. Dorda and M. Pepper, Phys. Rev. Lett. 45 (1980) 494.

[5] D.C. Tsui, H.L. Störmer and A.C. Gossard, Phys. Rev. B25 (1982) 1405.

[6] R.B. Laughlin, Phys. Rev. Lett. B23 (1981) 5632.

[7] S. Kivelson, C. Kallin, D. Arovas and J.R. Schrieffer, Phys. Rev. Lett. 56 (1986) 873.

[8] W.P. Su and J.R. Schrieffer, Phys. Rev. Lett. 46 (1981) 738.

[9] For a review of solitons and fractional charge in quasi one-dimensional conductors, see Highlights of Condensed Matter Theory, Int. School of Physics, Enrico Fermi, Course LXXXIX eds F. Bassani, F. Fermi and M.P. Tozzi (North-Holland, Amsterdam, 1985) p. 300.

[10] R.E. Peierls, Quantum Theory of Solids (London, 1955) p. 108.

[11] W.P. Su and J.R. Schrieffer, Proc. Nat. Acad. Sci (USA) 77 (1980) 5626.

[12] D.K. Campbell, A.R. Bishop and K. Fesser, Phys. Rev. B26 (1982) 6862.

[13] H. Takayama, Y.R. Lin-Liu and K. Maki, Phys. Rev. B21 (1980) 2388.

[14] A.J. Heeger, Philos. Trans. Soc. London A314 (1985) 17.

[15] S. Kivelson, in: Solitons, Modern Problems in Condensed Matter Sciences, Vol. 17, eds V.E. Zakharov, V.L. Pokrovskii and S.E. Trullinger (North-Holland, Amsterdam, 1986) ch. 6.

[16] W.P. Su, Handbook on Conducting Polymers, ed. T. Skotheim, (Marcel Dekker, New York, 1985).

[17] S. Kivelson and J.R. Schrieffer, Phys. Rev. B24 (1982) 6447;
J.S. Bell and R. Rajaraman, Phys. Lett. 116B (1982) 151.

[18] H. Thomann, L.R. Dalton, Y. Tomkiewicz and N.S. Shiren, Phys. Rev. Lett. 50 (1983) 533.

[19] A.J. Heeger and J.R. Schrieffer, Solid. State Commun. 48 (1983) 207.

[20] T. Ando, Y. Matsumoto and Y. Uemmura, J. Phys. Soc. Jpn 39 (1975) 279.

[21] F.D.M. Haldane, Phys. Rev. Lett. 51 (1983) 605.

[22] B.I. Halperin, Phys. Rev. Lett. 62 (1984) 1583.

[23] D. Arovas, J.R. Schrieffer and F. Wilczek, Phys. Rev. Lett. 53 (1984) 722.

[24] N.V. Berry, Proc. Soc. London Ser. A392 (1984) 45.

[25] B. Simon, Phys. Rev. Lett. 51 (1983) 2167.

[26] D. Arovas, J.R. Schrieffer, F. Wilczek, A. Zee, Nucl. Phys. B250 [FS13] (1985) 117.

[27] K. Maki and X. Zotos, Phys. Rev. B28 (1983) 4349.

[28] S.T. Chui and J.D. Weeks, Phys. Rev. B14 (1976) 4978.

[29] W.Y. Shih and D. Stroud, Phys. Rev. B32 (1985) 158.

Discussion, session chairman W. Kohn

Anderson: Is it obvious to you that the two representations of the fractional quantum Hall effect state, namely that of Laughlin and of KKAS are different? Or do they describe the same state in different ways?

Schrieffer: At present we have not been able to find a direct link between these two candidates for the ground state. Laughlin's is based on a fluid state while KKAS starts with a Wigner-crystal state. Thus, if one calculates the density correlation-function for the two states one would expect to find strong crystalline order, at least for short-range correlations in KKAS but not in the Laughlin fluid phase. Collective ring exchange weakens the crystallinity and long-wavelength fluctuation will smear out long-range crystalline behaviour.

Kohn: In your lecture you did not mention the fact that, associated with the quantum Hall effect, there is the occurrence of essentially vanishing resistance. Can you comment on this aspect of the quantum Hall effect within the framework of your theory?

Schrieffer: We are in the midst of attempting to understand this problem. In essence, the cusp nature of the energy leads to the incompressibility of the system at the densities v_i. This in turn apparently produces a gap in the collective excitation spectrum, leading to activated resistance, as is observed.

Kohn: What about the effect of imperfections?

Schrieffer: We have not put these effects in at present. We believe it will turn out that, like in superconductors, the flow simply adjusts to the impurity and goes around it.

The Lesson of Quantum Theory, edited by J. de Boer, E. Dal and O. Ulfbeck

The Study of the Nucleus as a Theme in Contemporary Physics

Ben R. Mottelson

Nordita
Copenhagen, Denmark

Contents

Introduction

Niels Bohr was just eleven years old when Becquerel discovered the first hint of the existence of the atomic nucleus in those faintly glowing ashes from the ancient cosmic fireworks that created the elements of which our solid earth and our living bodies are made; fifteen years later Rutherford was to exploit the natural radioactivity as a marvelous extension of the human sense organs able to resolve the atomic systems into their open, planetary electronic structures surrounding the dense, small, enigmatic atomic nucleus. This report will be an attempt to characterize and review the shifting questions and the physical problems that motivated them in the studies that have led to a growing understanding of this unexpected new form of matter occurring in atomic nuclei. I hope also to be able to indicate how some of these historically important issues have reappeared, transformed, as central issues in current research. In attempting to retell some of this history I cannot avoid a concern with the presumptuousness of doing this in front of an audience that includes some of the most important leaders and contributors to these developments. It may be a partial extenuation for me to admit that I see this as a chance to report how the historical tradition has been transmitted and understood by this particular member of the younger generation, and to strongly encourage you of the heroic generation to correct me and to bring your own witness where my interpretation seems to you to be inappropriate.

To indicate the broad structure of the development I find it useful to recognize in the history of nuclear physics three periods distinguished by the rather different character of the central question being asked.

(1) *The discovery of the nucleus and its constituents* (1886–1935). The achievement of this period was the identification of the "nuclear problem" as involving a composite non-relativistic quantal system built out of neutrons and protons, held together by a new force of nature—the "strong" interaction.

(2) *Defining the nuclear paradigm* (1935–1952). The developments of this period led to the recognition of nuclear structure as based on independent-particle motion capable of supporting a rich variety of collective dynamics.

(3) *Discovering the feel of the nuclear stuff* (1948–present).

The main focus of the present report is period 2, but I shall attempt to put the issues in a broader perspective by including some of the background from period 1 and the further development occurring in period 3.

1. The discovery of the nucleus and its constituents

Rutherfords discovery revealed the nuclei as a new constituent of matter.

To begin with, the different nuclei appeared as "elementary particles", in the sense that there did not appear to be any more *a priori* reasons for the existence of any one of these nuclei than there was for the existence of the electron. However, as is the way with "elementary particles" it gradually became apparent that the nuclei formed a large, but strongly ordered, family, and there accumulated compelling evidence for the view that the nuclei are composite systems built out of more elementary constituents.

(i) Radioactivity itself revealed the possibility of transition from one member state to another of the nuclear family. In particular, α-decay suggested the possibility of α-particles as potential constituents of nuclei.

(ii) The quantization of nuclear charge (Moseley 1913) and approximate quantization of nuclear mass [Prout's hypothesis (1815), enormously strengthened by the work of Aston (1920)] suggested that nuclei are composed of a discrete number of fundamental building blocks.

(iii) The discovery of induced nuclear reactions (Rutherford 1919) and artificial radioactivity (Joliot and Curie 1934) directly exhibited the possibility of changing and exchanging the elementary building blocks in nuclear processes.

Despite these significant clues, the construction of nuclei out of the then known particles, electron and proton, posed profound problems and, indeed, seemed to link the nuclear problem with the unsolved problems of relativistic quantum theory. The sense of confusion and mystery at this time is strikingly expressed in Niels Bohr's Faraday lecture (held in 1930 and published in 1932):

> "Still, just as the account of those aspects of atomic constitution essential for the explanation of the ordinary physical and chemical properties of matter implies a renunciation of the classical ideal of causality, the features of atomic stability, still deeper-lying, responsible for the existence and the properties of atomic nuclei, may force us to renounce the very idea of energy balance. I shall not enter further into such speculations and their possible bearing on the much debated question of the source of stellar energy. I have touched upon them here mainly to emphasize that in atomic theory, notwithstanding all the recent progress, we must still be prepared for new surprises."

We can now see that this situation was almost inevitable until the neutron had been discovered. Telescoping a marvelous scientific adventure into a mere telegraphic report we may remember that this discovery came from:

(i) *Rutherford* (1920) predicted the existence of the neutron (in his Bakerian lecture) by arguing that if heavy nuclei could form tightly bound states with electrons as revealed in the difference between the atomic number A and the positive charge number Z, one could very well expect a single proton to unite with a single electron to produce a neutral and very unusual nuclear system. He felt, also, the need for such neutral nuclear systems in order to account for the building up of the heavy elements. This vision by Rutherford appears to be the first successful prediction of an elementary particle.

(ii) *Chadwick* joined Rutherford (1920–1932) in a wide-ranging research program aimed at producing and exhibiting the expected neutron.

(iii) *Bothe and Becker* (1930) observed a penetrating radiation produced in Be + He reactions and interpreted this as a high-energy γ-ray. Joliot and Curie (1932) observed that the new radiation produces energetic recoils when passed through paraffin, but continue to interpret the radiation as a high-energy γ-ray.

(iv) *Chadwick* (1932) compared the recoils in H, He, and in N, to determine the mass of the new radiation and found $M_{rad} \approx M_{prot}$, and thus the neutron is discovered at last!

After the discovery of the neutron Heisenberg (1932), Majorana (1933) and Wigner (1933) took the first steps to pursue the consequences of this discovery with respect to the nuclear problem. Their program can be briefly summarized:

(i) Nuclei are composite systems built out of neutrons and protons. This picture provided an immediate interpretation of the integer quantization of nuclear charge and mass:

Z = number of protons,

A = number of protons + number of neutrons.

(ii) The nuclear binding required a new force of nature (which we now recognize as the first example of the "strong" interaction). A number of significant features of this interaction could be derived from the available systematics of nuclear binding energies:

–*saturation* (binding proportional to A);

–*strong force* (nuclear binding is of order 10^6 stronger than atomic binding);

–*charge symmetry* (from $A \approx 2Z$, with $A - 2Z$ increasing with Z as a result of Coulomb repulsion);

–*finite range* $\sim 2 \times 10^{-13}$ cm (from comparison of ^{2}H and ^{4}He binding).

The final resolution of the questions of period 1 had still to wait two years until Fermi (1934a) developed the theory of β-decay exploiting the freedom provided by the quantal formalism to have the electron and neutron created at the instant of the decay process. At last the nuclear dynamics could be totally freed from the terrible consequences of trying to think of bound electrons inside the nucleus.

Before leaving the achievements of period 1, and looking at the description of nuclei considered as built out of neutrons and protons, I would like to remind you that already at this juncture there began to appear, at first obscurely, but with constantly growing insistence, significant results that indicated the limitations of this picture and pointed toward the composite nature of the neutrons and protons themselves. Of course, the very existence of two states of the nucleon, the neutron and proton, can be seen (at least today) as a strong hint of internal structure, and then the discovery of the anomalous magnetic moment of the proton (Estermann and Stern 1933, Frisch and Stern 1933) should have removed all doubt about the elementarity of these particles. Finally Yukawa's invocation of massive quanta as the mediators of the strong interaction (Yukawa 1935) provided an energy scale setting the limits beyond which the compositeness of the nucleons would have to be seriously taken into account.

The question of the proper place of these "additional" degrees of freedom in the problems of nuclear structure has been a recurring theme and is currently a focus of active interest. Let me remind you of a rather extreme view considered by Niels Bohr (apparently sometime in the late 1930s). According to J.H.D. Jensen (1965), Bohr argued that

> "... since the field is strongly coupled to its sources, the hitherto existing picture of the 'compound nucleus' may still be much too naive. Perhaps, the only sensible concept is to consider the whole nucleus as an 'Urfeld' which is highly non-linear because of such strong couplings. When this field is quantized it must give (in addition to other conserved quantities, like angular momentum) integral charges Z, and energies (i.e. masses) that form a spectrum with values close to the integral numbers A, on which the 'excitation energy' bands are superposed. The assumption that inside the nucleus there exist Z protons and $A - Z$ neutrons, such as we encounter them as free particles in appropriate experiments, would then hardly make any sense." *

As I mentioned, these questions, slightly reformulated, are under active current investigation. Let me attempt a capsule assessment of the present status of these issues:

(i) The exchange of mesons between nucleons implies modifications in the electromagnetic and weak decay-properties of nucleons in the nucleus as compared with that of a collection of free nucleons. These modifications are relatively small in low-energy transitions, typically of order 10%, but in favourable cases they have been quantitatively identified. ** Note that this figure of 10% also represents the accuracy of the non-relativistic approximation in nuclei, $(v/c)^2$, as well as the ratio of the π-meson to nucleon mass, but I do not know a satisfactory general argument establishing a connection between these numbers.

(ii) Recent experimental studies of collective spin-excitations in nuclei (the so-called Gamow–Teller resonances) have revealed rather narrow and well-defined collective vibrational modes excited in high-energy proton–neutron (pn) reactions. [Goodman (1984) and fig. 1.] The absolute cross-sections for excitating these

---

* It is likely that it is this picture that is being referred to in letters by Rutherford in which he talks of Bohr's view of the nucleus as a "mush of particles of unknown kind, the vibrations of which can be deduced on quantum ideas" [Rutherford's letter to Born (1936), published in "Niels Bohr, Collected Works", Vol. 9 (Peierls, 1985)]. Indeed Bohr himself, in his compound-nucleus article in Nature (1936) refers obliquely to the possibility of this picture.

** An especially well-studied case is provided by the neutron–proton capture reaction [Riska and Brown (1972); for a review of the status of exchange effects in heavier nuclei see Yamazaki (1979) and Arima and Hyuga (1979)].

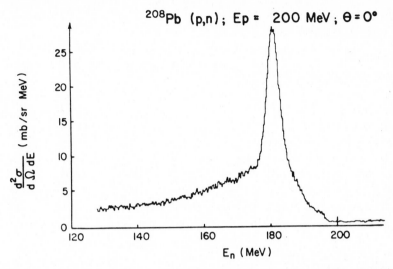

Fig. 1. Excitation of nuclear collective spin–isospin resonance. The figure exhibits the yield of the proton–neutron (p n) reaction at 0° for 200 MeV protons incident on ^{208}Pb (Goodman 1984). The strong peak with $Q \approx -20$ MeV, corresponds to a collective excitation produced by the "Gamow–Teller" operator $\tau_+ \sigma$.

resonances are about a factor of two less than predicted, assuming a simple mean-field description based on neutrons and protons. A significant part of this missing strength can be attributed to the effect of the spin-dependent nuclear mean-field acting on the spins of the quarks within the nucleons causing excitation of the Δ-resonance. It has not yet been possible to quantitatively determine the magnitude of this effect because of uncertainty concerning the line shape of the resonance. However, the nuclear-physics tools at present available should make it possible to settle this question and thus establish a quantitative measure of the role of the Δ-degree of freedom in this particular nuclear process.

(iii) A much more profound effect of additional degrees of freedom in nuclear matter would follow from conjectures perhaps suggested by quantum chromo-dynamics and bag models. The interpretation of quark confinement as an effect of the QCD vacuum acting as a medium from which color is excluded, has led to the suggestion that at sufficiently high energy-density (high temperature and/or baryon density) nuclear matter will exhibit a phase with unconfined color [see the review by Jacob and Van (1982)]. Here we would indeed encounter a phase of matter resembling that in Bohr's vision quoted above. The attempts to make quantative estimates of the energy density necessary in order to produce this new form of matter are still rather uncertain but indicate something like a doubling of the energy density as compared with the equilibrium state of nuclear matter. A possible environment for realizing such energy densities may be provided by collisions between heavy nuclei involving bombarding energies in the range 10 to 100 GeV/nucleon. There are intensive efforts to explore possibilities for creating matter under these conditions and to attempt to find diagnostic signals which would make it possible to probe the equation of state describing this regime.

2. Defining the nuclear paradigm

After the discovery of the neutron and Fermi's formulation of β-decay, it became possible to begin considering the dynamical patterns and structures formed by the neutrons and protons of the nuclei. The subsequent developments were strongly driven by the experimental discoveries that were constantly revealing new features of the nuclear systems. The beginning of the period saw the first nuclear reactions produced by artificially accelerated particles, as well as the use of the recently discovered neutron as a projectile capable of penetrating to even the heaviest nuclei and causing reactions. These neutron reactions, especially developed and exploited by Fermi and his collaborators, were uniquely important in focussing attention on the many-body aspects of the nuclear problem. Let me again resort to a telegraphic style to remind you of the bare outlines of the development:

(i) Early 1934, Fermi and his collaborators begin systematic neutron irradiation of all elements of the Periodic System, and find new radioactivities in most of them (Fermi 1934b, Fermi et al. 1934a).

(ii) October 1934, discovery of added effectiveness of slow neutrons (Fermi et al. 1934b).

(iii) Through the year 1935, theoretical analysis of neutron reactions on basis of particle motion in a static-potential model (Fermi and Rasetti 1935, Bethe 1935, Perrin and Elsasser 1935, Beck and Horsley 1935).
Main results:

- $\sigma_{n\gamma} \sim \dfrac{1}{v}$ at low neutron-energy,
- Short residence time of neutron in nucleus implies monotonic cross-sections in energy region below ~ 1 MeV,
- $\sigma_{cap} < \sigma_{scat}$ in all cases.

It is of some importance for assessing the frame of mind at this time to attempt to understand the motivation and degree of conviction with which the static-potential model was being used. It appears that the model was mainly motivated by its successes in describing collisions of electrons with atoms and, recognizing the great differences between atoms and nuclei, the model was being used without any great conviction; for example, near the end of his article Bethe writes:

> "It is not likely that the approximation made in this paper, i.e. taking the nucleus as a rigid body and representing it by a potential field acting on the neutron is really adequate … Anyway it is the only practicable approximation in many cases …"

(iv) Also in 1935, large capture cross-sections observed for some elements* and discovery of sharp resonances (Tillmann and Moon 1935, Bjerge and Westcott 1935) called "selective absorbtion", being in violent disagreement with the theory in (iii) and provoke the formulation of "compound nucleus" (Bohr 1936).

There exist published reports by J. Wheeler (1979) and by O.R. Frisch (1967) colorfully recounting discussions at the Niels Bohr Institute during the time the

---

* I am endebted to Professor Amaldi for pointing out to me that it was Dunning et al. (1935) who first established $\sigma_{abs} \gg \sigma_{scat}$ for slow neutrons on Cd.

Fig. 2. Bohr's picture, visualizing the formation of a compound nucleus by the capture of a neutron.

compound nucleus was being formulated. I shall not repeat these accounts here but would like to go directly to an examination of the content of Bohr's analysis.

The core of Bohr's thinking is the recognition that the densely packed nuclear system being studied in the neutron reactions, forces one to place the collective many-body features of the nuclear dynamics at the center of attention. To illustrate these ideas I do not know of any better figures than those prepared by Niels Bohr in connection with lectures which he gave at that time and which were published in the same issue of "Nature" (as a new item) that contains his famous article *. The first (fig. 2) draws attention to the far-reaching consequences for the course of a nuclear reaction, of the assumption of a short mean free path for nucleons in the nucleus.

If we imagine the balls removed from the central region of the figure, the ball entering from the right will be accelerated as it enters the central depression, but just this acceleration ensures that, after running across to the opposite side, the ball will have enough energy to surmount the barrier on that side and run out of the nuclear region.

A very different dynamical evolution results if we restore the balls to the central region. Now, the entering nucleon will soon collide with one of the nucleons of the target and, sharing its energy with the struck nucleon, will no longer be able to leave the confining potential. Being reflected back it will collide and share its remaining energy with still other nucleons; these struck nucleons will also collide and ultimately the total energy will be distributed among all the nucleons in a distribution of the type described by the equilibrium distribution of the kinetic theory of gases. In this situation the only possibility for one of the nucleons to escape from the central region requires the occurrence of a fluctuation in which almost all of the energy is again concentrated on a single particle, which will then be able to surmount the confining potential. The unlikelyhood of such an extreme fluctuation implies that the duration of the reaction phase is enormously increased (as compared with the first situation considered with only a static potential acting).

This increase of reaction time makes it possible to explain both the observed large ratio of capture to scattering cross-sections for slow neutrons as well as the

---

* Apparently the original draft of these figures was executed by O.R. Frisch for Niels Bohr (Frisch 1979).

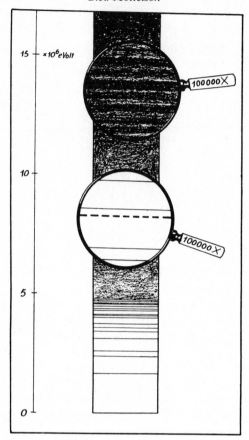

Fig. 3. Bohr's sketch of a schematic nuclear level spectrum. The dashed line indicates the neutron-binding
energy.

narrowness of the selective absorption bands. Perhaps even more important, the
intermediate stage representing a kind of thermal equilibrium from which the final
decay represents a rare fluctuation, ensures that the relative probability of different
final states will be governed by statistical laws and is independent of the mode of
formation of the compound system.

Figure 3 shows Bohr's sketch of a schematic nuclear level spectrum. The study of
radioactivity had shown that the lowest states in heavy nuclei have excitation
energies on the order of a fraction of 1 MeV, and Bohr assumed that these
excitations represent some sort of collective vibration of the whole nucleus. With
increasing excitation energy an increasing number of different vibrational modes
can be excited and the different possibilities for partitioning the total excitation
energy between these different modes leads to an enormous increase in the total
number of excited states. All of these quantum states can be resonantly excited by
an incoming neutron, thus accounting for the dense spacing and narrowness of the
levels observed in the selective absorption phenomena. The dotted line in the
magnifying glass at about 10 MeV indicates the neutron separation energy, but the

level scheme above and below this line are not significantly different; indeed, the neutron-escape probability is much less than the γ-emission probability for levels slightly above this energy as a result of the extreme improbability of the fluctuation required to concentrate all of the excitation energy on a single particle. Only at higher energies will the neutron-emission probability contribute appreciably to the width of the individual levels and lead eventually to a smearing out of the spectrum (indicated in the upper magnifying glass at about 15 MeV). Bohr contrasts this picture of densely-spaced many-particle levels in the nucleus with the spectrum of atoms excited in collisions with electrons, where the incident electron will at most collide with one of the atomic electrons causing it to change its binding state from one orbit to another.

The profound reordering of the picture of nuclear dynamics implied by Bohr's ideas was, apparently, rapidly and widely accepted in the nuclear physics community; within months the literature is dominated by papers applying, testing and extending the ideas of the compound nucleus.

In view of the subsequent history, it is an interesting and relevant question to ask whether Bohr's vigorous and effective contribution to the development of nuclear physics at this time had also an adverse element in preventing an earlier appreciation of the significance of independent-particle motion in the nucleus. The tentative use of an independent-particle picture has been mentioned above but after Bohr's paper such approaches were subjected to a much more critical attitude. The independent-particle starting point was further developed, especially by Feenberg and Wigner (1937) and by Rose and Bethe (1937) as a basis for the analysis of the configurations of light nuclei *, but as Maria Goepert Mayer (1964) says in her Nobel Lecture:

> "[the model] failed in predicting the properties of heavy nuclei, and somehow, the theory of individual orbits in the nucleus went out of fashion."

We may ask, did this going out of fashion delay the understanding of nuclear properties? To what extent was it a psychological question connected with Bohr's enormous prestige? As Victor Weisskopf remarked to me once when discussing this question, "You know, it wasn't easy to disagree with Niels Bohr".

It is in the nature of these questions that the answers can only be tentative and partial, but my impression is that the direction of the development of nuclear physics at this time was strongly bound to the available experimental tools and the limited number of facts about the nucleus that were then accessible. The discoveries that were being made focused attention mainly on a variety of reaction processes for which the compound nucleus was the uniquely appropriate and powerful concept. One can ask, what properties would one have understood better by invoking individual orbits? What data could have been used to test that idea? Only at a much later stage with the accumulation of more detailed and systematic knowledge on

---

* In the literature of that time it is stated again and again [see Wigner (1933), Bethe (1935, 1936)] that the independent-particle picture will not be applicable to heavy nuclei but might be appropriate for lighter systems. I am unable to discover, or understand, the basis for this expectation of a difference in the dynamics of light and heavy nuclei.

nuclear masses, spins, moments and excitation spectra could there be a proper assessment of the role of single-particle motion. Having said this, however, I think it is also relevant to notice that when that time came, the decisive contributions were made by scientists who were in a significant sense outsiders to the main development of the field; Hans Jensen in the scientifically isolated conditions of postwar Germany, and Maria Mayer, a chemical physicist newcomer to the field of nuclear physics, could look at this new data with uniquely creative vision. It might appear that by this time, ten years after the formulation of the idea of the compound nucleus, the successes of this idea had induced a certain orthodoxy such that most of the established figures were inhibited in reading the message contained in the burgeoning new facts about the nuclei.

I would also like to express the opinion that the close connection between experimental initiative and theory building, to which I referred above, has continued to be characteristic of the most fruitful developments in nuclear structure—one is almost tempted to say of fruitful developments in all those parts of physics dealing with systems with many degrees of freedom.

But now I would like to return to the early period after the formulation of the compound-nucleus idea and, again in a telegraphic style, remind you of the impressive series of developments in which this idea was extended, and successfully applied to the interpretation of the growing body of knowledge about nuclei (see table 1). For the first ten years after its formulation the compound nucleus served brilliantly as a basis for relating and interpreting the experiments that were gradually probing more and more deeply into the facets of nuclear structure. I do not know of any significant criticisms during that time of the assumptions of the compound nucleus or challenges to its explanatory power. Especially in the study and interpretation of the many phenomena associated with the fission reaction, the compound nucleus, coupled with the analogy of nuclear matter to that of a liquid drop, provided a marvelously successful conceptual basis.

Then, as is well known, there came a second major reordering of the picture of nuclear structure as it was recognized that a wide variety of nuclear systematics (mainly referring to binding energies, but also extending to the data on nuclear spins, magnetic moments, the occurrence of isomerism, etc.) testified to the existence of nuclear shell structure, i.e. independent-particle orbits as a basis for the nuclear ground states (Mayer 1948). This discovery carried a strong sense of paradox that is preserved in the early reference to the closed shells as "magic numbers" (an expression coined by Wigner). The paradox, of course, resulted from the fact that independent-particle motion seemed to be incompatible with the ideas of the compound nucleus.

At first, it was suggested that the shell structure might, in some way, be confined to the ground state while the compound-nucleus ideas would describe the excited states of the nucleus. But then Barschall (1952) pointed out that the neutron cross-sections (averaged over individual resonances) for incident neutrons of energy 0–3 MeV showed systematic variations (see fig. 4) that were in striking disagreement with the universal and monotonic pattern expected if the mean free path for the neutron in the nucleus would have been very short ("black nucleus"). This data were then interpreted by Feshbach, Porter and Weisskopf (1953) in terms of an

Table 1
Major developments bearing on compound nucleus (1936–1948).

Development	Parameters	References
Resonance formula	$\sigma_{n\gamma}(E) = \pi\lambda^2 \dfrac{\Gamma_\gamma\Gamma_n}{\left(E - E_0\right)^2 + \left(\frac{1}{2}\Gamma_{\text{tot}}\right)^2}$	Breit and Wigner (1936)
Level density and thermodynamic concepts	entropy: $\alpha \ln \rho$ temperature: $\dfrac{1}{T} = \dfrac{1}{\rho}\dfrac{\mathrm{d}\rho}{\mathrm{d}E}$	Bethe (1936) Bohr and Kalckar (1937)
Nuclear decay as evaporation	reciprocity arguments	Weisskopf (1937)
Cross-sections for "black" nucleus		Bethe (1940) Feshbach, Peaslee and Weisskopf (1947)
Semi-empirical mass formula	bulk energies (volume, surface, symmetry) pairing energy	Weizsäcker (1935)
Collective vibration of nucleus	shape oscillations density fluctuations electric dipole mode	Bohr and Kalckar (1937) Migdal (1944) Baldwin and Klaiber (1947) Goldhaber and Teller (1948)
Fission: The compound nucleus' finest hour!		Hahn and Strassmann (1939) Meitner and Frisch (1939) Bohr and Wheeler (1939)

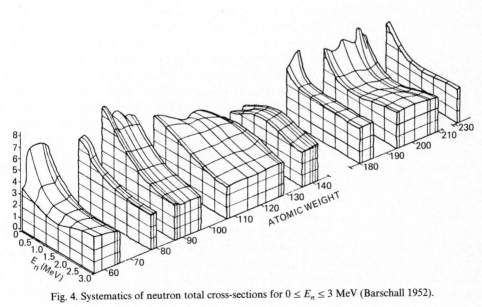

Fig. 4. Systematics of neutron total cross-sections for $0 \leq E_n \leq 3$ MeV (Barschall 1952).

Table 2
Time scales in nuclear reactions.

Aspect	Time scale
Traversal time, τ_0	$\tau_0 = \dfrac{2R_0}{v_{in}} \sim 10^{-22}$ s
Collision time, t_{col} (= mean free path/v_{in})	$t_{col} = \dfrac{\hbar}{w} \sim 6 \times 10^{-22}$ s w = absorption potential ~ 1 MeV
Single-particle residence time, t_{in}	$t_{in} = \tau_0/T$ T = transmission coefficient of nuclear surface $\sim (\dfrac{\lambda_{out}}{\lambda_{in}})$ $\sim 10^{-4}$ slow neutron $t_{in} \sim 10^{-18}$ s
Physical pictures	"black nucleus": $t_{col} < \tau_0$ shell structure: $t_{col} > \tau_0$ compound nucleus: $t_{col} < t_{in}$

"optical" potential in which the neutron mean free path for absorption was ~ 20 fm. Thus, the assumption that the mean free path was short compared to nuclear dimensions, believed to be a cornerstone of the compound nucleus, was shown to be wrong! But still the compound nucleus has survived and continues to be the basis for interpreting a large part of the data on nuclear reactions. The resolution of this paradox is provided by a more careful examination of the characteristic times involved in different nuclear processes. (See table 2.)

The necessary conditions for the occurrence of shell structure (and for systematics of the type pointed out by Barschall) is $t_{col} > \tau_0$, but the condition for formation of the compound nucleus is $t_{col} < t_{in}$ and thus both of these conditions are well satisfied. It is the strong reflection of slow neutrons at the nuclear surface that extends the residence time so effectively and makes the subsequent history of the reaction very sensitive to the rather weak coupling of the projectile to the complicated motion of the compound nucleus. The weakness of this coupling is revealed only in the somewhat detailed features of nuclear reactions, such as the relative narrowness of strength functions and other phenomena that measure the residual features of single-particle motion surviving in the compound nucleus somewhat like the smile that still remains after the disappearance of the Cheshire cat.

I would like to emphasize that the residence time of slow neutrons in the nucleus exceeds the traversal time by such a large factor that the compound nucleus would continue to be the crucial concept in the analysis of neutron reactions, even if the mean free path for energy exchange would have been appreciably longer than the observed value; the co-existence of independent-particle motion and the many-body phenomena of the compound nucleus is thus *not* an uncanny accident hinging on a fine balance in the parameters of the nuclear interactions, but appears to be a rather general feature that is expected in wide classes of quantal systems.

If we now look back over the development of nuclear physics in the period 1933–1952, we see, besides the great discoveries of different types of nuclear reactions and processes, a gradual clarification of the nature of that fascinating new form of matter encountered in nuclei. A deep understanding of the dynamics of this matter could not be built until one had settled on the correct starting point: Is one to start from something like the localized highly correlated picture of a solid, or from the delocalized orbits of particles quantized in the total volume of the nucleus? The question is, of course, intimately linked to the strength of the nuclear forces (measured in units of the Fermi energy which is a measure of the energy required to localize particles at the equilibrium density). From this point of view one may feel that from the start there were strong arguments to believe that the forces are rather weak—in the two-body system there is only one very weakly bound state for $T=0$ and no bound state at all for $T=1$—and thus unable to produce the localization necessary for a quantum solid. We must, however, remember that in assessing this question today we are exploiting the results of a long development in which the analysis of nuclear matter could be compared with a variety of quantal systems encountered in condensed-matter physics and that even with this advantage the answers are not very simple (see, for example, the necessary uncertainty in discussing the deconfinement transitions for quarks and gluons, as well as the question of a possible solid phase in the interior of neutron stars). We are here forcefully reminded that despite the impressive development of the powers of formal analysis, the important many-body problems of nature have repeatedly revealed the deep-seated limitations of straightforward reductionism. Each rung of the quantum ladder has revealed marvelous structures, the interpretation of which has required the invention of appropriate concepts which are almost never discovered as a result of purely formal analysis of the interactions between the constituents.

3. Discovering the feel of the nuclear stuff

The recognition of single-particle motion in the average nuclear potential provided a basis for developing a very detailed understanding of the nuclear dynamics, an understanding that reveals a fascinating tension between the concepts relating to independent-particle motion and those referring to collective features associated with the organized dynamics of many nucleons. I shall not attempt here to even enumerate in any systematic manner the rich variety of phenomena that have been revealed by these studies. Rather, I shall complete the present report with a few remarks on the further evolution of the compound-nucleus idea in connection with the statistical theory of quantal spectra, a development that will have to serve as a single illustrative example exhibiting some of the features of style and perspective characteristic of the third historical period of nuclear studies.

The experimental impetus for this development is again the neutron resonances which played such an important role in the original inspiration of the compound-nucleus idea. It is impossible for me to think about these resonances without a sense of awe at the profound generosity of nature in providing a window in the nuclear spectra at a point where the level densities are about a million times greater than those of the fundamental modes; where the quantal levels are still beautifully sharp in relation to their separation, and where the slow neutrons provide an exquisitly matched tool with which to resolve and measure the properties of each resonance. The effective exploitation of this tool has provided complete spectra comprising hundreds of individually resolved and measured neutron resonances, while corresponding developments in charged-particle spectroscopy have led to the measurement of similar spectra for proton resonances. It was Wigner (1955) who initiated thinking about this material in terms of random matrices.

The idea is to use statistical arguments in order to characterize the wavefunctions and spectra describing the quantal spectrum of the compound nucleus. The compound-nucleus idea implies that the quantal states are complicated mixtures involving all the available degrees of freedom of the many-body system (something like ergodic motion in classical mechanics). Wigner suggested that significant features of these spectra might be modeled by considering, for some region of the spectrum, an expansion of the Hamiltonian matrix on an arbitrarily chosen finite set of basic states. The strong mixing of different degrees of freedom and the randomness of the compound nucleus is expressed by chosing the elements of the Hamiltonian matrix independently and randomly from an appropriate ensemble. We may then ask

Table 3

The random-matrix model.

1. Object of study (Wigner 1955):
   (i)   an ensemble of real orthogonal $N \times N$ matrices (symmetry of H)
   (ii) invariance of ensemble under orthogonal transformation (independence of choice of basis)
   (iii) matrix elements independent random variables, (expressing "randomness" and strong coupling)
      $P(H) = \text{norm}\{\exp(-C \, \text{Tr} \, H^2)\}$ $C = $ constant related to level density.

2. Transform to variables:
   $E_i, \quad i = 1, \ldots, N,$ the eigenvalues
   $X_i = \frac{1}{2}N(N-1),$ "other" variables describing the eigenfunctions
   $P(E_1 \ldots E_N) = \text{norm.} \ \Pi \, | E_i - E_j | \exp(-C \Sigma E_i^2).$

3. Note (Dyson 1962) that the probability distribution of the eigenvalues is identical with the partition function for N particles moving in $1-d$ and interacting with
   (i) an average confining potential $U = -\sum_i x_i^2,$

   (ii) a repulsive two-body force

   $V_{12} = -\ln |x_1 - x_2|.$

4. The analogy in 3 provides a physical picture for unstanding the "repulsion" of levels:
   (i) nearest-neighbor spacings S, approximately described by the "Wigner distribution":

   $$P(S) = \frac{\pi}{2D^2} S \cdot \exp\left(-\frac{\pi}{4}\left(\frac{S}{D}\right)^2\right).$$

   (ii) suppression of long-range fluctuations (screening)
      $N(L) \equiv$ number of levels in interval L
      $\sigma_L^2 \quad \equiv$ mean square fluctuation in $N(L)$
      $\quad = \frac{2}{\pi^2} \ln(\frac{L}{D}) + \text{const.}$

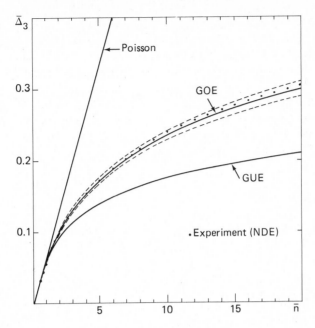

Fig. 5. Level statistic $\bar{\Delta}_3(\bar{n})$. The quantity plotted is the mean square fluctuation in the number of nuclear levels included in an energy interval of a length such that $\bar{n}$ levels would be expected in the average. The experimental data on neutron resonances is compared with the prediction based on the eigenvalues of random orthogonal matrices (Haq et al. 1982).

whether there are significant features in the eigenvectors and the eigenvalues which reflect the strong coupling of the different parts, but are otherwise universal in the sense of being the same for almost all of the matrices generated by this process.

It turns out that the answer to this question is: yes; indeed, as shown by Thomas and Porter, Mehta, Dyson, and French and coworkers, the *fluctuations* in level widths and spacings are just these universal properties [see table 3 and the review article by Brody et al. (1981)]. The extensive evidence from nuclear resonances referred to above has in recent years been shown to agree in striking detail with the prediction concerning these fluctuations based on random matrices (see fig. 5) and thus to confirm the applicability of this characterization of quantal states of the compound nucleus in the regions to which it has been applied. [These ideas have also been invoked in the interpretation of experiments on laser excitation of polyatomic molecules (Abramson et al. 1985, Sundberg et al. 1985) and in the discussion of electronic properties of small metallic particles (Gorkov and Eliashberg 1965).]

While the original formulation of this model was based on random matrices, current developments have made it possible to relate these characteristic features of quantal chaotic motion to more physical models [first to a model of electron motion in a disordered medium (Efetov 1983) and quite recently to direct semi-classical quantization of the classical chaotic motion based on the unstable periodic orbits (Berry 1985)].

The current questions are concerned with issues such as: How can one characterize the transition between the low-energy spectrum with its many conserved quantum numbers (classically multiply-periodic motion) and the compound-nucleus region, exhibiting quantal chaos? And how can one characterize the limitations on the random-matrix model that are associated with the existence of a finite relaxation-time for the nuclear configurations? These issues of chaos in quantal systems are a fascinating chapter in the continuing efforts to digest the significance of the quantal concept. For the nuclear physicist the compound nucleus provides a powerful inspiration in the struggle to understand this issue.

References

Abramson E., R.W. Field, D. Imre, K.K. Innes and J.L. Kinsey, 1985, J. Chem. Phys. **83**, 453.
Arima, A., and H. Hyuga, 1979, in: Mesons in Nuclei II, eds M. Rho and D. Wilkinson (North-Holland, Amsterdam) p. 683.
Aston, F.W., 1920, Philos. Mag. **39**, 449.
Baldwin, G.C., and G.S. Klaiber, 1947, Phys. Rev. **71**, 3.
Barschall, H.H., 1952, Phys. Rev. **86**, 431.
Beck, G., and L.H. Horsley, 1935, Phys. Rev. **47**, 510.
Berry, M.V., 1985, Proc. R. Soc. London **A400**, 229.
Bethe, H.A., 1935, Phys. Rev. **47**, 747.
Bethe, H.A., 1936, Phys. Rev. **50**, 332.
Bethe, H.A., 1940, Phys. Rev. **57**, 1125.
Bjerge, T., and C.H. Westcott, 1935, Proc. R. Soc. London **A150**, 709.
Bohr, N., 1932, J. Chem. Soc. London, p. 349.
Bohr, N., 1936, Nature **137**, 344.
Bohr, N., and F. Kalckar, 1937, Kgl. Dansk. Vid. Selsk. Mat.-Fys. Medd. **14** no. 10.
Bohr, N., and J.A. Wheeler, 1939, Phys. Rev. **56**, 426.
Bothe, W., and H. Becker, 1930, Z. Phys. **66**, 289.
Breit, G., and E. Wigner, 1936, Phys. Rev. **49**, 519.
Brody, T.A., J. Flores, J.B. French, P.A. Mello, A. Pandey and S.S.M. Wong, 1981, Rev. Mod. Phys. **53**, 385.
Chadwick, J., 1932, Proc. R. Soc. London **A136**, 692.
Curie, I. and F. Joliot, 1932, C.R. Hebd. Séances Acad. Sci. Paris **194**, 273.
Dunning, J.R., G.B. Pegram, F.A. Fink and D.P. Mitchell, 1935, Phys. Rev. **48**, 265.
Dyson, F.J., 1962, J. Math. Phys. **3**, 140, 157, 166, 1191, 1199.
Efetov, K.B., 1983, Adv. Phys. **32**, 53.
Estermann, I., and O. Stern, 1933, Z. Phys. **85**, 17.
Feenberg, E., and E. Wigner, 1937, Phys. Rev. **51**, 95.
Fermi, E., 1934a, Z. Phys. **88**, 161.
Fermi, E., 1934b, Ric. Scient. **5**(1), 283.
Fermi, E., and F. Rasetti, 1935, Nuovo Cimento **12**, 201.
Fermi, E., E. Amaldi, O. D'Agostino, F. Rasetti and E. Segré, 1934a, Proc. R. Soc. London **A146**, 483.
Fermi, E., E. Amaldi, B. Pontocorvo, F. Rasetti and E. Segré, 1934b, Ric. Scient. **5**(2), 282.
Feshbach, H., D.C. Peaslee and V.F. Weisskopf, 1947, Phys. Rev. **71**, 145.
Feshbach, H., C.E. Porter, and V.F. Weisskopf, 1953, Phys. Rev. **90**, 166.
Frisch, O.R., 1967, in: N. Bohr, His Life and Work as seen by his Friends and Colleagues. ed. S. Rozental (North-Holland, Amsterdam) p. 137.
Frisch, O.R., 1979, What little I remember (Cambridge University Press, Cambridge).
Frisch, O.R., and O. Stern, 1933, Z. Phys. **85**, 4.
Goldhaber, M., and E. Teller, 1948, Phys. Rev. **74**, 1046.

Goodman, C.D., 1984, Prog. Part. Nucl. Phys. **11**, 475.

Gorkov, L.P., and G.M. Eliashberg, 1965, Zh. Eksp. Theor. Fiz. **48**, 1407 [Sov. Phys. JETP **21**, 940].

Hahn, O., and F. Strassmann, 1939, Naturwissenschaften **27**, 11.

Haq, R.U., A. Pandey and O. Bohigas, 1982, Phys. Rev. Lett. **48**, 1086.

Heisenberg W., 1932 Z. Phys. **77**, 1; **78**, 156; **80**, 587.

Jacob, M., and J.T.T. Van, 1982, Phys. Rep. **8**, 321.

Jensen, J.H.D., 1965, Science **147**, 1419.

Joliot, F., and I. Curie, 1934, C.R. Hebd. Séances Acad. Sci. Paris **198**, 254.

Majorana E., 1933, Z. Phys. **82**, 137.

Mayer, M.G., 1948, Phys. Rev. **74**, 235.

Mayer, M.G., 1964, Science **145**, 999.

Meitner, L., and O.R. Frisch, 1939, Nature **143**, 239.

Migdal, A., 1944, J. Phys. (Moscow) **8**, 331.

Moseley, H.G.J., 1913, Philos. Mag. **26**, 1024.

Peierls, R. (ed.), 1985, Niels Bohr, Collected Works, Vol. 9 (North-Holland, Amsterdam) p. 21.

Perrin, F., and W.M. Elsasser, (1935) C.R. Hebd. Séances Acad. Sci. Paris **200**, 450; J. Phys. Radium **6**, 194.

Prout, W., 1815, published anonymously in: Annals of Philosophy **vi**, 321; **vii**, 111.

Riska, D.O., and G.E. Brown, 1972, Phys. Lett. **38B**, 193.

Rose, M. and H. Bethe, 1937, Phys. Rev. **51**, 205.

Rutherford, E., 1919, Philos. Mag. **37**, 581.

Rutherford, E., 1920, Proc. R. Soc. London **A97**, 374.

Sundberg, R.L., E. Abramson, J.L. Kinsey and R.W. Field, 1985, J. Chem. Phys. **83**, 466.

Tillman, J.R., and P.B. Moon, 1935, Nature **136**, 66.

von Weizsäcker, C.F., 1935, Z. Phys. **96**, 431.

Weisskopf, V.F., 1937, Phys. Rev. **52**, 295.

Wheeler, J., 1979, in: Nuclear Physics in Retrospect, ed. R.H. Strewer (Univ. of Minnesota Press, Minneapolis, MN) p. 213.

Wigner, E., 1933, Phys. Rev. **43**, 252.

Wigner, E., 1955, Ann. Math. **65**, 548.

Yamazaki, T., 1979, in: Mesons in Nuclei II, eds M. Rho and D. Wilkinson (North-Holland, Amsterdam) p. 651.

Yukawa, H., 1935, Proc. Phys. Soc. Jpn **17**, 48.

Discussion, session chairman S. Belyaev

Weisskopf: I was very much impressed by your presentation of a period which I experienced here in this place in the most delightful and most exciting way. I know that you and I have discussed the question before whether Niels Bohr has retarded the development of the shell model or not. Of course, one should never say something negative about a person at his 100th birthday, and I am far from saying anything negative about a man who formed my life and my thinking. However, in some ways you have actually supported my remark, which I have made several times, that the tremendous personality of Bohr has steered our thinking in certain directions. You made yourself the remark that the shell model was actually introduced by outsiders. Now, perhaps this is not quite true. The shell model was brought to Chicago by Enrico Fermi, whom you can hardly consider as an outsider in any part of physics, and he actually induced and encouraged Maria Goeppert-Mayer to investigate these phenomena. It is true that at that time there was a lot of experimental material available to support the shell model, but I think that magic-number effects were already known before, but were not exploited. This may have been caused by the tremendous—I would not say influence of Niels Bohr—but by the tremendous success of the compound-nucleus picture, which opened up so many new perspectives, including fission, that you call the finest hour of that picture. I would call it the most tragic hour.

Peierls: We should not exaggerate the responsibility of Bohr, through his authority, in delaying the study of the shell model. I must admit that I belonged to those who were convinced that the shell model could not work. This view started with the success of Bohrs compound-nucleus picture, but we then looked very seriously and quantitatively at properties on which the validity of the shell model would depend, and convinced ourselves about its impossibility. Our arguments were misleading for a number of subtle reasons, but they did not rely on Bohrs authority. I simply want to say that many of us shared the responsibility for maintaining Bohrs original view longer than it should have been.

Kohn: You indicated that the criteria for nuclear single-particle motion (the shell model) and of the compound-nucleus picture could be explained by ratios between characteristic times, and that these ratios are both of the order of 10^4. What happens with these estimates for really small nuclei where, as far as I know, the compound-nucleus model is not very useful?

Mottelson: There still is enormous difference between the wavelength of the neutron outside the system, where its energy may be on the order of an eV, and the wavelength it has inside the nucleus. So there are still strong reflection effects for slow particles entering also light systems. That corresponding factor is also part of the description of nucleon capture by light nuclei.

Amaldi: I am really very much impressed by the capacity of Mottelson to summarize in three quarters of an hour the essential developments of such a long period. I would, however, like to make two minor remarks of historical nature. You have correctly said that one of the facts that led to the development of the compound-system model by Bohr was the fact that the large capture cross-section was not accompanied by a large scattering cross-section. It should be mentioned that this was proved experimentally by a group at Columbia University. The people were J.R. Dunning, G.B. Pegram, G.B. Fink and D.P. Mitchell, who published a paper in Physical Review in the summer of 1935. You also correctly mentioned the paper by Bjerge and Westcott (1935), and by Tillman and Moon (1935). They were the first to observe that the neutron-capture cross-section does not show the same dependence upon velocity in the different elements as would be necessary if the $1/v$ law had general validity. The fact that the cross-section was changing rapidly was then shown by Fermi and me some months later. In the winter of 1935–1936 we measured the width of the resonances of a few nuclei, and got values close to 0.1 eV, corresponding to a lifetime of 10^{-14} seconds. This value agreed perfectly with the estimate given by Bohr in his paper on the compound nucleus, published in Nature. This made a great impression on everybody.

Weisskopf: Just a short remark about history. The concept of nuclear temperature and evaporation which was ascribed in some extent to me, actually should be ascribed also to Landau. Landau discussed the nuclear temperature first, and I learned about it from a paper by him.

Bjørnholm: How come that Bohr completely ignored the Pauli exclusion principle when proposing the compound-nucleus concept? Would you care to comment on how Bohr was reconciling the idea of a short mean free path with the idea that fermions should have a long mean free path inside the nucleus?

Mottelson: That is an interesting question. Apparently that kind of thinking was not understood—or was not used—in the period before the war, as far as I can gather. There are some notes by Niels Bohr after the discovery of shell structure in the late forties. I believe the notes are dated 1947 or 1948, in which he is trying to face up to the evidence for a long mean free path. He does not use that argument in those notes. He talks about a quantal non-localisation of the particles. But about six months later, in 1948 or 1949, he does refer to the argument that the Pauli principle will effectively prevent the correlations which would be involved in the short-range interaction.

Amaldi: I am sorry to speak again, but I should say something different from my dear friend Weisskopf. The first persons who spoke about nuclear temperature were André Debierne, A pupil and collaborator of Marie Sklodowska Curie and Henri Poincaré, in 1911–1912. Debierne published in part alone, in part in collaboration with Marie Sklodowska Curie, some interesting papers concerning the fact that the "atoms" of radioactive bodies "disintegrate at random". On various occasions, in particular at the end of a lecture that Debierne gave in January 1912 in front of the Société Francaise de Physique, he arrived at the conclusion that inside the atom there is an element of disorder, which causes the atom to pass through a great number of different states, in a very short instant of time, but that such an element of disorder is different from thermal agitation. As an example he suggested that the constituents of atoms are endowed with disordered movements similar to those of molecules of a gas inside a container. Commenting these views of Debierne, Poincaré noticed that this element of disorder should be described by statistical laws, and therefore by a "thermodynamics appropriate to the internal part of the atom", implying that one should define a temperature for the interior of the atom, which is not in thermal equilibrium with the external part. A short presentation of these ideas was also given by Marie Sklodowska Curie at the Solvay Conference of 1913. I would like to emphasize that all these ideas of Debierne and Poincaré were developed and presented in 1911–1912, i.e. before the existence of atomic nuclei was universally recognized. This is why these authors were speaking about atoms and not about nuclei.

The Lesson of Quantum Theory, edited by J. de Boer, E. Dal and O. Ulfbeck

"Complementarity" between Energy and Temperature

Jens Lindhard

Institute of Physics, University of Aarhus
Aarhus, Denmark

Contents

Abstract

Niels Bohr's general conception of "complementarity" between energy and temperature was previously taken up quantitatively only by L. Rosenfeld. In the present chapter it is attempted to reconsider the problem. A detailed discussion is made on statistical equilibria, primarily canonical and microcanonical ensembles, as well as their connection to measurements and fluctuations of energy and temperature. In particular, by formal methods and by direct inspection it is shown how a temperature distribution is obtained for an energy fixation. An "uncertainty" relation is obtained for energy and temperature, of a somewhat different kind than the uncertainty relations in quantum theory. A similar relation is found to connect particle number and chemical potential. But the quantities pressure and volume do not show this behaviour, because pressure, for fixed volume, has no fluctuations in an equilibrium ensemble.

1. Introduction

The concept of complementarity was introduced by Niels Bohr in his analysis of the salient features of quantum phenomena. But within physics he also used it when comparing classical mechanics with thermodynamics. He wrote only little about it.

Still, in his Faraday lecture (Bohr 1932) one whole page is devoted to the mechanical-thermodynamic complementarity; it is formulated in rather general terms and written less lucidly than he usually did.

Many of you will have heard him emphasize the superiority of Gibbs' conceptions as compared to those of Boltzmann. He said that Gibbs' ensemble was the proper kind of theoretical approach, at first extremely abstract, but then when one brought together two ensembles with the same modulus, they turned out to reproduce exactly experimental findings and the basic concept of temperature equilibrium. Heisenberg tells vividly about this in an interview from 1963, quoted in Niels Bohr, Collected Works, Vol. 6 (Kalckar 1985 pp. 324–326).

Heisenberg has also attempted to formulate Bohr's views in his memoirs: "Der Teil und das Ganze" (Heisenberg 1969), where he reconstructs a discussion with Bohr, Kramers and Klein. Heisenberg leaves the impression—an impression that many others have had from Bohr—that a "complementary" conception of the relation between mechanics and thermodynamics was important to Bohr long before he introduced the concept, and the word, in quantum theory. Heisenberg discusses explicitly the complementarity between energy and temperature for a molecule in a cup of tea. But there is no quantitative discussion.

The only quantitative attempt that I know of was published by Rosenfeld (1962). I shall come back to that below, since it will be the starting point of my analysis.

A discussion of complementarity between energy and temperature should have several implications. The quantitative aspects of it may possibly be expressed in terms of "uncertainty" relations between energy and temperature, as will be discussed in the following. But there is also the question of whether energy and temperature belong to different experimental arrangements, and how idealized measurements are performed. Next, we might be able to learn what entropy increases are associated with measurement. This was treated by Szilárd some sixty years ago in his familiar example of a molecule in a box, where a shutter can divide the box in two parts (Szilárd 1929). Szilárd's conclusion about entropy increase in the measuring process was discussed by Bohr in correspondence with Pauli and with Stern (cf. Kalckar 1985 pp. 326–330, 449–456, 467–473). When describing the measuring process in quantum theory, and in later years especially, Niels Bohr emphasized the importance of irreversibility in measurements. The problem of Szilárd is not the subject of the present chapter, however, and it will only be touched upon briefly.

It should be added that a discussion of complementarity between energy and temperature and of their measurement, all within classical physics, may be a useful background for the understanding of quantal phenomena. In point of fact, Heisenberg (1969) concludes the above-mentioned discussion by noting the different attitudes of Bohr and Einstein:

"Wir konnten nun gut verstehen, warum für Niels der grundsätzliche Unterschied zwischen den statistischen Gesetzen der Wärmelehre und denen der Quantenmechanik viel weniger bedeutsam war als für Einstein. Niels empfand die Komplementarität als einen zentralen Zug der Naturbeschreibung, der in der alten statistischen Wärmelehre, insbesondere in der durch Gibbs gegebenen Fassung, schon immer vorhanden, aber nicht genügend beachtet worden war; während Einstein immer noch von der Vorstellungswelt der Newtonschen Mechanik oder der Maxwellschen Feldtheorie ausging und die komplementären Züge in der statistische Thermodynamik gar nicht bemerkt hatte."

2. The assertion of Rosenfeld

Consider the simplest thermodynamic system, where only one parameter is varied, i.e. within classical thermodynamics we are concerned with energy E and entropy S, and with the derived quantities, temperature T and specific heat C,

$$E = E(S), \qquad dE = T\,dS, \qquad \frac{dE}{dT} = C, \tag{2.1}$$

where the specific heat $C = C(T)$ can be used to characterize the properties of the system.

For a system of this kind, L. Rosenfeld (1962) applied fluctuation theory of equilibrium statistical mechanics and obtained the following connection between the fluctuations of energy and temperature:

$$\delta E \cdot \delta T = kT^2, \quad \text{or} \quad \delta E \cdot \delta\beta = 1, \quad \beta = \frac{1}{kT}, \tag{2.2}$$

where the fluctuations are given by the averages

$$(\delta E)^2 = \overline{(E - \overline{E})^2}, \qquad (\delta T)^2 = \overline{(T - \overline{T})^2}. \tag{2.3}$$

Rosenfeld emphasizes that, whereas δE depends on the size of the system, the result (2.2) is independent of the size. He says that the reciprocal relationship between energy and temperature is closely analogous to the uncertainty relations in quantum theory.

Now, there appears to be something quite strange in this result of Rosenfeld. In fact, the canonical ensemble one conceives as having an exact temperature T, and a finite energy fluctuation, in disagreement with eq. (2.2). Similarly, if we have a system with vanishing δE it is hard to imagine that the fluctuation δT is unlimited large.

At this stage it is proper to introduce the canonical distribution of energy E, for a system with differential phase volume, or density of states, $\rho(E)\,dE$. The total differential probability is

$$W(E)\,dE = P(E)\,\rho(E)\,dE = K\,\exp\!\left(-\frac{E}{kT}\right)\rho(E)\,dE, \tag{2.4}$$

where T is the temperature, and where the normalization constant K is associated with the free energy, $\log K = F/kT$.

In the following, I use mostly Gaussian approximations, since they are sufficiently accurate for my purpose. We can expand the density $\rho(E)$ around the point of most probable energy, E_p,

$$\rho(E) \cong \rho(E_p) \exp\left[\frac{E - E_p}{kT} - \frac{(E - E_p)^2}{2\sigma_c^2}\right], \tag{2.5}$$

where E_p and σ_c are determined by, respectively,

$$\frac{\mathrm{d}}{\mathrm{d}E_p} \log \rho(E_p) = \frac{1}{kT}, \quad \text{and} \quad \frac{\mathrm{d}^2}{\mathrm{d}E_p^2} \log \rho(E_p) = -\frac{1}{\sigma_c^2}. \tag{2.6}$$

It follows that, in this approximation, eq. (2.4) becomes

$$W(E) = \frac{1}{\sigma_c\sqrt{2\pi}} \exp\left[-\frac{(E - E_p)^2}{2\sigma_c^2}\right], \tag{2.7}$$

and the energy square fluctuation is determined by the specific heat according to eq. (2.6),

$$\sigma_c^2 = kT^2 C. \tag{2.8}$$

How Rosenfeld derived eq. (2.2) is not completely clear. But he refers to the discussion of fluctuations by Landau and Lifshitz (1958). They consider a small subsystem and derive fluctuations, like Einstein, by connecting probability to entropy and for the remainder use classical thermodynamics.

Now, if in the above canonical fluctuation (2.8) we write $(\delta E)^2 = \sigma_c^2$, and furthermore introduce a formal temperature, changing with E such that

$$\delta T = \delta E / \frac{\mathrm{d}E}{\mathrm{d}T} = \delta E / C,$$

we might replace one δE by δT and arrive at Rosenfeld's formula (2.2). But this replacement contains two errors. First, the temperature does not fluctuate in eq. (2.4) because it is a canonical distribution; second, if we let the formal temperature fluctuate with E, the fluctuations δT and δE in eq. (2.2) are not independent, so that it is not an uncertainty relation, where the fluctuations must be independent.

In order to clarify the situation, I shall proceed in small steps, looking first at the simplest cases, explaining each "Gedankenexperiment" and the connected formalism.

3. Measurements of canonical energy distribution

Canonical and microcanonical ensembles are distributions where complete equilibrium has been obtained within the available phase space, and where thus time does not exist. One can compare equilibria before and after a process has occurred,

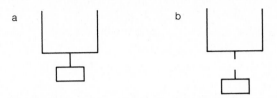

Fig. 1. (a) Canonical equilibrium; (b) isolation of a small system.

however. This is quite like the assumptions in basic thermodynamics. My task is to look into the concepts of energy and temperature, as well as their measurement, for equilibrium ensembles.

The canonical ensemble is an idealized case, like a plane wave in quantum theory. Within Gedankenexperiments, it may be realized with arbitrary accuracy as the phase space distribution of a small system in equilibrium with a very large one. Together, the two form a total system which is isolated and may be supposed to be microcanonical, i.e. with a rather sharply defined total energy. Again, the isolated system is an idealized concept: we can isolate with high perfection, but not completely.

Let me compare one aspect of measurements in quantum theory and in statistical ensembles. If, for the quantal case as well as for the canonical distribution, one desires to measure probability density at a single point ($\bar{r}_0$ or E_m), the fixation of the variable in question requires a drastic intervention and a change of the physical system. This is the type of measurement with which we are concerned for the present. But it should not be forgotten that there are also less drastic measurements, where *averages* of the probability distributions are measured (e.g. average energy), and where the system itself may be left in an essentially unchanged state; in the case of wave functions it could be elastic scattering of an external particle on the system (form factor), and for canonical distributions an example is afforded by measurements of pressure, as we shall see in section 8.

With this in mind we see that the obvious way in which to measure the canonical distribution in energy for the small system is, first, to isolate it from the large system, as illustrated in fig. 1. The small system then becomes a microcanonical

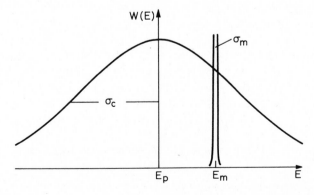

Fig. 2. Microcanonical ensemble after measurement in a canonical ensemble.

system, which is characterized by an energy centre E_m, and a very small width, $\sigma_m \ll \sigma_c$. Second, the value of E_m is measured. The outcome of the experiment will then be some value of E_m, in the neighbourhood of E_p (cf. fig. 2). The distribution of E_m must be given by the Gaussian (2.7), where E is to be replaced by E_m and where the width is the canonical one, σ_c. This is like any proper probability distribution, such as a stationary wave function in space, where $|\varphi(\bar{r})|^2$ gives the probability density for the coordinate $\bar{r}$, when we make a measurement.

We can therefore state that the uncertainties are

$$(\delta E)^2 = \sigma_c^2 = kT^2C, \qquad \delta T = 0, \text{ canonical ensemble.} \tag{3.1}$$

3.1. Entropy change and microcanonical width σ_m

In the above arises, as a side issue, the question of entropy change by reduction from a canonical to a microcanonical ensemble. It is necessary to show that, for the present purposes, this entropy change is not of importance, and neither is the magnitude of the microcanonical width σ_m. But they are of interest in a study of the measuring process itself.

For simplicity, I suppose that a microcanonical ensemble is Gaussian, or

$$W_m(E) = P_m(E)\,\rho(E) = \frac{1}{\sigma_m\sqrt{2\pi}} \exp\left[-\frac{(E - E_m)^2}{2\sigma_m^2} \right]. \tag{3.2}$$

Let us here consider only the change in entropy due to the change of width of the distribution; the consequence of E_m being different from E_p is dealt with in section 5. We therefore assume that $E_m = E_p$, and since the entropy is given by

$$S = -k\int dE\, W(E) \log P(E), \tag{3.3}$$

the entropy change becomes

$$\delta S_{\text{width}} = S_m - S_c = -k \log\frac{\sigma_c}{\sigma_m} + \tfrac{1}{2}k\left(1 - \frac{\sigma_m^2}{\sigma_c^2}\right), \tag{3.4}$$

the formula being valid for any value of σ_m, with a maximum equal to zero for $\sigma_m = \sigma_c$. For $\sigma_m \ll \sigma_c$, it can be seen that eq. (3.4) corresponds to simple expectations, because the available volume in phase space has been reduced by a factor σ_m/σ_c, and $\delta S_{\text{width}}/k$ is essentially the logarithm of that factor.

The total entropy change is slightly different from eq. (3.4) because, as follows from eq. (5.1), the shift of E_m with respect to E_p gives on the average a small change $-k/2$, so that the total entropy change becomes, since $\sigma_m \ll \sigma_c$

$$\delta S \approx -k \log\frac{\sigma_c}{\sigma_m} = -\frac{k}{2}\log\frac{3N}{2} - k \log\frac{kT}{\sigma_m}, \tag{3.5}$$

where, for definiteness, the value of σ_c for a free gas of N particles has been introduced. It should also be noted that the energy shift $E_m - E_p$ gives rise to an exchange of entropy $\sim \pm kN^{1/2}$ between the small system and the large one.

It follows then that, for large N, both of the terms on the right-hand side of eq. (3.5) are negligible compared to the leading entropy terms, proportional to N, or to $N^{1/2}$. By narrowing down one degree of freedom (energy), we cannot essentially affect the entropy of a system with many effective degrees of freedom. In this connection it is noteworthy that, because of the logarithm, an upper limit of as little as $\sim 100\, k$ exists for the value of $|\delta S|$ for any system whatever.

But there is another aspect of the moderate entropy decrease. If we have to suppose, with Szilárd (1929), that the entropy of the measuring apparatus plus system can never decrease, the total process of isolation and measurement must involve an entropy increase at least compensating for the decrease (3.5). This interesting question, however, is outside the main purpose in the present discussion.

4. Temperature determined from energy measurements

In the other sections I approximate all distributions by Gaussians. In order to make a few rigorous deductions, I shall now use the accurate distributions.

The preceding section was concerned with a familiar situation within problems of statistics. With a known parameter $(\beta = 1/kT)$ of the probability distribution, we can observe the various outcomes $(E_1, E_2, \ldots)$. There are simple basic rules for probabilities, including a product rule for probabilities of independent events, or successive measurements.

In the further discussion it is important that the canonical distribution has special properties. Suppose that the density $\rho(E)$ is composed of densities of two independent systems, $\rho_1(E_1)$ and $\rho_2(E_2)$. It holds then that the canonical distribution $\exp(-\beta E)\,\rho(E)\,dE$ is a product of two distributions, $\exp(-\beta E_1)\,\rho_1(E_1)\,dE_1$ and $\exp(-\beta E_2)\,\rho_2(E_2)\,dE_2$, where next $E_1 + E_2 = E$. The total density is given by

$$\rho(E) = \int_0^E dE_1\, \rho_1(E_1)\, \rho_2(E - E_1). \tag{4.1}$$

If we now ask for the probability distribution $W_{E,\beta}(E_1)$ of E_1 for given E and β, we observe that the exponential factor $\exp(-\beta E_1 - \beta E_2) = \exp(-\beta E)$ is independent of E_1, and therefore

$$W_{\beta,E}(E_1)\, dE_1 = \frac{\rho_1(E_1)\, \rho_2(E - E_1)}{\rho(E)}\, dE_1. \tag{4.2}$$

This result is remarkable in that the probability for E_1 is independent of β, so that E_1 has become a redundant variable. It follows that if we make a number of measurements by means of the small system, the only relevant energy is the sum of the measured energies $E = E_1 + E_2 + E_3 + \cdots$, together with the formal total density $\rho(E)$, obtained by successive integrations of the type of eq. (4.1). When the number of measurements goes to infinity, the relative width of the canonical distribution of E tends to zero, and we obtain the limit of classical thermodynamics.

Our central problem consists in the inversion of the above situation (Lindhard 1974): Suppose that we know the results $(E_1, E_2, \ldots)$ of one or more measurements, and ask what statement can be made about the unknown parameter (β) of the distribution. To this end we already found a characteristic property of the canonical distribution, in that only the energy sum E and the integrated density $\rho(E)$ are relevant. Our problem is reduced to an inversion of a distribution of the simple type $W_\beta(E) = -K \exp(-\beta E) \rho(E)$. Since this is a normalized mass distribution along the E-axis changing monotonously with β, the corresponding normalized mass distribution along the β-axis becomes

$$\Pi_E(\beta) \, \mathrm{d}\beta = \mathrm{d}\beta \, \frac{\partial}{\partial \beta} \int_0^E \mathrm{d}E' \, W_\beta(E'). \tag{4.3}$$

The result (4.3) is the unique solution of inversion. Thus, it is obvious that the formula has the necessary property of a repeated inversion leading back to the original distribution. It is also easy to show that it is the only possible solution.

If we introduce the canonical distribution in eq. (4.3), we find for the inversion

$$\Pi_E(\beta) \, \mathrm{d}\beta = \mathrm{d}\beta \, \frac{\partial}{\partial \beta} \int_0^E K \, \mathrm{e}^{-\beta E'} \rho(E') \, \mathrm{d}E'. \tag{4.4}$$

In the particular case of a gas of N free particles, simple scaling prevails, and the two distributions are of the same kind, i.e.

$$W_\beta(E) \, \mathrm{d}E = \frac{1}{\Gamma(\tfrac{3}{2}N)} \beta^{\frac{3}{2}N} \, \mathrm{e}^{-\beta E} E^{\frac{3}{2}N-1} \, \mathrm{d}E, \tag{4.5}$$

$$\Pi_E(\beta) \, \mathrm{d}\beta = \frac{1}{\Gamma(\tfrac{3}{2}N)} E^{\frac{3}{2}N} \, \mathrm{e}^{-E\beta} \beta^{\frac{3}{2}N-1} \, \mathrm{d}\beta. \tag{4.6}$$

Note that when n measurements are made with a gas of N particles, we replace N by nN in eqs. (4.5) and (4.6), as is seen from the composition rule for densities, eq. (4.1).

By means of the distributions (4.4) or (4.6) we have obtained statements about the unknown parameter β of the heat source, when measurements of total energy E are made. The statements are not unlike usual probability distributions, but their contents are of a more abstract kind. The two distributions may be conceived, however, in another way than to give an estimate of an unknown source parameter. In fact, we have merely a total microcanonical system of energy E, irrespective of the way in which it is achieved. Therefore, eqs. (4.4) and (4.6) also represent the distribution of β, or of temperature, for a microcanonical system.

In eqs. (4.4)–(4.6), it is easy to obtain the Gaussian approximations. In point of fact, if we had started from a Gaussian distribution $W_\beta(E)$ with small relative width, the problem of inversion would have been trivial, and would have had the same results.

5. *Indeterminacy in T for microcanonical ensemble*

In place of the formal derivation in the preceding section, let us consider the basic and conceptually simple way of determining the temperature of a system, and thereby find the indeterminacy in T. In fact, suppose that we have a very large system with well-defined temperature T, bring the small microcanonical system in contact with it, and demand that there is no essential change by contact. If this is the case, the temperature of the small system was equal to T. This experiment can also be illustrated by fig. 2, where the narrow peak then represents the small system before contact, and the broad Gaussian indicates its subsequent canonical equilibrium.

The measure of the lack of equilibrium by contact must be the magnitude of the irreversible change of entropy. I have already given the change of entropy, δS_{width}, due to the change in width of the distribution [cf. eq. (3.3)]. We can disregard this unavoidable and constant term. Consider therefore the entropy change connected with the shift of the most probable energy from E_m to E_p. There is an energy transfer $E_m - E_p$ to the large system, for which the temperature remains constant. For the small system we can also use classical thermodynamics [eq. (2.1)], but its temperature changes slightly during the process. In all we obtain an entropy change

$$\delta S_{shift} = \frac{E_m - E_p}{T} + \int_{E_m}^{E_p} \frac{dE'}{T'}$$

$$= \frac{E_m - E_p}{T} + \int_{E_m}^{E_p} dE' \left\{ \frac{1}{T} - \frac{T' - T}{T^2} \right\} = \frac{\left(E_m - E_p \right)^2}{2CT^2}. \tag{5.1}$$

It is not surprising that eq. (5.1) corresponds to minus the exponent in the Gaussian (2.7), representing the canonical distribution, because we are concerned with the same process in the opposite direction.

The increase of entropy, eq. (5.1), remains less than $k/2$ when $(E_m - E_p)^2 < \sigma_c^2$, and when this condition is fulfilled there is effective temperature equilibrium with the large system. Since here the change of energy corresponds to a change of temperature, $dE = C \, dT$, we obtain the following uncertainties for the microcanonical ensemble

$$C^2(\delta T)^2 = \sigma_c^2 = kT^2C, \qquad \delta E \approx 0, \text{ microcanonical ensemble.} \tag{5.2}$$

This simple estimate is in agreement with the precise description (4.4), where the distribution of temperature was obtained.

6. *General fluctuation*

So far, I have merely discussed the two limiting cases of fluctuations, represented by the microcanonical and canonical ensemble. But when we examine these cases, we

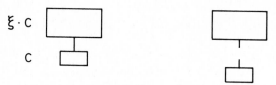

Fig. 3. Equilibrium and isolation.

find that it is not difficult to construct intermediate situations too. In fact, suppose that the small system, with heat capacity C, is in equilibrium with another system having heat capacity ξC. If $\xi = 0$, the small system is microcanonical, and if $\xi \to \infty$ it becomes canonical. The process of measurement, by isolation and subsequent energy determination of the small system, is illustrated in fig. 3 and is quite analogous to the canonical equilibrium in fig. 1.

The combined system is microcanonical with total energy E_{tot}. The density of states is $\rho(E)$ for the small system, and $\rho_0(E_{tot} - E)$ for the other one. From this we find the probability distribution and make a Gaussian expansion around the most probable energy E_p,

$$W(E) = K \rho_0(E_{tot} - E) \, \rho(E)$$

$$\simeq K' \exp\left(-\frac{E}{kT} - \frac{(E - E_p)^2}{2\sigma_c^2 \xi}\right) \exp\left(\frac{E - E_p}{kT} - \frac{(E - E_p)^2}{2\sigma_c^2}\right)$$

$$= \frac{1}{\sigma\sqrt{2\pi}} \exp\left(-\frac{(E - E_p)^2}{2\sigma^2}\right), \tag{6.1}$$

where

$$\sigma^2 = \sigma_c^2 \frac{\xi}{1 + \xi}, \tag{6.2}$$

σ_c being the canonical width [see eq. (2.8)]. This means that the energy fluctuation in equilibrium is somewhat smaller than the canonical one:

$$(\delta E)^2 = \sigma_c^2 \frac{\xi}{1 + \xi}. \tag{6.3}$$

It remains to find the indeterminacy δT in temperature for the equilibrium. But since the total system is microcanonical, with heat capacity $(1 + \xi)C$, we can use the fluctuation eq. (4.2) with the heat capacity changed by a factor $(1 + \xi)$, i.e.

$$C^2(\delta T)^2 = \sigma_c^2 \frac{1}{1 + \xi}. \tag{6.4}$$

This is the temperature indeterminacy for the total system, and hence also for its subsystems. By a more cursory argument, we can also arrive at eq. (6.4) directly from the property (6.3) of the small system. In fact, if the result (6.3) for the fluctuation is supposed to arise from an ensemble mixture of a canonical and a microcanonical system, the former must have a probability $\xi/(1 + \xi)$ and the latter therefore a probability $1/(1 + \xi)$. The temperature fluctuation arises from the microcanonical system only, for which it is given by eq. (4.2), and the probability factor $1/(1 + \xi)$ thus leads to eq. (6.4).

Combining eqs. (6.3) and (6.4) we arrive at the general fluctuation formula

$$(\delta E)^2 + C^2 (\delta T)^2 = kT^2 C, \tag{6.5}$$

for a system with heat capacity C. It is implicitly assumed, because of the Gaussian approximation, that the fluctuations are small in a relative sense. This means that k/C is small, so that the effective number of particles participating is $N_{eff} \gg 1$.

The result (6.5) was obtained for Gaussian distributions with widths between zero and σ_c, corresponding to the range of possibilities obtainable for systems in equilibrium. If we imagine other distributions $W(E)$, the left-hand side of eq. (6.5) cannot become less; it can only increase. Simple examples are a Gaussian with width greater than σ_c, or a non-Gaussian distribution with a square fluctuation equal to σ_c^2, where the left-hand side of eq. (6.5) would exceed the right-hand side. We therefore get a more general result when replacing " $=$ " by " $\geq$ " in eq. (6.5).

We have hereby obtained a quantitative expression for complementarity between energy and temperature. The result is not quite like the uncertainty relations in quantum theory. In fact, it also follows from eq. (6.5) that if we form the product $\delta E\, \delta T$ corresponding to Rosenfeld's formula (2.2), it will not have any particular physical significance, being between zero and an upper limit $kT^2/2$, and with inequality sign in eq. (6.5) there is not even an upper limit.

Let me finally exemplify and extend the result (6.5). Consider a gas of N free particles, where $C = 3NkT/2$, and rewrite the resulting equation in the form

$$\frac{(\delta E)^2}{E^2} + \frac{(\delta\beta)^2}{\beta^2} \geq \tfrac{2}{3}N,$$

so that it holds for more general probability distributions $W(E)$, as well as for small values of N, where eqs. (4.5) and (4.6) are applicable.

7. Particle number and grand potential

It is natural to ask whether connections similar to those in eq. (6.5) exist for other sets of thermodynamic variables too. An obvious possibility is afforded by the grand canonical ensemble, where the particle number and the chemical potential play similar roles as energy and temperature, respectively. For the grand ensemble the number of particles N becomes a free variable and, in analogy to $\exp(-E/kT)$, there appears a probability factor $\exp(N\mu/kT)$, μ being the chemical potential. The analogy to $\rho(E)$ is a weight factor decreasing with increasing N, for large N. In the simplest case—a gas of free particles—the weight factor becomes $\lambda^N/N!$, and the probability of N particles is then given by the Poisson distribution, the average number $\overline{N}$ then being proportional to $\exp(\mu/kT)$.

In general, the square fluctuation of the number of particles in the grand ensemble is

$$\sigma_g^2 = kT \frac{\partial}{\partial \mu} \overline{N}. \tag{7.1}$$

One might now, using Gaussian approximations for the distribution, go through the derivations corresponding to sections 2–6. The final result, corresponding to eq. (6.5), is found easily by

$$(\delta N)^2 + \left(\frac{\partial \overline{N}}{\partial \mu}\right)^2 (\delta \mu)^2 = kT \frac{\partial}{\partial \mu} \overline{N}. \tag{7.2}$$

The Gaussian approximations imply again that relative fluctuations, such as $\delta N/\overline{N}$, are small. In the case of a gas of free particles, eq. (7.2) becomes

$$\frac{(\delta N)^2}{\overline{N}^2} + \frac{(\delta \mu)^2}{(kT)^2} = \frac{1}{\overline{N}}. \tag{7.3}$$

For completeness, it should be mentioned that there are two exceptions to the complete analogy with results in the previous sections, both due to the particle number being discrete, in contrast to the energy. They are of no significance when N is large. First, the decrease in entropy δS_{width}, when the grand ensemble is replaced by a definite number of particles, is

$$\delta S_{\text{width}} = -k \log \sigma_g, \tag{7.4}$$

which quantity becomes $-(k/2) \log \overline{N}$, for a gas.

The second exception to the analogy is concerned with the exact inversion $W_\mu(N) \to W_N(\mu)$, where the attempt to represent a continuum variable (μ) by means of a discrete one (N) introduces a peculiar latitude (cf. Lindhard 1974, §5).

8. The question of pressure and volume

It might seem as if also pressure and volume were a pair of variables which could be of interest in the present context. It should be realized, however, that pressure is quite a peculiar quantity. Since it corresponds to work divided by volume change, it can hardly be well-defined unless the work is performed infinitely slowly. This is because a volume change in a finite time interval contains an ambiguous velocity of a piston, depending on its area, the velocity possibly competing with molecular velocities. The concept of pressure then applies precisely for systems in equilibrium, like canonical and microcanonical ensembles. Pressure can be measured directly in either ensemble by an adiabatic process. But that process contains an unlimited number of collisions, and if we consider the corresponding work for a given volume change δV, the average work $\overline{\delta W} = \bar{p}\delta V$ will be composed of an unlimited number of equivalent terms and therefore the work will have a vanishing fluctuation. Because pressure depends on the number of collisions, and not on the number of particles in the system, it is without fluctuations, and we have no connection to the present fluctuation problems.

This result is also obtained by closer scrutiny of current estimates of pressure fluctuations of canonical distributions, cf. the review by Münster (1959), Wergeland (1962) and Klein (1960). In such treatments it is explicitly, or implicitly, assumed that there are no pressure fluctuations in a microcanonical ensemble. The fluctuations in pressure are claimed to arise as a consequence of energy fluctuations, and estimated to be $(\delta p/p)^2 = (\delta E/E)^2 = \frac{2}{3}N$ in a canonical gas of N particles. However, when a microcanonical ensemble has no fluctuations in pressure, and a canonical ensemble can be a subsystem of a microcanonical one, the canonical ensemble cannot either have fluctuations, since a measurement of pressure for the canonical subsystem can also be a measurement for the total system. The cited pressure fluctuations therefore do not belong to a direct measurement of pressure for a canonical distribution, but rather to a series of experiments corresponding to figs. 1 and 2, where the system is isolated, after which pressure (and energy) is measured; the resulting pressures will then follow the variation of the energies and acquire their fluctuation.

This discussion shows, again, that it is important to specify clearly the detailed experimental background, when one makes theoretical estimates of some physical quantity belonging to a system in statistical mechanics.

Acknowledgements

In a previous paper, in collaboration with Jørgen Kalckar (Lindhard and Kalckar 1982), the relationship of energy and temperature was treated, and preliminary results were obtained. I am much indebted to Jørgen Kalckar for numerous discussions, and for encouragement and criticism, from which I have profited much. I also wish to express my indebtedness to J.U. Andersen and E. Eilertsen for many critical comments on the manuscript.

I am particularly grateful to Susann Toldi for careful and competent preparation of the paper.

References

Bohr, N., 1932, Faraday Lecture, in: J. Chem. Soc. London, p. 349–384; cf. p. 376–377.
Heisenberg, W., 1969, Der Teil und das Ganze (R. Piper & Co. Verlag, München) ch. 9.
Kalckar, J. (ed.), 1985, Niels Bohr, Collected Works, Vol. 6, Foundations of Quantum Physics I (North-Holland, Amsterdam).
Klein, M.J., 1960, Physica **26**, 1073.

Landau, L.D., and E.M. Lifshitz, 1958, Statistical Physics (Pergamon Press, London, New York).
Lindhard, J., 1974, Mat.-Fys. Medd. Dan. Vidensk. Selsk. **30**, No. 1.
Lindhard, J., and J. Kalckar, 1982, Fysisk Tidsskrift **80**, 60.
Münster, A., 1959, Handbuch der Physik, Vol. III/2 (Springer Verlag, Heidelberg) p. 176.
Rosenfeld, L., 1962, Proc. Enrico Fermi School of Physics, Vol. 14 (Academic Press, New York) p. 1.
Szilárd, L., 1929, Z. Phys. **53**, 840.
Wergeland, H., 1962, in: Fundamental Problems in Statistical Mechanics, ed. E.G.D. Cohen (North-Holland, Amsterdam) p. 33.

Discussion, session chairman S. Belyaev

Kubo: You discussed a small system which is in contact with a heat bath. Its distribution of energy is a canonical distribution, determined by the temperature of the bath. So the energy of this system fluctuates, and the fluctuations are given in terms of the bath temperature. The temperature of that small system is something different from the temperature of the heat bath, and you can directly interpret the energy fluctuations as temperature fluctuations. Introducing the heat capacity, you can easily get the result $\Delta E \cdot \Delta T = kT^2$. I think that this is the simplest interpretation of this relation.

Lindhard: This was the Rosenfeld result, with a product $\Delta E \cdot \Delta T$, but that is a case of dependent fluctuations of E and T. It is not at all fluctuations of the kind that are involved in uncertainty relations in quantum mechanics. The two statistical fluctuations ΔE and ΔT have to be independent of each other in order to be of interest.

The Lesson of Quantum Theory, edited by J. de Boer, E. Dal and O. Ulfbeck
© Elsevier Science Publishers B.V., 1986

Radiation by Uniformly Moving Sources

V. L. Ginzburg

P.N. Lebedev Physical Institute
Moscow, USSR

Contents

Abstract

Uniformly and rectilinearly moving charges and other sources with zero eigenfrequency do not radiate in vacuo. In the case of motion in a medium, however, the absence of radiation is rather the exception. Radiation is absent if in a non-moving, homogeneous and time-independent medium a charge moves uniformly at a velocity v, smaller than the phase velocity c_{ph} for all waves that can propagate in this medium. If $v > c_{ph}$, the Vavilov–Cherenkov radiation may occur. However, radiation is also possible for $v < c_{ph}$. This is a transition radiation; it occurs if a source with a zero eigenfrequency moves at constant velocity in an inhomogeneous and (or) nonstationary medium (or near such a medium). The simplest form of transition radiation occurs when a charge crosses the boundary between two media. If the properties of a medium change periodically, the transition radiation is somewhat specific (resonance transition radiation or transition scattering). In particular, transition scattering occurs when a permittivity wave falls on a non-moving charge. Transition scattering is closely connected with transition bremsstrahlung radiation. Transition processes are essential for plasma physics.

We also touch upon the quantum interpretation of the Vavilov–Cherenkov effect and the classical and the quantum theory of the Doppler effect in a medium. The Vavilov–Cherenkov effect, transition radiation and transition scattering have analogies beyond the limits of electrodynamics to be described below.

1. Introduction

There exists a rather large section of electrodynamics—and, in particular, optics—which deals with the radiation of uniformly and rectilinearly moving sources and, first of all, charges. The effects to be discussed below have analogies beyond the limits of electrodynamics (acoustics, chromodynamics and, properly speaking, any field theory).

We shall deal with the radiation of a source moving uniformly in (or near) a medium. Most important are sources with zero eigenfrequency, i.e. static in the reference frame where they are at rest. The sources may be charges, various permanent dipoles, etc. The corresponding radiation is the Vavilov–Cherenkov * radiation or transition radiation in its different forms. If a source has a non-zero eigenfrequency and moves uniformly and rectilinearly, we can observe the Doppler effect. For motion in vacuo the Doppler effect is well-known, but in a medium it may be more complicated (in textbooks this case is usually not considered).

We have chosen the radiation of uniformly moving charges as the subject of the present chapter for three reasons. First, the corresponding class of problems seems to be rather interesting both from the general physical point of view and for the analysis of a number of concrete effects and their application. Second, I have been engaged in the study of these problems from the very beginning of my scientific activities. Third, there exists a fairly close connection with the studies of Niels Bohr who published his famous papers [1,2] devoted to the motion of charged particles through matter, in 1913 and 1915.

A consistent application of electrodynamics of continuous media to the motion of fast charged particles through matter is presented in a book by Landau and Lifshitz [10, ch. 14] (note, however, that for different dipoles and higher multipoles the situation is more complicated than for charges; this question is considered below, in section 5). For charges the role of the medium is essential only for far collisions with an impact parameter $\rho \gg a$, where for a condensed medium $a \sim 3 \times 10^{-10}$ m is the atomic dimension and for a plasma $a \sim N^{-1/3}$ is the mean distance between particles. Under such conditions, say, a radiation of waves with a wavelength $\lambda \gg a$ can be considered on the basis of the equations of electrodynamics of continuous media. Assuming for simplicity the medium to be isotropic and neglecting spatial dispersion, we can take into account all the properties of the medium by introducing the dielectric constant $\epsilon(\omega)$ and the magnetic permeability $\mu(\omega)$.

So, we shall deal with a transparent medium with a refractive index $n(\omega) = \sqrt{\epsilon(\omega)\mu(\omega)}$ and disregard absorption. In the case of transition radiation, however, we also consider an absorbing or nontransparent medium (for a nonabsorbing but nontransparent medium ϵ and μ are real quantities, but $\epsilon\mu < 0$). Here we only present the most important results and make some remarks. References may be found in some original papers [3–5,7,9] and books [8,11,12].

---

* In the Western literature this effect is called the Cherenkov radiation. We (my colleagues, acquainted with the history of the question, and the author; see [6]) believe, however, that only the term "Vavilov–Cherenkov effect" is correct.

2. Vavilov–Cherenkov effect for a charge

If a charge moves uniformly, not in vacuo, but in a medium, the absence of radiation is the exception rather than the rule. In fact, a charge emits no radiation only if its velocity v is less than the phase velocity c_{ph} for all of the electromagnetic waves which can propagate in a given medium. If $v > c_{ph}$, the Vavilov–Cherenkov radiation occurs. Moreover, even if $v < c_{ph}$ the absence of radiation refers only to a medium at rest, which is everywhere uniform and, besides, time-independent. In a non-uniform and/or non-stationary medium a uniformly moving charge radiates also if $v < c_{ph}$: this is the transition radiation. It is interesting that even a charge at rest can radiate in a medium, although this is a rather exotic case (see section 8).

Sommerfeld considered in some detail the following problem: a charge (charged pellet) moves in vacuo at a velocity $v =$ constant and we must find its electromagnetic field. For $v < c$ there is no radiation (a stationary problem is considered). For $v > c$ there occurs radiation and its front forms a conic surface with an angle θ_0 between the normal to it (the wavevector $\mathbf{k}$) and the velocity $\mathbf{v}$. In this case (see fig. 1):

$$\cos \theta_0 = \frac{c}{v}. \tag{1}$$

One may say that condition (1) has a kinematic character—it is the condition for interference of secondary waves excited by the charge along its path. Rewritten in the form

$$\cos \theta_0 = \frac{c_{ph}}{v}, \tag{2}$$

condition (1) holds for any kind of wave and for any static source; but it can be satisfied only if

$$v > c_{ph}. \tag{3}$$

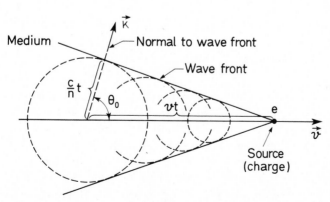

Fig. 1. Formation of Vavilov–Cherenkov radiation; $(c/n)t$ is the light path for the time t, vt is the length passed by the particle during the same time.

Sommerfeld had in a sense bad luck: within a year, we write 1905, the special theory of relativity appeared. According to this theory, the particle momentum is given by

$$p = \frac{mv}{\sqrt{1 - v^2/c^2}}$$

and it is impossible to accelerate it from small velocities to a velocity $v > c$. Moreover, it would seem that the causality requirements also do not allow particles to move with a velocity $v > c$, since they could be used as superluminal signals.

Nobody thought of the simple idea of transferring Sommerfeld's results to the case of a charge moving through a medium. True, even before the appearance of Sommerfeld's papers Heaviside [13] understood correctly the possible role of the medium. However, at that time his papers did not attract due attention. History took a different run. In 1934. S.I. Vavilov and P.A. Cherenkov observed radiation, the nature of which was explained in 1937 by Frank and Tamm [4] (for the history of the discovery see ref. [6]).

Vavilov–Cherenkov radiation is the radiation of a uniformly moving charge in a transparent medium with refractive index $n(\omega)$ and, hence, with phase velocity $c_{ph} = c/n(\omega)$. Conditions (2), (3) thus become

$$\cos \theta_0 = \frac{c}{n(\omega)\, v}, \tag{4}$$

and

$$v \geqslant \frac{c}{n(\omega)}. \tag{5}$$

Frank and Tamm obtained for the energy emitted by a particle of charge e per unit time (that is, along a pathlength v) the following expression:

$$\frac{dW}{dt} = \frac{e^2 v}{c^2} \int_{\frac{c}{n(\omega)\, v} \leqslant 1} \mu(\omega) \left(1 - \frac{c^2}{n^2(\omega)\, v^2}\right) \omega \, d\omega$$

$$= \frac{e^2 v}{c^2} \int \mu(\omega) \left[\sin^2 \theta_0(\omega)\right] \omega \, d\omega. \tag{6}$$

Vavilov–Cherenkov radiation now occupies a prominent place in physics, and a huge number of papers is devoted to it, including books and review articles (see refs [8,11,12] and the literature cited therein). Not a smaller role is played by the Landau damping, i.e. the damping of longitudinal (plasma) waves in a collisionless plasma. Landau [14] concluded that such a damping should exist when, in 1946, he was working on the solution of the initial value problem concerning the propagation of longitudinal perturbations in a collisionless plasma, using the kinetic equation. The collisionless damping which then occurs and which also appears in a number of problems of plasma physics and plasmalike media (for instance, in the case of the "solid-state plasma"—the electron liquid in metals), can be interpreted (if one considers it from a physical point of view) in different ways. One of them is the following: the condition for collisionless wave absorption by electrons in the plasma, which has the form:

$$\omega = kv, \tag{7}$$

is simply the condition for emission (2) for waves (in this case the longitudinal plasma waves) with a phase velocity

$$c_{ph} = \frac{\omega}{k} = \frac{c}{n_\ell}, \qquad n_\ell(\omega) = \frac{ck}{\omega}, \tag{8}$$

where $n_\ell(\omega)$ is the refractive index for the longitudinal waves considered. The collisionless Landau absorption is thus closely connected with the inverse Vavilov–Cherenkov effect for plasma waves (if recoil is neglected, the kinematic conditions for absorption and for emission of waves are the same). When we are dealing with an "external" wave (in this case a longitudinal wave propagating in the plasma) we must consider its interaction not only with a single particle, but with an ensemble of them. As a result it is necessary to take into account not only the absorption of the waves, but also the stimulated emission.

Up to recently it seemed rather obvious that Vavilov–Cherenkov radiation would be impossible in vacuo and also in media with a refractive index $n(\omega) < 1$ (in particular, in an isotropic plasma under conditions where the well-known formula $n(\omega) = \sqrt{1 - \omega_p^2/\omega^2}$ is valid). In fact, however, such a conclusion is incorrect or too rash (see [8, ch. 8]). Quite realistic radiation sources can move at a velocity $v > c$ (particle beams incident upon a metal plate), and there appears a possibility to observe the Vavilov–Cherenkov effect both in vacuo (under normal conditions, however, only if a boundary is present) and also in an isotropic plasma.

The Vavilov–Cherenkov effect is possible in vacuo also far from any boundaries if a strong constant magnetic field B exists which is comparable with the well-known critical value $B_c = m^2c^3/e\hbar = 4.4 \times 10^{13}$ Gauss (here e and m are the charge and the mass of the electron, respectively). As was noted already in the beginning of the 1930s, vacuum in a strong field behaves like a bi-refringent medium. In some cases the refractive index for weak electromagnetic waves propagating in a strong magnetic field is $n_i > 1$ and, hence, a uniformly moving charged particle can emit Vavilov–Cherenkov waves. Here we should not be confused by the fact that (unless we consider motion strictly in the direction of the strong magnetic field) a charged particle can be deflected by the magnetic field. The fact is that we can, in principle, maintain a constant velocity v of a particle by some external means (sources). Besides, one can formally assume the particle mass m to be arbitrarily large; its velocity will then be constant.

I wish to make yet one more methodical remark.

Frank and Tamm derived eq. (6) in 1937 by evaluating the electromagnetic energy flux S through a cylindrical surface surrounding the trajectory of the particle. The present author obtained the same formula in 1939 by evaluating the change in energy $dW_{e\ell m}/dt$ of the electromagnetic field per unit time in the entire space (see [15] and [8, ch. 1]). Finally, the same result (see [9,10]) can be obtained by evaluating the work done by the field on the particle per unit time, i.e. the quantity $ev E$, where the field E is calculated at the position of the charge (eE is the radiative friction force; the other parts of the field do not contribute to the corresponding expressions).

One would perhaps expect that all three methods would give identical results. Indeed, this is the case with the Vavilov–Cherenkov effect. But in the general case for non-stationary charges in vacuo (and in a medium) and, for example, for transition radiation, the quantities S, $dW_{e\ell m}/dt$ and $ev E$ may be different (for details see refs [8,12]).

3. The quantum theory of the Vavilov–Cherenkov effect

Let us now turn to the quantum interpretation of the Vavilov–Cherenkov effect. In general, the Vavilov–Cherenkov effect is described in the framework of the classical theory and quantum corrections are not important. It seems to me, however, that from a methodical (and, if you like from a physical) point of view the quantum approach is useful and interesting.

Let us restrict ourselves to obtain condition (4) for emission and to its quantum generalization (of course, quantum theory enables us to derive eq. (6) with its quantum corrections [16]).

How can one explain, in quantum–mechanical language, the absence of emission by a charge or another static source moving uniformly in vacuo? To do this, it is sufficient to use the energy and momentum conservation laws:

$$E_0 = E_1 + \hbar\omega, \qquad E_{0,1} = \sqrt{m^2 c^4 + c^2 p_{0,1}^2}\,, \tag{9}$$

$$p_0 = p_1 + \hbar k, \qquad \hbar k = \frac{\hbar\omega}{c}\,, \tag{10}$$

where $E_{0,1}$ and $p_{0,1}$ are the energy and the momentum of the charge (source) of rest mass m before (index 0) and after (index 1) the emission of a photon of energy $\hbar\omega$ and momentum $\hbar k = (\hbar\omega/c)(k/k)$ (ω is the radiation frequency). One can verify that it is impossible (and this is also clear from eq. (13) with $n = 1$) to satisfy relations (9) and (10) for $\omega > 0$.

In order to consider radiation by a source in a medium, one must know the energy and the momentum of the radiation (the expression for the energy $E = \sqrt{m^2 c^4 + c^2 p^2}$ of the source is, evidently, not changed). It is not that simple to do this fully consistently, but on an intuitive level the answer is clear at once. Indeed, the presence of the non-moving and time-independent medium does not affect the frequency ω at all, and the wavelength in the medium $\lambda = \lambda_0/n(\lambda)$, where $\lambda_0 = 2\pi c/\omega$ is the wavelength in vacuo; in other words, in the medium the wavenumber is

$$k = \frac{2\pi}{\lambda} = \frac{\omega}{c} n(\omega).$$

Using this substitution, we get, instead of eq. (10),

$$p_0 = p_1 + \hbar k, \qquad k = \frac{\hbar \omega \, n(\omega)}{c}, \qquad p_{0,1} = \frac{m v_{0,1}}{\sqrt{1 - v_{0,1}^2/c^2}}. \tag{11}$$

A simultaneous solution of eqs. (9) and (11) leads to the result

$$\cos \theta_0 = \frac{c}{n(\omega) v_0} \left(1 + \frac{\hbar \omega (n^2 - 1)}{2mc^2} \sqrt{1 - \frac{v_0^2}{c^2}} \right), \tag{12}$$

$$\hbar \omega = \frac{2(mc/n)(v_0 \cos \theta_0 - c/n)}{\sqrt{1 - v_0^2/c^2} \, (1 - 1/n^2)}, \tag{13}$$

where θ_0 is the angle between v_0 and k. If

$$\frac{\hbar \omega}{mc^2} \ll 1, \tag{14}$$

[or for a somewhat more general inequality resulting from eq. (12)], which corresponds to the classical limit, eq. (12) changes to eq. (4), as one expects. The classical limit corresponds to neglecting the recoil when a "photon in the medium" with momentum $\hbar k$ is emitted. It is also clear from eq. (13) that $\omega > 0$ and $\cos \theta_0 < 1$, i.e. emission is possible, only for $v_0 > c/n(\omega)$ [see eq. (5)]. In the classical limit, where the result [in this case expression (4)] does not contain Planck's constant $\hbar$, the quantum calculation has merely a methodical character: it may turn out to be convenient, but it is not unavoidable. The energy and momentum conservation laws can be formulated in the classical region by taking into account the connection between the emitted energy $W_{e\ell m}$ and the change in momentum G of the radiation and of the medium. In accordance with eq. (11) we get

$$G = \frac{W_{e\ell m} n}{c} \cdot \frac{k}{k}. \tag{15}$$

Further, for a freely moving particle with sufficiently small changes in energy and momentum,

$$\Delta E - E_1 - E_0 = v \Delta p = v(p_1 - p_0);$$

indeed,

$$\frac{dE}{dp} = \frac{d}{dp} \sqrt{m^2 c^4 + c^2 p^2} = \frac{c^2 p}{E} = v,$$

furthermore one can put $v_0 \approx v_1 \approx v$, and from the conservation laws (9) and (11), and replacing $\hbar\omega$ by $W_{e\ell m}$, we obtain

$$\Delta E = W_{e\ell m} = v\Delta p = \frac{W_{e\ell m}n}{c}\left(\frac{kv}{k}\right).$$

The energies $W_{e\ell m}$ cancel out, and we are thus led to the classical condition for emission [see eq. (4)]

$$\frac{nv}{c}\cos\theta_0 = 1$$

Relation (15), or $k = \hbar\omega n/c$ [see eq. (11)] corresponds to writing the energy-momentum tensor in a medium in the Minkowskii form. In fact, however, the energy-momentum tensor of the field in a medium has the form proposed by Abraham (we mean the simplest case of a non-dispersive medium) which is reflected in the existence of the Abraham force upon the medium with a density

$$f^A = \frac{n^2 - 1}{4\pi c}\frac{\partial}{\partial t}[EH].$$

One can show that expression (15) is valid also when we use the Abraham tensor, if we are interested in the total momentum of both radiation and medium (for details see ref. [8], ch. 12; the connection with the phonon momentum in a solid body is also treated there). But it is just that quantity which occurs in the conservation law (11) or its classical analogue. The necessity to use expressions (11) and (15) for $\hbar k$ and G follows also from the classical eq. (4) for the angle of the Vavilov–Cherenkov radiation.

4. Vavilov–Cherenkov radiation in the case of motion in channels and gaps

Energy losses due to Vavilov–Cherenkov radiation make up a part of the total, so-called ionization losses. Vavilov–Cherenkov radiation in a transparent medium can go far from the source trajectory. The possibility to eliminate practically all other losses seems nonetheless very interesting. To this end a radiating particle must move in an empty channel or gap in a medium. In this case the losses due to near collisions are absent altogether, polarization losses are strongly suppressed (see also [17]), and the Vavilov–Cherenkov radiation on waves with wavelength λ changes little if $\lambda \gg r$, where r is the radius of the channel, or the width of the gap. This fact was mentioned by L.I. Mandelstam in 1940 (when he spoke at Cherenkov's thesis defence). The transverse electromagnetic field of a charge in a direction perpendicular to the source trajectory is formed in a region on the order of λ. If $r \gg a$, the influence of a channel or a gap can be accounted for by macroscopic equations with the usual boundary conditions. Such calculations [18] show, for example, that in a medium with $n = 1.5$ a charge with $v \to c$ in an empty channel, for which $r/\lambda \sim 0.1$, radiates only by 10–20% weaker than it would in a continuous medium. Even in optics, not to mention longer waves, $r/\lambda \sim 0.1$ for a channel of radius $r \sim 5 \times 10^{-6}$ cm $\gg a \sim 3 \times 10^{-8}$ cm. The possibility to use a channel or a gap is of particular importance, not only for a moving charge, but also for an atom or another complex "system" (see section 6) which would be destroyed in a continuous medium. For suppressing ionization losses the trajectory of a charge or another radiator can simply be placed, instead of into a channel or a gap, in a vacuum outside the medium, but sufficiently close to it (in this case the Vavilov–Cherenkov radiation is small unless $d/\lambda \ll 1$, where d is the distance from the trajectory to the boundary between the medium and the vacuum).

The corresponding electrodynamic problems require rather cubersome calculations [18–20]. For methodical considerations it is instructive to note that in some cases it is useful to use the reciprocity theorem (see, for example, [10, section 89]). The application of this theorem makes it possible to establish that sufficiently thin channels or gaps do not influence the Vavilov-Cherenkov radiation caused by a charge [21] (see also [8, ch. 7]). By means of the same reciprocity theorem one finds that for dipoles and other multipoles even thin channels or gaps do, in general, affect the Vavilov–Cherenkov radiation. This problem is discussed in section 5.

The reciprocity theorem is also useful in solving other problems concerning radiators in a medium. For instance, let an oscillator (a dipole of frequency ω_0) be placed in the centre of an empty spherical cavity of radius $r \ll \lambda_0 = 2\pi c/\omega_0$ in a medium with permittivity $\epsilon(\omega)$. Then, on the basis of the known solution of the electrostatic problem concerning the field in a spherical cavity, the reciprocity theorem immediately suggests that the oscillator radiation field for the oscillator in a cavity differs by a factor of $\frac{3}{2}\epsilon(\omega_0)/(\epsilon(\omega_0) + 1)$ from the radiation field in the case of continuous medium.

5. Vavilov–Cherenkov radiation for electric, magnetic and toroidal dipoles

Condition (4) for the opening of the cone for Vavilov–Cherenkov radiation has a kinematical or interferential character. The opening of the cone is therefore the same for all radiators—charges, dipoles, etc. (for anisotropic medium, see [22,23]). The intensity of radiation, its distribution and polarization along the cone depend on the character of the radiator.

Here we shall consider the Vavilov–Cherenkov radiation for the case of different dipoles. Although this problem has been discussed already for a long time [16,21,24,25], unclear points arose from time to time, which only recently have become clarified [26,27].

Let us write the field equations by means of which one can calculate the

Vavilov–Cherenkov radiation using the macroscopical method:

$$\text{curl } \boldsymbol{H} = \frac{1}{c} \frac{\partial \epsilon \boldsymbol{E}}{\partial t} + \frac{4\pi}{c} \boldsymbol{j}, \tag{16}$$

$$\text{curl } \boldsymbol{E} = -\frac{1}{c} \frac{\partial \mu \boldsymbol{H}}{\partial t}, \tag{17}$$

$$\text{div } \epsilon \boldsymbol{E} = 4\pi\rho, \qquad \text{div } \mu \boldsymbol{H} = 0. \tag{18}$$

We deal here with an everywhere non-moving medium whose properties are described by the permittivity ϵ and the permeability μ (ϵ and μ in eqs. (16)–(18) must be considered as operators, but for the problems considered one may at first put $\epsilon = \text{const}$, $\mu = \text{const}$, and in the final result substitute $\epsilon(\omega)$ and $\mu(\omega)$; see [8]). Besides, in the case of the Vavilov–Cherenkov effect the medium is assumed to be homogeneous.

If a source moves uniformly at velocity $\boldsymbol{v}$ and has a point charge e, an electric moment $\boldsymbol{p}$ and a magnetic moment $\boldsymbol{m}$, then

$$\boldsymbol{j} = e\boldsymbol{v}\delta(\boldsymbol{r} - \boldsymbol{v}t) + \frac{\partial}{\partial t}\{\boldsymbol{p}\delta(\boldsymbol{r} - \boldsymbol{v}t)\} + c\,\text{curl}\{\boldsymbol{m}\delta(\boldsymbol{r} - \boldsymbol{v}t)\} \tag{19}$$

Here, $\boldsymbol{p}$ and $\boldsymbol{m}$ are moments in the laboratory frame of reference (by definition it coincides with a non-moving medium). In the rest frame of the source the values $\boldsymbol{p}'$ and $\boldsymbol{m}'$ are different (see below). If there exists only a charge, the solution of eqs. (16) and (17) leads to eq. (6). We will not write here all the equations for the case of dipoles (see refs [25,26]), but we will give those which are connected with unclear points. For a magnetic dipole $\boldsymbol{m}_\perp$ perpendicular to $\boldsymbol{v}$, we have

$$\left(\frac{\mathrm{d}W}{\mathrm{d}t}\right)_{m_\perp} = \frac{m_\perp^2}{2vc^2} \int \epsilon\mu^2 \left[2\left(1 - \frac{1}{\epsilon\mu}\right)^2 - \left(1 - \frac{v^2}{\epsilon\mu c^2}\right)\left(1 - \frac{c^2}{\epsilon\mu v^2}\right) \right] \omega^3 \, \mathrm{d}\omega. \tag{20}$$

The dipole considered here is purely magnetic in the rest frame: in the laboratory frame it possesses also an electric dipole moment $\boldsymbol{p}_\perp = (1/c)[\boldsymbol{v}\boldsymbol{m}_\perp]$ (for clarity we should recall that in this case $\boldsymbol{m}' = \boldsymbol{m}$, $\boldsymbol{p}' = 0$). For an electric dipole $\boldsymbol{p}_\perp$ perpendicular to $\boldsymbol{v}$ (in this case $\boldsymbol{m}' = 0$, $\boldsymbol{m}_\perp = -(1/c)[\boldsymbol{v}\boldsymbol{p}_\perp]$) we have

$$\left(\frac{\mathrm{d}W}{\mathrm{d}t}\right)_{p_\perp} = \frac{p_\perp^2 v}{2c^4} \int \epsilon\mu^2 \left(1 - \frac{c^2}{\epsilon\mu v^2}\right)^2 \omega^3 \, \mathrm{d}\omega. \tag{21}$$

The magnetic dipole considered above [see eqs. (19) and (20)] is a usual "current" magnetic dipole. But in principle, there may exist also magnetic dipoles of other types (let us call them "true" magnetic dipoles) which are formed by two magnetic monopoles $+g$ and $-g$ [the moment of such a dipole is $\tilde{m} = gd$, where $\boldsymbol{d} = \boldsymbol{r}_2 - \boldsymbol{r}_1$, $\boldsymbol{r}_2$ and $\boldsymbol{r}_1$ being positions of the monopoles $+g$ and g (see fig. 2); of course, for a point dipole $d \to 0$, $gd = \tilde{m}$].

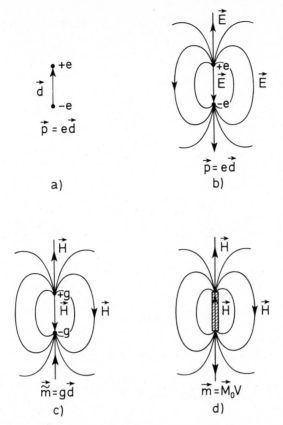

Fig. 2. (a, b) Electric dipole p, (c) "true" magnetic dipole $\tilde{m}$ and (d) "current" magnetic dipole m. The current magnetic dipole is schematically shown here as a rod with magnetization M_0 in a volume V; another possible model is a ring with a current.

To calculate the fields of magnetic monopoles *, dipoles, etc. in a medium, we can in a certain approximation, which we just use here, assume $j = 0$, $\rho = 0$ in eqs. (16)–(18), but add to the right-hand side of eq. (17) the term $-(4\pi/c)\boldsymbol{j}_\mathrm{m}$, where $\boldsymbol{j}_\mathrm{m}$ is the current density of the magnetic monopoles (besides, the second of eqs. (18) takes the form div $\mu\boldsymbol{H} = 4\pi\rho_\mathrm{m}$, where ρ_m is the magnetic charge density).

If the solution of the problem for an electric charge is known, the solution for a magnetic monopole is obtained from the duality principle (for more details see [26]). Thus, for the Vavilov–Cherenkov radiation of a magnetic monopole, where $\boldsymbol{j}_\mathrm{m} = g\boldsymbol{v}\delta(\boldsymbol{r} - \boldsymbol{v}t)$ we obtain from eq. (6):

$$\left(\frac{\mathrm{d}W}{\mathrm{d}t}\right)_g = \frac{g^2 v}{c^2} \int \epsilon \left(1 - \frac{c^2}{\epsilon\mu v^2}\right) \omega \, \mathrm{d}\omega. \tag{22a}$$

---

* For magnetic monopoles the electrodynamics of continuous media is, in general, more complicated than indicated here [39], but this is not especially important for us.

In the case of a "true" magnetic dipole $\tilde{m}_{\perp}$, which is a direct analogue of an electric dipole $p_{\perp}$, we similarly obtain from eq. (21):

$$\left(\frac{dW}{dt}\right)_{\tilde{m}_{\perp}} = \frac{\tilde{m}_{\perp}^2 v}{2c^4} \int \mu \epsilon^2 \left(1 - \frac{c^2}{\epsilon \mu v^2}\right)^2 \omega^3 \, d\omega. \tag{22b}$$

Expression (22b) differs from eq. (20) which refers to the "current" dipole $m_{\perp}$. This result [25] seems paradoxical, because the fields of a current and a true magnetic dipole are quite similar, at least outside the dipoles. But in fact these fields are not quite equivalent. For instance the field of a resting true dipole $\tilde{m}$ is as follows:

$$H = \frac{3(\tilde{m}r)r - r^2\tilde{m}}{\mu r^5}, \quad E = 0, \tag{23}$$

which is analogous to the field of an electric dipole

$$E = \frac{3(pr)r - r^2p}{\epsilon r^5}, \quad H = 0. \tag{24}$$

At the same time the field of a current dipole is

$$H = \frac{3(mr)r - r^2m}{r^5} + 4\pi m \delta(r), \quad E = 0. \tag{25}$$

The absence of the factor $1/\mu$ in the first term of this expression as compared with eq. (23) is associated with different definitions of the moments m and $\tilde{m}$ [26]. The term $4\pi m \delta(r)$ in eq. (25) reflects the difference between the dipoles, which is clear from fig. 2 (for a current dipole the field lines are closed; eq. (25) is the solution of eq. (16) with $j = c \, \text{curl}\{m\delta(r)\}$).

If monopoles move in a medium with rather thin empty channels and gaps, the Vavilov–Cherenkov radiation remains unchanged (see section 4). In the case of dipoles, however, an arbitrarily thin channel or a gap, generally speaking, changes the field and the energy radiated by the source.

The presence of thin channels and gaps does not affect dipoles parallel to the velocity v (and, thus, to the channel axis or to the gap plane). This can be seen by means of the reciprocity theorem [21].

An electric dipole $p_{\perp}$ perpendicular to v radiates in a thin cylindrical channel $[2\epsilon/(\epsilon + 1)]^2$ times more than in a continuous medium; for a gap the amplification is characterized by a factor ϵ^2 (above we have put $\mu = 1$ and assumed that in the laboratory frame of reference the magnetic moment $m = 0$; see [21] and [8, ch. 7]). This fact shows that a narrow surrounding is important for radiation by a dipole. Moreover, it was shown in [28] that a "current" dipole moment radiates like a "true" one, if the medium inside it is not at rest but is moving at the same velocity v as the dipole itself. I should confess, nonetheless, that I have not understood the

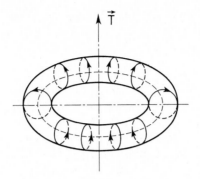

Fig. 3. Toroid with current. T is the toroidal dipole moment.

situation quite well until recently. The consideration of the Vavilov–Cherenkov radiation for a toroidal dipole moment [26,27] has provided a good insight into this matter. An example of a toroidal dipole is a "toroid", i.e. a toroid-like solenoid with current (fig. 3). If such a system is not charged ($\rho = 0$), it does not possess an electric dipole and higher multipole moments. Further, if an azimuthal current is absent (to this end the winding must be, for example, double—it must first go in one and then in the other direction), the system does not possess a magnetic moment either. But inside the toroid the field $H \neq 0$ and the system has a toroidal dipole moment. Its general definition is as follows:

$$T = \frac{1}{10c} \int \{ (rj)r - 2r^2 j \} \, d^3\tau, \tag{26}$$

where j is the current density.

If a point toroidal dipole is at rest, the density of the toroidal moment $\mathcal{T} = T\delta(r)$ and

$$j = c \text{ curl curl} \{ T\delta(r) \}. \tag{27}$$

If a toroidal dipole is moving in a vacuum at constant velocity v, then, as follows from Lorentz transformations, outside the dipole the fields H and E are as before equal to zero. Inside the dipole the field $H \neq 0$ and the field $E = -(1/c)[vH]$.

The fields of a toroidal dipole moving uniformly in a medium cannot, of course, be found by means of Lorentz transformations. The calculation on the basis of eqs. (16) and (17) with the current eq. (27) shows that outside such a toroidal dipole fields are present [27], and if condition (5) holds, the Vavilov–Cherenkov radiation appears with a power

$$\left(\frac{dW}{dt} \right)_{T_\parallel} = \frac{T^2}{c^4 v} \int \mu (\epsilon\mu - 1)^2 \left(1 - \frac{c^2}{\epsilon\mu v^2} \right) \omega^5 \, d\omega. \tag{28}$$

The dipole T is here assumed to be directed along the velocity v and, what is essential, in the rest frame there exists only a toroidal dipole [i.e. the current density in the rest frame is given by eq. (27)]. If in the laboratory frame of reference

$$j = c \text{ curl curl}\{T_v\delta(r - vt)\},$$

then in the rest frame there also exists a quadrupole electric moment, and we obtain for $(dW/dt)_{T_{v,\parallel}}$ an expression [26,27] which differs from eq. (28) by the replacement of $(\epsilon\mu - 1)^2$ by $(\epsilon\mu)^2$ and, of course, by the replacement of T by T_v. By virtue of what has been said one should expect that if a toroidal dipole moves in an empty channel, the Vavilov–Cherenkov radiation completely vanishes. The calculation, naturally, confirms this conclusion [20,30]. For an empty channel or a gap the calculation is, however, not needed—the result is obvious because the field in vacuum, outside a toroidal dipole is equal to zero. But the calculation is necessary if the channel is filled with a medium with permittivity ϵ_0 and permeability μ_0. Then for a thin channel we obtain expression (28) with the replacement of $(\epsilon\mu - 1)^2$ by $(\epsilon_0\mu_0 - 1)^2$, which vanishes as $\epsilon_0\mu_0 \to 1$ and, of course, as $\epsilon_0 \to 1$, and $\mu_0 \to 1$.

What does all this mean? The whole point is that in the case considered the medium fills up the toroidal dipole inside which there is a field. This field affects, that is, polarizes the medium, and this effect lasts also after the dipole has passed. In other words, the dipole leaves a "trace". This picture is especially good for a plasma. Particles (electrons, ions) passing through a dipole are deflected inside it by the field, and therefore the plasma behind the dipole is perturbed. For a toroidal dipole this effect manifests itself, so to say, in a pure form. This also occurs with magnetic moments, and, due to different fields inside a "current" dipole and a "true" dipole [see eqs. (23, (25)] the corresponding "traces" in the medium are also different. According to this conclusion, a "current" dipole and a "true" magnetic dipole moving in an empty channel or in a gap must radiate similarly [21].

How will different dipoles radiate when they move in a continuous medium? Since the radiation of dipoles is already influenced by the motion in thin channels, and while the fields inside the dipoles are essential, it is clear that not only "far" collisions, but also "near" collisions are responsible for the radiation. If the dipoles are macroscopic, then the passage of a medium (for example, plasma) through the dipoles may play some role. But for microscopic (point) dipoles the role of "near" collisions may be considered only in the framework of the microtheory or with an account of spatial dispersion. This also concerns the problems dealing with particle radiation in the channelling process. If we omit channelling of charged particles, the corresponding problems are of no real importance (at least at the present time) simply because of the smallness of the dipole moments of different particles (neutrons, etc.) Radiation by toroidal dipoles moving in a medium may apparently be only of methodical interest. In solid-state theory crystals with a toroidal moment of nonzero density are, however, specific magnetic substances [32].

6. The classical and quantum theories of the Doppler effect in a medium

The quantum theory of the Vavilov–Cherenkov effect in the classical limit (14) does not give anything new, apart from an understanding of the role played by the conservation laws. It is therefore interesting that in more complicated cases quantum theory enables us to reveal interesting points even in the classical limit. To illustrate this, we consider the Doppler effect in a medium.

First, we recall the classical situation using as an example an oscillator with frequency ω_{00} (this is a frequency in a frame of reference in which the oscillator as a whole is at rest). If the oscillator moves in vacuo with a constant velocity v (in the laboratory frame of reference), then in this frame of reference the frequency of the waves emitted by it equals

$$\omega(\theta) = \frac{\omega_{00}\sqrt{1 - v^2/c^2}}{1 - (v/c)\cos\theta} = \frac{\omega_0}{1 - (v/c)\cos\theta}, \tag{29}$$

where θ is the angle between the wavevector (in the direction of observation) and v, and ω_0 is the oscillator frequency in the laboratory frame.

Now let there be a transparent medium [with a refractive index $n(\omega)$], which is at rest in the laboratory frame of reference. We should not be disturbed by the fact that the motion of the source in a medium may lead to large energy losses and, which is important, to the destruction of the source itself (say, of an excited atom). Indeed, as pointed out in section 4, to eliminate losses and destructive collisions, one can make an empty gap or an empty channel in the medium or direct a beam of atoms near the boundary between the medium and the vacuum.

If a medium is present, eq. (29) is replaced by

$$\omega(\theta) = \frac{\omega_{00}\sqrt{1 - v^2/c^2}}{\left|1 - \dfrac{v}{c}n(\omega)\cos\theta\right|} = \frac{\omega_0}{\left|1 - \dfrac{v}{c}n(\omega)\cos\theta\right|}. \tag{30}$$

One can obtain eq. (30) from eq. (29) by using the general rule consisting in the replacement of v/c by vn/c (in the expression $\sqrt{1 - v^2/c^2}$ one should not, of course, make this substitution, as it does not concern the emission process). However, one can obtain eq. (30) also by solving the problem of an oscillator moving in a medium. It is nontrivial that absolute values occur in eq. (30). If the motion is subluminal ($v < c/n$) or if in the case of a superluminal motion the emission proceeds outside the cone [eq. (4)], i.e.

$$\frac{v}{c}n(\omega)\cos\theta < 1, \tag{31}$$

we are dealing with the usual, normal Doppler effect. The so-called complex Doppler effect due to dispersion (i.e. due to the dependence of n on ω) is also possible in this case (the complex Doppler effect as well as eq. (30) with the modulus were first considered by Frank in 1942 [24]).

If the motion is superluminal, then under the condition

$$\frac{v}{c}n(\omega)\cos\theta > 1 \tag{32}$$

[i.e. when there is emission into the cone (eq. 4) which is often referred to as the

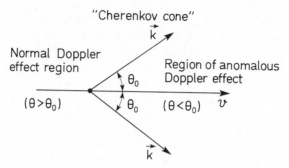

Fig. 4. Regions of normal and anomalous Doppler effect.

"Cherenkov cone"; see fig. 4] eq. (30) without the modulus would give negative values of the frequency ω. From this it is clear that it is necessary to introduce the absolute signs.

Under condition (32) the Doppler effect is called anomalous. If dispersion is taken into account, the whole picture is rather complicated. Since here we are interested in another aspect of the question, we shall neglect dispersion. In this case it follows from eq. (30) with $n(\omega) = n = $ const. that, on the Cherenkov cone itself [for $(vn/c) \cos \theta = (vn/c) \cos \theta_0 = 1$; see eq. (4)] the frequency $\omega(\theta_0) = \infty$ and $\omega(\theta) \to \infty$ as $\theta \to \theta_0$ on both sides of the cone. It is impossible to say anything more on the basis of eq. (30), and the difference between the normal and the anomalous Doppler effects does not seem to be a profound one.

We now turn to a quantum derivation of the formula for the Doppler effect in a medium [33]. To do this, one should use the conservation laws (9) and (11), but replacing the particle energies $E_{0,1}$ for a charge (or, more generally, for a source without internal degrees of freedom) by the expression $E_{0,1} = \sqrt{(m + m_{0,1})^2 c^4 + c^2 p_{0,1}^2}$, where $(m + m_0)c^2 = mc^2 + W_0$ is the total energy of the system (atom) in the lower state, 0, and $(m + m_1)c^2 = mc^2 + W_1$ is the total energy in the upper state, 1.

For simplicity we consider here a two-level system and refer to the state with a larger energy as the upper state [that is, $W_1 > W_0$, and the frequency of the radiation by an atom at rest is $\omega_{00} = (W_1 - W_0)/\hbar$].

If we now apply the conservation laws in the classical limit (14) and the relation $\Delta E = E_1 - E_0 = v \Delta p$, we arrive at eq. (30). A more general calculation, which takes into account recoil, is presented in ref. [33]. It is, however, not on account of the quantum corrections that this is of importance, but because of the fact that, tracing down the signs one can observe an important point which is, of course, completely hidden in the classical derivation of eq. (30): In the region of the normal Doppler effect [eq. (31)] the emission at frequency ω corresponds to a transition of the atom from the upper state 1 to the lower state 0 (the direction of the transition is determined from the requirement that the energy $\hbar\omega$ of the emitted quantum should be positive, that is from the requirement $\omega > 0$). In the case of an anomalous Doppler effect [eq. (32)], i.e. if a quantum of energy $\hbar\omega$ is emitted into the Cherenkov cone, the atom must transit from below (state 0) upwards (to state 1)

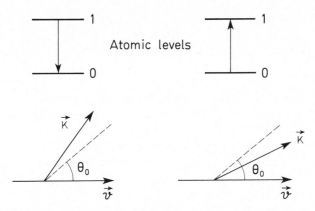

Fig. 5. Transitions between levels 0 and 1 in the case of a normal and an anomalous Doppler effect.

(For some more details see also fig. 5 and refs. [33,34].) There is no contradiction here—the energy that goes to the excitation of the radiating system (the atom), as well as the radiation energy $\hbar\omega$ itself, is derived from the kinetic energy of translational motion.

Therefore, in the case of superluminal motion $v > c/n$, where the anomalous Doppler effect is possible, the radiating atom, even if initially not excited will get excited with the simultaneous emission of quanta inside the Cherenkov cone. Making a transition to a lower state, the excited atom emits quanta at angles $\theta > \theta_0$, i.e. outside the Cherenkov cone. In this connection see also [40].

I think it would have been rather difficult to establish this unusual picture without quantum calculation. We can confirm the result and develop the theory further using the classical calculations of the radiative friction force acting upon a superluminal oscillator. In accordance with the above conclusion, the waves emitted outside the Cherenkov cone lead to a damping of the oscillator vibrations, whereas the waves emitted inside the cone (the anomalous Doppler effect) pump up the oscillator vibrations (see ref. [8, ch. 7] and the literature cited therein).

7. Transition radiation at the boundary between two media

If the medium is inhomogeneous and (or) changes in time or if such a medium lies near the trajectory of a source, transition radiation occurs. This radiation originates when a charge (or another source with zero eigenfrequency) moves uniformly and rectilinearly under non-uniform conditions—in an inhomogeneous or a time-dependent medium (or near such media). Transition radiation in general may coexist and interfere with the Cherenkov radiation and with the radiation due to charge acceleration (i.e. with bremsstrahlung, synchrotron radiation, etc.). But for a deeper insight into the physics of the case, we consider the transition radiation alone.

Let a charge move at a constant velocity

$$v < \frac{c}{n}, \tag{33}$$

such that Vavilov–Cherenkov radiation does not occur. If, besides, we are dealing with a vacuum ($n = 1$), there is no radiation at all. For the radiation to appear in a vacuum, a charge (or a multipole) must be accelerated or, in other words, the parameter v/c, which characterizes the radiation, must change. If the medium is transparent, this parameter has the form $v/c_{ph} = vn(\omega)/c$. This is equal to the ratio of the particle velocity v to the phase velocity of light, $c_{ph} = c/n(\omega)$.

The radiation occurring when the parameter vn/c changes due to changes in n along or close to the trajectory of a source with $v = $ const is called transition radiation. To be precise, in the general case of an absorbing medium the role of the refractive index n is played by $\sqrt{\epsilon} = n + i\kappa$, where ϵ is the complex dielectric permittivity of the medium (for simplicity we assume here, and below, that the medium is nonmagnetic, i.e. $\mu = 1$).

The simplest problem of this kind is a charge crossing the boundary between two media considered by I.M. Frank and the present author in 1944 [35]. Transition radiation is an even simpler effect than Vavilov–Cherenkov radiation. The reason for the possibility of transition radiation to be revealed so late is the same as in the case of the Vavilov–Cherenkov effect.

It is useful to recall the most obvious explanation of the reason for the occurrence of transition radiation when a charge crosses the boundary between media. It is well known that the electromagnetic field in the first medium (in the medium in which the charge moves at a given moment of time) can be represented as the field of the charge itself and the field of its "mirror image" moving in the second medium towards the charge. When the charge and its image cross the boundary, they partially "annihilate" as it were "from the point of view" of the first medium, and "reconstruct themselves"; and this leads to the radiation. Especially simple is the case of a charge impinging normally on an ideal mirror. When crossing the boundary of the mirror, the charge and its image $-e$ "annihilate" each other completely or, rather, stop at the boundary (in the sense that the radiation occurring in a vacuum is the same as the radiation by an incident charge e and its image $-e$ which simultaneously stop at the boundary; c.f. fig. 6).

To find the emitted energy W in this simplest case, we need not solve the rather cumbersome boundary problem, but we can use a simple formula for radiation by charges which suddenly change their velocity:

$$W(\omega, \theta, \varphi) = \frac{1}{4\pi^2 c^3}\left\{\sum_i e_i\left(\frac{[v_{i2}s]}{1 - (v_{i2}s)/c} - \frac{[v_{i1}s]}{1 - (v_{i1}s)/c}\right)\right\}^2. \tag{34}$$

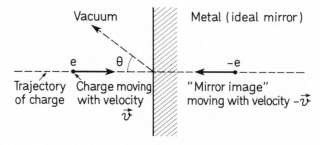

Fig. 6. Transition radiation of a charge e crossing the boundary between vacuum and a metal.

Here e_i is the charge of the ith particle, whose velocity changes sharply from v_{i1} to v_{i2}, and $s = k/k$ is the direction of the radiation wavevector characterized by the angles θ and φ; the total frequency-dependent energy density of the radiation is $W(\omega) = \int W(\omega, \theta, \varphi) \sin\theta \, d\theta \, d\varphi$, and the total energy $W = \int W(\omega) \, d\omega$.

If in a vacuum (in the absence of boundaries, etc.) one charge $e_1 = e$ stops abruptly or accelerates rapidly from the rest state to velocity v, then [see eq. (34)],

$$W(\omega, \theta, \varphi) = \frac{e^2 v^2 \sin^2\theta}{4\pi^2 c^3 [1 - (v/c)\cos\theta]^2},$$

$$W(\omega) = \frac{e^2}{\pi c}\left(\frac{1}{v/c}\ln\frac{1 + v/c}{1 - v/c} - 2\right). \tag{35}$$

In the case of transition radiation on an ideal mirror, one assumes in eq. (34) that the charge $e_1 = e$ with velocity v and the charge $e_2 = -e$ with velocity $-v$ stop abruptly at the boundary (see fig. 6). As a result, one observes radiation in medium 1 (in vacuo) with energy

$$W_1(\omega, \theta) = \frac{e^2 v^2 \sin^2\theta}{\pi^2 c^3 \left[1 - (v/c)^2 \cos^2\theta\right]^2},$$

$$W_1(\omega) = \frac{e^2}{\pi c}\left[\frac{1 + (v/c)^2}{2v/c}\ln\frac{1 + v/c}{1 - v/c} - 1\right]. \tag{36}$$

Here θ is the angle between k and $-v$, as shown in fig. 6. In the nonrelativistic case (i.e. for $v \ll c$)

$$W_1(\omega, \theta) = \frac{e^2 v^2 \sin^2\theta}{\pi^2 c^3}, \qquad W_1(\omega) = \frac{4e^2 v^2}{3\pi c^3}. \tag{37}$$

In the ultrarelativistic case ($v \to c$)

$$W_1(\omega) = \frac{e^2}{\pi c}\ln\frac{2}{1 - v/c} = 2\frac{e^2}{\pi c}\ln\frac{2E}{mc^2},$$

$$E = \frac{mc^2}{\sqrt{1 - v^2/c^2}} \gg mc^2, \tag{38}$$

which coincides with the radiation of a single particle in the same limit [see eq. (35)]. The latter is clear since, when $v \to c$, the radiation is directed along the charge velocity, and therefore the radiation of the charge e "going into metal" is not observed (in the medium 1—in vacuo—one observes only the radiation of the mirror image, of the charge $-e$, with a velocity $-v$). In the nonrelativistic limit the energy (37) is four times larger than the energy (35) for a single charge radiating into the backward hemisphere (into the vacuum), because for a nonrelativistic velocity the fields of the charge e and of its mirror image $-e$ add up, i.e. they are doubled.

The transition radiation discussed here occurs if a boundary between two media with different "electrical" parameters (e.g. dielectric permittivity, refractive index) is crossed. However, initially attention was concentrated on the incidence of a charge on a metal (which may not be a perfect mirror) and, hence, on transition radiation —mainly optical—in the backward direction, which is observed in vacuum. For relativistic particles of sufficiently high energy it is, however, quite realistic that a particle may well pass through a medium and go into a vacuum. This problem is equivalent to the previous one, and the corresponding formula for the radiation intensity is simply derived by a replacement of the velocity v by $-v$ (see below). At the same time, in the calculation of fields there is no symmetry in these cases and the radiation intensities are different when we replace v by $-v$, and under certain conditions the differences are large. For forward radiation and, in particular, when a particle leaves the medium and enters the vacuum, the radiation spectrum has higher frequencies. In a condensed medium the transition radiation of relativistic particles can stretch out into the X-ray part of the spectrum. It is not expedient to give here the solution of the corresponding boundary problems (see refs. [10–12] and the literature cited therein).

The final formula for the case where medium 1 is a vacuum and medium 2 is described by the complex permittivity ϵ [35], is given by

$$W_1(\omega, \theta)$$

$$= \frac{e^2 v^2 \sin^2\theta \cos^2\theta \left|(\epsilon - 1)\left[1 - v^2/c^2 + (v/c)\sqrt{\epsilon - \sin^2\theta}\right]\right|^2}{\pi^2 c^3 \left|\left[1 - (v^2/c^2)\cos^2\theta\right]\left[1 + (v/c)\sqrt{\epsilon - \sin^2\theta}\right]\left(\epsilon\cos\theta + \sqrt{\epsilon - \sin^2\theta}\right)\right|^2}.$$

$$(39)$$

For an ideal mirror one can assume that $|\epsilon| \to \infty$ and eq. (39) goes over into eq. (36), as should be the case. The expression for $W_2(\omega, \theta)$, referring to the case where the charge e leaves a medium of permittivity ϵ (medium 1) for a vacuum (medium 2) with velocity v, is derived from eq. (39) by a replacement of v by $-v$; besides, the angle θ is now the angle between k and v [but not beteen k and $-v$, as in eq. (39)]. The replacement of v by $-v$ in eq. (39) is not at all innocent—in the denominator there appears a factor $[1 - (v/c)\sqrt{\epsilon - \sin^2\theta}]$, instead of $[1 + (v/c)\sqrt{\epsilon - \sin^2\theta}]$. This is just the reason why there appear higher frequencies in the radiation spectrum when a particle leaves the medium (we must take into account that ϵ approaches unity for high frequencies). As a result the total intensity (integrated over all angles and frequencies) also increases; in the simplest case it turns out to be proportional to

$$\frac{E}{mc^2} = \frac{1}{\sqrt{1 - v^2/c^2}}$$

(see section 10 below; E is the total energy of a radiating charge of mass m). This important fact was clarified in 1959 by Barsukov and Garibyan [36]. This opened up much wider perspectives for the creation of efficient "transition counters" intended for determination of relativistic particle energies.

The appearance of a possibility for a practical application to high-energy physics, stimulated interest to transition radiation. Altogether, transition radiation, this rather simple and clear effect in the field of classical electrodynamics, had hardly attracted any attention for about 15 years; now it is very popular, although mainly in the context of developing and applying transition counters. The latter problem has been mentioned even in a number of review articles, apart from a large number of papers (for references see [12,38]).

8. Transition radiation as a more general phenomenon; formation zone

Transition radiation in the broad sense is also of value from a general physical point of view. It develops certain ideas and a "language", and thereby facilitates further developments. The situation here is similar to the one with the Vavilov–Cherenkov effect used in Cherenkov counters.

Our attention has so far been concentrated on the transition radiation which occurs when one or several boundaries between media are crossed. In the latter case we deal either with an ordered sequence of boundaries, i.e. a system with a definite period, or with randomly distributed boundaries (inhomogeneities). Another trend which has developed is based on the fact that any radiation is formed not in a point, but in some region (the formation zone) whose dimension is determined by the wavelength λ, but its dimension can also be appreciably larger. This is the reason why the Vavilov–Cherenkov effect occurs when a particle moves in a vacuum but near a medium (in a channel, a gap or near a boundary between media). Quite similarly, transition radiation (which in this case is called diffraction radiation) occurs when a source (charge) moving uniformly in a vacuum (or in a uniform medium) passes close to some obstacles—metallic or dielectric globules, diaphragms, a diffraction grating, etc. Apart from the above general remark, the occurrence of this transition radiation can also be explained on the basis of the method of images.

For relativistic particles, when we consider radiation in the direction of their velocity, the formation zone generally increases with an increasing particle energy. For instance, in vacuo the size of the formation zone L_f in the direction of the velocity for a given radiation wavelength λ increases proportionally to

$$\left(\frac{E}{mc^2}\right)^2 = \left(1 - \frac{v^2}{c^2}\right)^{-1},$$

where E is the total charge (source) energy and it is assumed that $E \gg mc^2$.

The concept of the radiation formation zone and its size L_f, and the concept of the radiation formation time $t_f = L_f/v$, are of great importance not only in electrodynamics, but also in high-energy physics (see [11,24] and [37, section 93]).
The derivation of the expressions for L_f and t_f for a source moving at velocity v in a transparent medium with a refractive index $n(\omega)$ and emitting waves at an angle θ

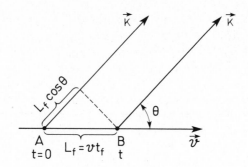

Fig. 7. The length L_f of the radiation formation zone.

to v (fig. 7) seems worth considering here. Let at the moment $t = 0$ the source be at a point A and the phase of the wave emitted by it in direction k be equal to φ_A. We define the formation time t_f as the time after which the phase of the wave φ_B emitted at a point B in the same direction k differs from the phase of the wave φ_A emitted at point A by 2π. The phase factor of the wave has the form $\exp i\varphi = \exp i(\mathbf{k}\mathbf{r} - \omega t)$. As is clear from fig. 7,

$$|\varphi_A - \varphi_B| = |kL_f \cos\theta - \omega t_f| = \left|\frac{\omega n}{c} v t_f \cos\theta - \omega t_f\right|, \tag{40}$$

since the size of the formation zone is the path $L_f = v t_f$. It follows from eq. (40) that

$$L_f = v t_f = \frac{2\pi v}{\omega|1 - (v/c)\, n(\omega)\cos\theta|} = \frac{(v\, n(\omega)/c)\lambda}{|1 - (v/c)\, n(\omega)\cos\theta|}. \tag{41}$$

The meaning of the formation time t_f is especially obvious when we deal with forward radiation in a vacuum: in this case, within the formation time t_f the radiation is ahead of the particle by a wavelength λ [indeed, in a vacuum for $\theta = 0$, according to eq. (41),

$$t_f = \frac{\lambda}{c(1 - v/c)} \quad \text{and} \quad (c - v)t_f = \lambda = \frac{2\pi c}{\omega};$$

when we pass over to a medium, we should replace c by $c_{\text{ph}} = c/n$ (since we deal with phase relations) and from

$$\left(\frac{c}{n} - v\right)t_f = \lambda = \frac{2\pi c}{\omega n}$$

we obtain expression (41) for $\theta = 0$]. In a vacuum, for $\theta = 0$, the formation length

$$L_f = \frac{\lambda \cdot v/c}{1 - v/c},$$

and for $v \to c$, as has already been mentioned,

$$L_{\mathrm{f}} = 2\lambda \left(\frac{E}{mc^2} \right)^2, \quad E \gg mc^2. \tag{42}$$

For sufficiently high energies the dimension of the formation zone L_{f} and time t_{f} can increase and finally exceed greatly the wavelength λ and time λ/v. For instance, the intensity of transition radiation from two boundaries, say, boundaries of a plate of thickness d may be considered (the interference terms being neglected) as a sum of the radiation intensities from one boundary only, provided that $L_{\mathrm{f}} \ll d$. If $L_{\mathrm{f}} \gtrsim d$ and especially if $L_{\mathrm{f}} \gg d$, the radiation from a plate differs markedly from the radiation from two independent boundaries.

What has been said determines to a large extent the specificity of transition radiation by a pile of plates or, in general, by a regularly inhomogeneous medium (for details concerning this radiation, which was called the resonance transition radiation or transition scattering, see refs. [11,12]).

Another type of transition radiation occurs in a homogeneous but time-dependent medium. The essence of the matter is explained most easily in terms of the parameter vn/c. In order that a transition radiation can occur (for $v = \text{const.}$), the refractive index must change on the charge trajectory or near to it. But this change will also take place if the index n changes in time. This kind of transition radiation can occur even for a charge which is at rest relative to the medium. Indeed, if by applying a magnetic or an electric field, or by some other means, one changes the medium rather sharply from an optically isotropic state to an anisotropic state, the polarization of the medium surrounding the fixed charge changes, thereby losing its spherical symmetry. This change in the polarization entails the emission of electromagnetic waves.

Like the Vavilov–Cherenkov radiation, the transition radiation is also of a very general character in the sense that it takes place for various kinds of waves. As an example we mention transition radiation of acoustic waves, arising when a moving dislocation crosses a grain boundary in a polycrystalline body. Other problems connected with transition radiation are also of interest in acoustics (see also section 11, below).

Interesting and may be of importance in applications to pulsar magnetospheres is transition radiation in the presence of a strong magnetic field leading to nonlinear electrodynamic effects.

9. Transition scattering; transition bremsstrahlung

If transition radiation occurs when a charge moves in a medium with a periodically (say, sinusoidally) changing refractive index, it may be called not only transition radiation or resonance transition radiation, but also transition scattering. A dielectric permittivity (refractive index) wave, which can be a standing or a travelling wave, is in this case scattered by a moving charge generating electromagnetic (transition) radiation. But the term transition scattering used in this case instead of transition radiation would be irrelevant, if the effect did not take place also in the limiting case of a charge at rest.

We consider an isotropic medium characterized by a dielectric permittivity ϵ which depends on the density ρ of the medium only. If a longitudinal acoustic wave propagates in this medium, the density is $\rho = \rho^{(0)} + \rho^{(1)} \sin(k_0 r - \omega_0 t)$ and, by virtue of what has been said above,

$$\epsilon = \epsilon^{(0)} + \epsilon^{(1)} \sin(k_0 r - \omega_0 t), \tag{43}$$

where $\epsilon^{(1)}$ is the change in ϵ caused by the change in ρ (in the simplest case $\epsilon^{(1)} = \text{const.} \, \rho^{(1)}$). We have chosen a definite mechanism for the change in ϵ (in this case due to a change in the density ρ) to make our consideration concrete.

We now place in a medium a fixed (or, say, an infinitely heavy) charge e. Around this charge there appears an induction and a field

$$D(r, t) = \epsilon E(r, t), \quad D^{(0)} = \frac{er}{r^3}, \quad E^{(0)} = \frac{er}{\epsilon^{(0)} r^3}, \tag{44}$$

where the superscript (0) implies an "unperturbed" problem (the field of a charge without a permittivity wave). In the presence of a permittivity wave a varying polarization

$$\delta P = \frac{\delta D}{4\pi} = \frac{\epsilon^{(1)} E^{(0)}}{4\pi} \sin(k_0 r - \omega_0 t) \tag{45}$$

arises around the charge, in a first approximation (corresponding to the assumption $|\epsilon^{(1)}| \ll \epsilon^{(0)}$). This polarization which possesses no spherical symmetry for $k_0 \neq 0$, results in the appearance of an electromagnetic wave with a frequency ω_0, emanating from the charge (see fig. 8). The wave number of this wave is

$$k = \frac{2\pi}{\lambda} = \frac{\omega^{(0)}}{c} \sqrt{\epsilon^{(0)}} \, .$$

If, as we assumed, the permittivity wave is caused by the acoustic wave, $k \ll k_0 = \omega_0/U$, where U is the sound velocity (we assume here that $U \ll c/\sqrt{\epsilon^{(0)}}$).

The electromagnetic wave may be regarded as being scattered in the same sense as for other kinds of scattering, such as, for instance, the scattering of an electromagnetic wave by an electron at rest (in this case, we mean a rest state, only when the effect of an incident wave is disregarded). Transition scattering plays a prominent role in plasma physics (see section 10, below) and is on the whole a rather general phenomenon which occurs in a vacuum when an electromagnetic or a gravitational wave is incident on the region with a strong constant (static) or quasi-stationary electromagnetic field.

Closely related to the transition scattering process is transition bremsstrahlung. It occurs in a medium if a uniformly and rectilinearly moving charge e passes close to another charge e at rest. Transition bremsstrahlung is similar in its characteristics to

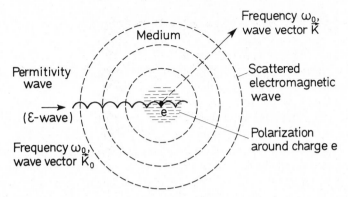

Fig. 8. Schematic picture characterizing the process of transition radiation formation on a non-moving (fixed) charge.

ordinary bremsstrahlung, although for its occurrence an acceleration (a change in the rectilinear trajectory or deceleration) is not necessary. The term "transition bremsstrahlung" is justified also because this radiation occurs in particle collisions. Moreover, transition bremsstrahlung of electrons is described by expressions which are very close to the corresponding formulae for ordinary bremsstrahlung. Further, transition bremsstrahlung interferes with ordinary bremsstrahlung. However, in contrast to ordinary bremsstrahlung, transition bremsstrahlung does not disappear in the limit of infinitely heavy colliding particles. The general theory of bremsstrahlung of particles in a medium should take into account also transition bremsstrahlung and its interference with ordinary bremsstrahlung (in the same way as the general theory of scattering must take into account transition scattering).

It is easy to understand the physical nature of transition bremsstrahlung if we bear in mind that the field E and the polarization $P = (\epsilon - 1)E/4\pi$ of a uniformly moving and, in particular, a resting charge can be expanded into waves with a wave vector k_0 and a frequency $\omega_0 = (k_0 v)$, where v is the charge velocity. These waves are connected to permittivity waves with the same values k_0 and ω_0. Such permittivity waves "dragged along" by a single charge may undergo transition scattering by another charge, as a result of which electromagnetic radiation, in this case transition bremsstrahlung, may be produced.

Transition bremsstrahlung may generate any "normal waves" (excitons, photons, phonons, and so on) which can propagate in the medium considered and is therefore a phenomenon of rather general character.

10. Transition radiation, transition scattering and transition bremsstrahlung in a plasma

Transition radiation, transition scattering and transition bremsstrahlung play a particularly important role for plasmas. In plasma physics one can go far by using a microscopical approach and avoid introducing the concept of transition scattering, but it may provide an insight into the situation and facilitate the development in this field [12].

In a rarified (in the collisionless limit) plasma the processes of transition scattering and transition bremsstrahlung may turn out to be particularly important. The above mentioned transition processes occur without any particle acceleration due to inhomogeneity of the medium along the trajectory of rectilinearly and uniformly moving charges. Collisions being neglected, the particle motion in a plasma is, in a first approximation, rectilinear and uniform if the action of macroscopic magnetic and electric fields are excluded. Various instabilities may appear in a collisionless plasma leading to a background of different kinds of waves. These waves are permittivity waves undergoing transition scattering. These permittivity waves are coupled with the most typical and most often observed high-frequency (Langmuir) wave in an isotropic plasma. Let us consider this example in more detail.

For a collisionless isotropic plasma the longitudinal dielectric permittivity has the form (see, for instance, [8, ch. 11]; the role of ions is assumed to be negligible small)

$$\epsilon_\ell(\omega, k) = 1 - \frac{\omega_p^2}{\omega^2} - 3\frac{\kappa T \omega_p^2 k^2}{m\omega^4},$$

$$\omega_p^2 = \frac{4\pi e^2 \mathcal{N}}{m}, \quad \omega^2 \gg \frac{\kappa T}{m} k^2, \tag{46}$$

where κ is Boltzmanns constant, T the temperature and $\mathcal{N}$ the electron density. The dispersion equation for longitudinal waves $\epsilon_\ell(\omega, k) = 0$ then leads to the following form for a longitudinal (plasma) wave

$$E = E_0 \cos(k_0 r - \omega_0 t), \quad [E_0 k] = 0, \quad E_0 k = E_0 k,$$

$$\omega_0^2 \approx \omega_p^2 + \frac{3\kappa T}{m} k^2. \tag{47}$$

Further,

$$\text{div } E = -E_0 k_0 \sin(k_0 r - \omega_0 t) = -4\pi e \mathcal{N}^{(1)}, \quad \mathcal{N} = \mathcal{N}^{(0)} + \mathcal{N}^{(1)}$$

(the electron charge is $-e$), from which it is clear that the longitudinal wave is coupled to the permittivity wave (the charge $-e\mathcal{N}^{(0)}$ is compensated by the ion charge)

$$\epsilon_\ell = \epsilon_\ell^{(1)} \sin(k_0 r - \omega_0 t),$$

$$\epsilon_\ell^{(1)} = -\frac{4\pi e^2 \mathcal{N}^{(1)}}{m\omega_0^2} = -\frac{e k_0 E_0}{m\omega_0^2} \tag{48}$$

Thus, plasma particles in a plasma wave are affected by the electric field E of the wave and at the same time a permittivity wave falls on them. The oscillations of the electrons in the field lead to Thomson scattering with the well-known cross-section

$$\sigma_T = \tfrac{8}{3}\pi r_e^2 = \tfrac{8}{3}\pi \left(\frac{e^2}{mc^2}\right)^2 = 6.65 \times 10^{-25} \text{ cm}^2.$$

[we are dealing with a total cross-section for non-polarized transverse radiation; it is assumed that the electron velocity $v \ll c/n(\omega)$.] Simultaneously occurring transition scattering interferes with Thomson scattering. For electrons, both effects are of the same order of magnitude. In this respect a plasma is a complicated object since one should take into account both spatial and frequency dispersion. The corresponding formulae are presented in ref. [12]. In the case of ions the role of Thomson scattering is small due to the large ion mass, whereas transition scattering prevails and is of the same order of magnitude as for electrons. The same can be said about transition bremsstrahlung—it is as important for ions as it is for electrons, whereas ordinary bremsstrahlung is significant only for electrons.

At boundaries between media, transition radiation in a plasma is usually not so interesting because plasma boundaries are smeared out. This remark is, however, rather conditional since all depends on the wavelength of waves we are dealing with. For sufficient long waves, even if there are no walls (to say nothing of the conditions where such walls do exist) the plasma boundary may turn out to be sharp enough for the appearance of a noticeable transition effect.

For sufficiently large frequencies $\omega^2 \gg \omega_s^2$ (where ω_s is the eigenfrequency of the medium) all media obey in a good approximation, the plasma formula

$$\epsilon(\omega) = 1 - \frac{\omega_{p,t}^2}{\omega^2}, \qquad \omega_{p,t}^2 = \frac{4\pi e^2 \mathcal{N}_t}{m}, \tag{49}$$

where $\mathcal{N}_t$ is the total electron concentration in the medium.

In the forward radiation of a relativistic particle leaving a plate, the energy is concentrated mainly in the X-ray range and, therefore, one can use eq. (49) to calculate the total energy W_2 for any medium. The result mentioned in section 6 is (see ref. [12])

$$W_2 = \frac{1}{3} \frac{e^2}{c} \omega_{p,t} \cdot \frac{E}{mc^2}, \tag{50}$$

the maximum of radiation corresponding to the frequency

$$\omega_m \sim \omega_{p,t} \cdot \frac{E}{mc^2}. \tag{51}$$

According to eq. (42) with $\lambda \sim \lambda_m$, the dimension of the formation zone is in this case,

$$\lambda_m \sim \frac{2\pi c}{\omega_m} \sim \frac{2\pi c}{\omega_{p,t}} \left(\frac{mc^2}{E}\right),$$

equal to

$$L_f \sim \frac{4\pi c}{\omega_{p,t}} \left(\frac{E}{mc^2} \right) \sim 10^{-5} \left(\frac{E}{mc^2} \right) \text{ cm,} \tag{52}$$

where we have taken into account that for an ordinary medium $\omega_{p,t} \sim 10^{16}-10^{17}$ sec^{-1} and $\lambda_{p,t} = 2\pi c/\omega_{p,t} \sim 10^{-5}-10^{-6}$ cm. For high energy particles, for example, for protons with $E \sim 10^{15}$ eV, $E/mc^2 \sim 10^6$ and $L_f \sim 10$ cm showing, how large the formation zone may be in some cases.

For the boundary between plasma and vacuum [i.e. in case eqs. (49) and (39) are used for $E \gg mc^2$] the total energy of backward radiation (in vacuum) is equal to

$$W_1 = \int_0^\infty W_1(\omega) \, d\omega = \frac{32}{15} \frac{e^2}{\pi c} \omega_{p,t} \ln \frac{E}{mc^2}, \tag{53}$$

and in the region of plasma transparency (i.e. for frequencies $\omega > \omega_{p,t}$) the energy is

$$W_1' = \int_{\omega_{p,t}}^\infty W_1(\omega) \, d\omega = \frac{2}{15} \frac{e^2}{\pi c} \omega_{p,t} \ln \frac{E}{mc^2}.$$

The causes of this increase of the energy W_1, which is slower than according to eq. (50), have already been analyzed in section 7 (see also [12]).

Even in the case of eq. (50) where one boundary is crossed, the probability of the appearance of a single transition quantum, is only on the order of $W_2/\hbar\omega_m \sim e^2/\hbar c \sim \frac{1}{137}$, according to eq. (51). That is why transition counters must have many dividing boundaries. The number of boundaries is in turn limited by the necessity to have layers (plates) with a thickness comparable with or exceeding the dimension of the formation zone L_f. Nonetheless, transition counters have their advantages and are being applied [38]. Transition counters, as well as other possible devices exploiting transition radiation and scattering of various types, may find application in experimental physics.

11. Concluding remarks

Motion of sources at a constant velocity is important for electrodynamics in continuous media and, in particular, in plasmas. New problems of this sort may arise and attract attention.

To illustrate concretely the fruitfulness of understanding the physics of transition processes, I would like to give an actual example. In 1973 a paper appeared which, within the framework of the general theory of relativity, considered a charge in the centre of mass of a binary (two identical, electrically neutral stars moving in a circle relative to their centre of mass). Such an absolutely non-moving charge emits electromagnetic waves. This result is, at first glance, unexpected. However, it becomes obvious if one knows what transition scattering is and if one bears in mind that in the general theory of relativity a gravitational field affects the electromagnetic properties of the vacuum: one can say that vacuum possesses an electric permittivity and a magnetic permeability which depend on the metric tensor g_{ik}. It is thus clear that moving stars modulate permittivity and permeability and, so to say, generate permittivity waves. As a result, transition scattering of these waves by a non-moving charge occurs, and there appears a "scattered" electromagnetic wave. The understanding of this fact enables us to treat such problems as the transformation (scattering) of a gravitational wave by a charge or, what is more realistic in astrophysics, by a magnetic dipole (a pulsar) [12].

References

[1] N. Bohr, Philos. Mag. 25 (1913) 10.
[2] N. Bohr, Philos. Mag. 30 (1915) 581.
[3] N. Bohr, Mat.-Fys. Medd. Dan. Vidensk. Selsk. 18 (1948) N8.
[4] I.M. Frank and I.E. Tamm, C.R. (Doklady) Acad. Sci. USSR, 14 (1937) 109.
[5] I.E. Tamm, J. Phys. USSR 1 (1939) 439.
[6] I.M. Frank, Usp. Fiz. Nauk. 143 (1984) 111.
[7] E. Fermi, Phys. Rev. 57 (1940) 485.
[8] V.L. Ginzburg, Theoretical Physics and Astrophysics (Pergamon Press, London, New York, 1979); a new supplemented issue of this book is to appear by VNU, The Netherlands.
[9] A. Bohr, Mat.-Fys. Medd. Dan. Vidensk. Selsk. 24 (1948) N19.
[10] L.D. Landau and E.M. Lifshitz, Electrodynamics of Continuous Media (Pergamon Press, London, New York 1984).
[11] M.L. Ter-Minakelyan, High-Energy Electromagnetic Processes in Condensed Media (Wiley, New York, 1972).
[12] V.L. Ginzburg and V.N. Tsytovich, Phys. Rep. 49 (1979) N1; Phys. Lett. 79A (1980) 16. The book by the same authors, Transition Radiation and Transition Scattering—Some Aspects of the Theory (in Russian) (Nauka, Moscow, 1984). The English translation is planned for publication by Plenum Press, New York.
[13] O. Heaviside, Electromagnetic Theory, Vol. 3 (The Electrician Publ. Co., London, 1912).
[14] L.D. Landau, J. Phys. USSR 10 (1946) 25; Collected Papers (Pergamon Press, London, New York, 1965) p. 445.
[15] V.L. Ginzburg, C.R. (Doklady) Acad. Sci. USSR 24 (1939) 131.
[16] V.L. Ginzburg, J. Phys. USSR 2 (1940) 441.
[17] V.L. Ginzburg, Sov. Phys. JETP 7 (1958) 1096.
[18] V.L. Ginzburg and I.M. Frank, Dokl. Akad. Nauk SSSR 56 (1947) 699.
[19] L.S. Bogdankevich and B.M. Bolotovsky, Zh. Exp. Teor. Phys. 32 (1957) 1421.
[20] V.N. Tsytovich, Izv. VUZ Radiofiz. 29 (1986) N5.
[21] V.L. Ginzburg and V.Ya. Eidman, Sov. Phys. JETP 8 (1959) 1055.
[22] V.L. Ginzburg, J. Phys. USSR 3 (1940) 95.
[23] V.M. Agranovich and V.L. Ginzburg, Crystal Optics with Spatial Dispersion, and Excitons (Springer-Verlag, Heidelberg, 1984).
[24] I.M. Frank, Izv. Akad. Nauk SSSR Ser. Fiz. 6 (1942) 3.
[25] I.M. Frank, in: Vavilov Memorial Volume (Acad. Sci. USSR, Moscow, 1952) p. 173; Usp. Fiz. Nauk 144 (1984) 251.
[26] V.L. Ginzburg, Radiophys. Quantum Electron. 27 (1985) 601.
[27] V.L. Ginzburg and V.N. Tsytovich, Sov. Phys. JETP 61 (1985) 48.
[28] V.L. Ginzburg, in: Vavilov Memorial Volume (Acad. Sci. USSR, Moscow, 1952) p. 193.
[29] V.M. Dubovik and L.A. Tosunyan, Fiz. Elem. Chastits At. Yadra 14 (1983) 193.
[30] V.L. Ginzburg, Izv. VUZ Radiofiz. 28 (1985) 1211.
[31] L.S. Bogdankevich, Sov. Phys. Tech. Phys. 4 (1960) 992.

[32] V.L. Ginzburg, A.A. Gorbatsevich, Yu.V. Kopaev and B.A. Volkov, Solid State Commun. 50 (1984) 339.
[33] V.L. 'Ginzburg and I.M. Frank, Dokl. Akad. Nauk SSSR 56 (1947) 583.
[34] I.M. Frank, Usp. Fiz. Nauk 129 (1979) 685.
[35] V.L. Ginzburg and I.M. Frank. Zh. Exp. Teor. Fiz., 16 (1946) 15.
     Short version of this paper: I.M. Frank and V.L. Ginzburg, J. Phys. USSR 9 (1945) 353.
[36] K.A. Barsukov, Zh. Eksp. Teor. Fiz. 37 (1959) 1106.
     G.M. Garibyan. Sov. Phys. JETP 10 (1960) 372.
[37] V.B. Berestetskii, E.M. Lifshitz and L.P. Pitaevskii, Quantum Electrodynamics (Pergamon Press, London, New York, 1982).
[38] C.W. Fabian and H.G. Fischer, Rep. Progr. Phys. 43 (1980) 1003.
     K. Kleinknecht. Phys. Rep. 84 (1982) N2, 87.
[39] D.A. Kirzhnits and V.V. Losgakov, JETP Lett. 42 (1985) 226.
[40] V.L. Ginzburg and V.P. Frolov, JETP Lett. 43 (1986) 265; Phys. Lett. A (in press).

Discussion, session chairman S. Belyaev

Rubbia: You have shown us situations in which a particle can emit photons. There exist also the inverse processes, in which charged particles can absorb photons, like for example inverse Vavilov–Cherenkov radiation. This kind of phenomenon can also provide us with novel ways of accelerating particles using coherent radiation.

Ginzburg: You are correct. I simply forgot to mention this.

The Lesson of Quantum Theory, edited by J. de Boer, E. Dal and O. Ulfbeck
© Elsevier Science Publishers B.V., 1986

Does Elementary Particle Physics Have a Future?

Sheldon L. Glashow * **

Harvard University
Cambridge, Massachusetts, USA

How pleased I am—having spent 27 months of the past 27 years in Denmark—to return after a lamentable lapse of 21 years! And how honored to speak at this celebration of the genius of Niels Bohr! What a marvelous century it has been, and how wonderful Copenhagen truly is!

The tantalizing discovery of the Balmer series took place the year before Bohr's birth, 101 years ago. Last year, the Nobel Prize was awarded to an Italian and a Dutchman for their discovery of the intermediate bosons W and Z—the penultimate predictions of today's theory. It was a fitting climax to the century of quantum mechanics, the century we now celebrate as Niels Bohr's.

Physicists deal with an incredible range of distances, from the inconceivably small Planck scale of 10^{-33} cm to the incomprehensible size of the visible universe, 61 powers of ten larger. Arranged sequentially upon this cosmic ruler are the many disciplines of science: particle, nuclear, and atomic physics, then chemistry, biology and geology, and finally astronomy and cosmology. All these fields are ultimately quantum mechanical, and quantum mechanics began with Bohr.

Typically, Bohr began not at the beginning of the ruler nor at the end, but at the muddle in the middle, the atom. He saw that the classical rules could not describe Rutherford's nuclear atom so he invented new ones which did. His rules evolved into a theory which explained the mysteries of the atom and the successes of the periodic theory. But, quantum mechanics is a greedy master which admits of no competition. Quantum rules must rule the whole ruler.

The quantum atom led to the quantum nucleus, which is built up of nucleons held quantum-mechanically together. Nucleons themselves are built up of quarks, whose quantum nature is demonstrated by the discovery of dozens of "stationary states" of the proton. The smaller things are, the more they are quantum mechanical. For this reason, it took physicists decades to recognize these states for what they are.

For objects the size of a mountain and larger, gravity is the dominant force and the relevance of quantum mechanics seems to fade. But, stars shine because of a

---

* Research supported in part by the National Science Foundation under Grant No. PHY82-15249.
** Discussion on p. 153.

complex interplay of all of the forces of nature in an essentially quantum-mechanical game. Quantum mechanically as well, the hot early universe of ten billion years ago cooked up many of the light nuclei now about us. Things work out so well that we know that the laws of quantum mechanics were the same then as they are now. Truly, *plus ça change, plus c'est la même chose*.

In the earliest moments of the formation of our universe, things were so very hot that particle physics and its bestiary of quarks and leptons reigned supreme throughout the universe. The rule has curled up upon itself—it is no ruler at all but a snake swallowing its own tail, Ouroboros. The physics of the microworld is the physics of the entire cosmos. The large and the small are one. Our earthly accelerators are at once microscopes of supernal resolution, and miniature replicas of the greatest accelerator of them all, the entire universe.

The unification of the small and the large is twofold. Gravity itself must be a quantum-mechanical theory: Bohr and Einstein must ultimately be reconciled. Quantum gravity is dominant only at times so early that the temperature of the universe approaches the Planck mass, or conversely, at the very smallest distance scale of the Planck–Compton wavelength. Once again the snake swallows its tail, but at energies and at distances far removed from any conceivable laboratory but the universe itself.

Bohr's modest domain of atomic sizes has been expanded and expanded by the arrogant reductionism of today's physical scientist to cover the whole shebang. Mysteries still confront us at the largest distance scales: What is the dark matter of the universe? Why do globular clusters seem to be older than the universe itself? How did the galaxies form? And so on. But, we have made remarkable progress at the smaller scales. We have what appears to be a correct, complete and consistent theory which describes all the known phenomena of the microworld: quantum chromodynamics and the electroweak theory.

On vacation upon the lovely isle of Jamaica I discovered that Jamaicans know but one variety of cheese. It is used on pizzas, in sandwiches, and in omelets. It is called standard cheese. So it is in particle physics. We have only one theory that works, and a very good theory it is. Quantum chromodynamics and the electroweak theory comprise our standard model of particle physics. It is used in cosmology, astrophysics, nuclear physics, and in particle physics. There is really no choice.

No known phenomena suggest structure beyond the standard model. No measured quantity contradicts the standard model. There are no internal contradictions, and there are no loose ends. Yet, the standard model appears to no one to be a satisfying conclusion to the search of the particle physicist. For one thing, quantum chromodynamics (QCD) is not yet as predictive as one might hope. It has not yielded the observed mass spectrum of the hadrons. Presumably, this is a computational question which will be resolved by further study and future development of computer systems. There are more fundamental puzzles outstanding.

Why is the gauge group what it is? Why are there three families of fundamental fermions? Is there a Higgs boson, or what? Aren't seventeen basic particles and seventeen arbitrarily tunable parameters far too many? How about a quantum theory of gravity?

Have you noticed that we have made the great leap sidewise? Once upon a time, we particle physicists could honestly claim to be studying the ultimate structure of ordinary matter, from atoms, to their nuclei, to the garden varieties of quarks. Things began to change about 40 years ago with the discovery of the muon. I. Rabi, like other New York physicists fond of Chinese food, is said to have said upon hearing about the muon, "Who ordered that?" It is among the very few questions that Rabi's student, Julian Schwinger, hasn't answered. Schwinger's own student, yours truly, doesn't yet know.

Today the muon has been joined by the tau lepton, strangeness, charm, top and bottom. We no longer study the structure of the atom, for two thirds of our particles have nothing whatever to do with mundane matter. They are exotica found only at large accelerators or in the debris of cosmic-ray collisions. They have no more to do with "ordinary physics" than do elements numbers 108 and 109 have to do with "ordinary chemistry". We seem to be following an endlessly difficult and expensive side issue. Have we lost the thread of relevance?

We have not at all lost our way. It is just that we are not yet there. The standard theory cannot be our final answer just because it cannot justify the great leap sidewise. Like my Harvard predecessor, Percy Bridgeman, I (and, surely, we) have a quite unjustifiable faith in ultimate simplicity, in the existence of a one and true theory: unjustifiable but always justified by the remarkable progress in our discipline. Once again, think of the century and the man we now honor. Today's side issue will one day be central. The muon will find its essential place in the sun, along with its curious brethren.

Yes, my children, elementary-particle physics does have a future. Yet, today it is threatened, and its exposure may be greater than ever before.

The greatest danger is the possibility of rejection by our parent society: because of its seeming irrelevance, and because of its considerable cost. The American high-energy physics community is uncharacteristically unanimous in its desire and perceived need for a large hadron collider. Budgets are tight, and the multi-billion-dollar SSC has not yet been funded. Will it ever be? The answer is not obvious. That such a machine is essential to the American technological renaissance is hardly an incontrovertible argument. Perhaps, as R.R. Wilson has argued, large accelerators are today's analog to the great cathedrals. This too is not a convincing argument, for we are not any more prone to monument building. We might try to argue that particle physics builds sound minds in strong bodies. Those lured into intellectual combat by the challenge of particle physics often make their mark in other and more useful fields, like Wally Gilbert's discovery of the repressor, or Allen Cormack's contribution to the CAT scanner, or Luis Alvarez, or George Charpak, or Max Delbrück, and so on. Ultimately our arguments should be more self-contained. We must, and I hope we will, support particle physics for its own sake. We must convince our governments of the importance of pursuing pure science, of our obligation to understand the world we are born into. The American people understand this kind of argument far better than do its elected representatives.

The high cost of particle physics is not an exclusively American problem. Not all the European countries enthusiastically support fundamental physics. Denmark, the land of Niels Bohr, is certainly not one of the big spenders, not even on a per capita basis. And, England has produced the Kendrew Commission report which seriously puts forth the suggestion of a fatal 25% cut in the entire CERN budget. Fortunately, it is unlikely that things will come to such a sorry pass, since the entire British contribution to CERN is a mere 16%, and not all of Europe is so disenchanted with big science. France has discovered, after all, that it *does* have need for its savants.

A second threat to our discipline is the recent divorce between particle experiment and particle theory. Perhaps it all began with quantum chromodynamics, an apparently correct theory underlying the quark structure of nucleons and the nuclear force itself. It is not merely *a* theory, but within a certain reasonable context, it is *the* unique theory. In principle, QCD offers a complete description and explanation of nuclear physics and of particle physics at accessible energies. While most questions are computationally impossible to answer fully, the theory has had very many qualitative (and, a few quantitative) confirmations. It is almost certainly "correct". QCD is not the threat I have in mind. It has *not* produced a divorce between experiment and theory—indeed, it has led to closer coordination and cooperation between experimenters and theorists. Yet, it has planted a seed that has blossomed elsewhere. It suggests and affirms the belief that elegance and uniqueness can be criteria for truth. I believe in these criteria. But, they must be reinforced by experiment. I agree with Lord Kelvin when he says,

> "When you can measure what you are speaking about and express it in numbers, you know something about it. And when you cannot express it in numbers, your knowledge is of a meagre and unsatisfactory kind. It may be the beginning of knowledge, but you have scarcely in your thought advanced to the stage of a science."

By this criterion, too, QCD is a science. But, can the same be said of the superstring and its ilk?

Quantum mechanics is contagious, and gravity must be framed within its context. Some of my theorist friends feel that they have come upon the unique quantum theory of gravity: a supersymmetric system of strings formulated within a ten-dimensional space–time. Most of the physics of the superstring lies at forever inaccessible energies, up around the Planck mass. Within its context, the theory is unique. It may even be finite and self-consistent. It seems capable of describing the low-energy phenomena which we can observe in the laboratory, but it is hard to prove that it really does. In principle, it predicts what particles exist. In principle, the number of tunable parameters is reduced to zero. In practice, however, it has made no verifiable prediction at all, and it may not do so for decades to come. The string theorist has turned towards an inner harmony. But, can it be argued that elegance, uniqueness, and beauty define truth? Has mathematics supplanted and transcended experiment which has become irrelevant? Will the mundane problems which I call physics, but which they call phenomenology, simply come out in the wash in some distant tomorrow? Is further experimental endeavor not only difficult and expensive, but *unnecessary* and *irrelevant*? Perhaps I have overstated the case made by string theorists in defense of their new version of medieval theology where angels are replaced by Calabi–Yau manifolds. The threat, however, is clear. For the first time ever, it is possible to see how our noble search could come to an end, and how Faith could replace Science once more. Personally, I am optimistic. String

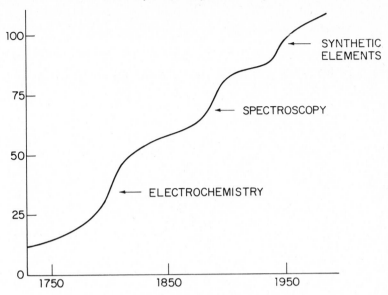

Fig. 1. The first population explosion. The growth of the number of known chemical elements over the past three centuries. The curve is roughly linear, with superimposed spurts of discovery resulting from technological developments. Elements 108 and 109 were produced in the 1980s. Will there be more?

theory may well dominate the next fifty years of fundamental theory, but only in the sense that Kaluza–Klein theory has dominated the past fifty years. Perhaps we should turn our attention to the past for guidance about the future.

The search for the ultimate constituents of matter has had a cyclic history passing from chaos, to order, to a new level of structure, and to chaos once more. We have passed through four such cycles in recent history: atoms, their nuclei, hadrons, and now quarks and leptons.

Recall the search for new chemical elements, which precedes Lavoisier and continues to this day. Figure 1 shows the growth of the number of known species with time. The most recent additions, Nos. 108 and 109, were produced and observed in Germany in the 1980s. The curve increases almost linearly over two-and-a-half centuries, showing spurts of discovery due to the new technologies of electrochemistry, spectroscopy, and more recently, artificial synthesis.

Midway in the history of the discovery of chemical elements, the Periodic Table was devised. The systematic order thus revealed was fully explained in terms of electrons, nuclei, and the quantum rules of Bohr. Nuclei were soon identified as composite structures themselves. The discovery of isotopes showed that there are far more nuclear species than there are chemical elements. Indeed, fig. 2 shows a plot of the 399 nuclear species whose half-life exceeds one year. (My colleague, Roy Glauber, had guessed 400.) Clearly, there are far too many nuclides for them to be elementary. The integral values of Z discovered by Moseley, and the almost integral values of A, represent the second level of order. Structure awaited the discovery of the neutron in the year of my birth.

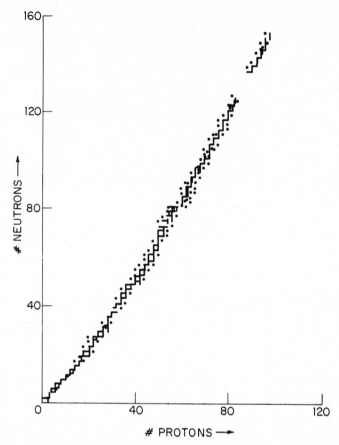

Fig. 2. The second population explosion. There are precisely 399 nuclear species whose half-lives exceed one year. Most of them lie on a connected "peninsula", but some, including uranium, form an isolated "island of stability". Is there a second such island awaiting discovery?

Things were beginning to look simple. In the late 1940s, Gamow wrote that the elementary particles were only three in number: nucleons, electrons, and neutrinos. These were the pointlike particles that were the basic building blocks of all matter. Alas, it was not to be so simple. Pions and strange particles were discovered. As large accelerators were deployed, there was a virtual population explosion among the nuclear particles, or hadrons. Figure 3 shows the time-evolution of the number of known hadron types. The curve rises approximately linearly from a few to more than a hundred in a time interval of only several decades. The curve is reminiscent of the explosion of known atomic species, but compressed in time by a factor of ten. As before, wiggles in the curve correspond to developments in experimental technology: bubble chambers and electron–positron colliders.

Once again, midway in the evolution of the curve, a new level of order is discovered: the eightfold way of Gell-Mann and Ne'eman. New particles with specific properties were predicted by the theory. The discovery of the famous Ω^-

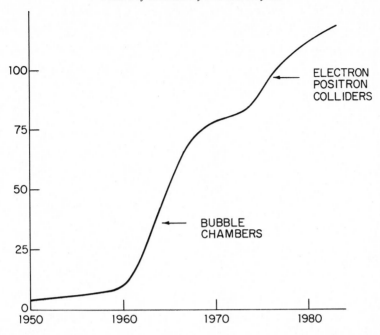

Fig. 3. The third population explosion. The growth of the number of known hadron multiplets over the past three decades. Again, the curve is roughly linear with spurts due to the deployment of new instruments. There are now about as many hadrons as there are chemical elements. Soon there will be more.

particle played the same role in this third cycle that the finding of scandium, germanium and gallium did for the first. Gell-Mann was no more mad than Mendeleev. For a third time, the appearance of systematic order led to the revelation of a new layer of structure. The hadrons were found to have all the attributes of composite systems: they were shown to be made up of quarks.

Today's candidate elementary particles have endured a quieter population explosion than their predecessors. The time evolution of the number of known quark species, leptons, and force particles is shown in fig. 4. There are, in our standard theory, just 17 building blocks, and 14 have been unambiguously discovered in the laboratory. Indirect evidence for the existence of the tau neutrino is compelling. Muted reports of the discovery of the top quark appear regularly in *Physics Today* and other journals of record. By far, the most important of the missing but predicted particles is the Higgs boson. If it is sufficiently light, it will be discovered at LEP. If its mass lies in an intermediate mass window, it will show up at the SSC or the LHC. The Higgs boson is the last great confirmation of the standard model that awaits discovery.

While the most recent population explosion, or descent into chaos, has been gentle, it is of a profoundly new and disturbing aspect. In earlier cycles, we were studying the nature of matter, quite ordinary matter such as is found on earth. Of our fundamental fermions, this is true for just two varieties of quarks, electrons, and perhaps the electron neutrino. None of the other quarks and leptons have a relevant

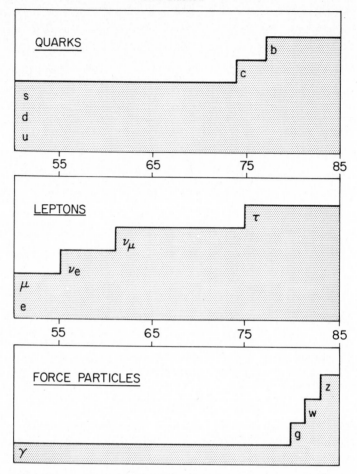

Fig. 4. Today's population explosion of fundamental particles. According to our canonical theory, only one particle of each type remains to be found. The top quark and the tau neutrino are relatively easy. The search for the Higgs boson is the outstanding challenge to defenders of the standard model. Will it be found? Will something unexpected be found?

role to play in the standard model. This great leap sidewise indicates that great progress remains to be made. The one-and-true theory that we seek is perhaps not the best, but it is surely the *only* possible world, and in it each and every particle will have an essential (an discernible!) *raison d'etre*.

The next level of order is surely revealed by today's periodic table of quarks and leptons, fig. 5. The fundamental fermions form families, each with the same recurrent pattern of strong and electroweak properties. The pattern is explained in terms of grand unification, but the reason for the seemingly superfluous replication of families remains obscure.

From the time of Bohr to the present time particle physics has progressed enormously. Many, indeed most, of the problems of yesteryear have been solved. To a large extent, our success is the result of experimental discovery. Generally, the

-1	-1/3	0	2/3
e	d	ν_e	u
μ 1938	s 1947	ν_μ 1961	c 1974
τ 1975	b 1976	ν_τ	t

MATTER PARTICLES

	1980	1983	1984	
γ	g	W	Z	HIGGS

FORCE PARTICLES

Fig. 5. The periodic table of quarks and leptons and the supplemental list of force particles. Two decades ago, the table suggested the possible existence of the charmed quark. Today, it hints at a deeper level of fundamental structure. Each row or family is a minimal anomaly-free unit. Why are there three families, or, are there more than three? Mendeleev's earlier table left no room for inert gases. Have we made such an omission again?

essential empirical basis to our theory is the result of patient, even plodding, endeavor. It is the accumulated knowledge due to very many scientists, all too few of whom will be remembered for a particularly startling or significant discovery. It has been a history, though, which is punctuated by dramatic and surprising events. Things like the unanticipated discoveries of X-rays or of radioactivity have taken place with remarkable frequency.

For example, in the 1930s, the neutron was discovered. So were the deuteron the positron, and the muon. And, let us not forget nuclear fission.

The 1940s saw the taming of the nuclear force, if such may be said of the use of the atomic bomb. The Lamb shift was measured, and it led to the surprising development of a consistent relativistic quantum theory. It was the decade of pions, and of the strange particle.

Parity-violation surprised almost everybody in the 1950s. So did the ever stranger properties of strange particles: associated production and neutral kaon behavior. Neutrinos were actually "seen" in the laboratory, and the first excited state of the nucleon was detected. An accelerator powerful enough to produce antimatter was commissioned. The discovery of the antiproton was a great accomplishment but hardly a surprise.

Schwinger was not surprised by the discovery of a second neutrino species in the 1960s, but most of us were. The population explosion of new hadron species grew out of hand. As if by magic, these new particles filled out complete supermultiplets of the SU(3) symmetry scheme. Deep inelastic electron scattering produced convincing evidence for the existence of pointlike "partons" in the proton, which, like the atom, turned out not to be a plum pudding after all. And who predicted that time-reversal symmetry would be violated?

So it went in the 1970s. A kooky theory purporting to unify weak interactions and electromagnetism was shown to be renormalizable, and wary experimenters were amazed to find that its neutral currents were for real. The discovery of the J/ψ particle was a big surprise, and even the existence of charmed particles surprised some of us. As if that were not enough, there were the tau lepton, the upsilon particle and its associated beauty particles.

The present decade is only half over. It has been a remarkably quiet decade so far. It was a surprise that CERN was able to mount a search for intermediate vector bosons in a timely and effective fashion. Any demonstration of international cooperation is surprising. However, the existence of W and Z bosons (like the existence of the antiproton) cannot really be thought of as something unexpected.

The lack of a fundamental but unanticipated discovery in this decade has not been for want of trying. Quite a number of surprises were *reported* during the 1980s. The trouble is that none of them seem really to be there. Perhaps a list of such non-discoveries will suffice:

(1) The magnetic monopole,
(2) Neutrino oscillations,
(3, 4) Neutrino masses (twice: Russian and Canadian),
(5) The zeta particle,
(6) No-neutrino double beta decay,
(7) Muons from Cygnus X-3 (still alive?),
(8) Proton decay,
(9) Forbidden decays like $\mu \to e\gamma$,
(10) Inexplicable "wrong sign dileptons",
(11) Free quarks,
(12) Anomalons,
(13) About a half-dozen varieties of anomalous events seen at the CERN collider and purporting to show that there is new physics beyond the standard model. None of these effects is established. Those who have done so much to confirm the standard model have not, as yet, succeeded in demolishing it. Don't they wish!

What is the meaning of this almost incredible list of failed but noble efforts? Has the era of great surprises in particle physics come to an end? Have we exhausted nature's bag of tricks? Do we already have enough clues in hand to build a theory of everything? Or, have we set into effect a self-sustaining prophecy wherein no new discoveries will lead to no new machines, and a guarantee that there can be no discovery tomorrow? These are dangerous times for particle physics.

There are two approaches to our current dilemma, the possession of a theory which is on the one hand too successful, but on the other, clearly incomplete. There is the pedestrian and the grandiose: the upwards path from mere experiment to theory, and the downwards path of pure positive thinking: the way of Bohr, and the way of Einstein. I think that there *is* a lesson to be learned from the past. Bohr's route has proven itself to be successful beyond any reasonable expectation. Einstein's path—the search for a complete and unified theory *now*—has proven to be a dismal failure.

Some day, if our species lives so long, Einstein's dream may be fulfilled. Of course there *is* a connection between gravitation and the other forces of nature. Michael Faraday, like Einstein, and like all of us, believed in the existence of such a relation. Unlike Einstein, he was a follower of the upwards path. Towards the end of his life, on July 19, 1850, after an unsuccessful search for an experimentally verifiable connection between the forces, he wrote:

> "Here end my trials for the present. The results are negative. They do not shake my strong feeling of the existence of a relation between gravity and electricity, though they give no proof that such a relation exists."

The Theory of Everything will come in its time if we let it. I am convinced that we still have a lot to learn about the phenomena of nature. One reason that Einstein failed in his quest is that he simply didn't know enough physics. Particle physics has not necessarily come to the end of the road. Astonishing experimental discoveries certainly remain to be made. If, and only if, we look, shall we find. The question is one of perseverance. Shall the scientific traditions established in the Renaissance survive in today's bizarrely materialistic society? The Way is clear, but what of the Will?

Discussion, session chairman H. Bethe

Bjørnholm: Could you comment on the question of the two additional families of quarks? Are they an indication of a substructure of the quarks?

Glashow: Let's remember that the Periodic System was discovered simultaneously by Mendeleev and Meyer and that they came to opposite conclusions. To Mendeleev the superfluous repetition of the Periodic System suggested that the atom had a substructure, while to Meyer it indicated the elementarity of the atom.

We have seen a very similar superfluous repetition in the case of the three identical families of quarks and leptons. To some people this suggests another layer of structure, and yet at the highest energies studied, there is absolutely no evidence of any structure. For molecules, atoms, nuclei and even the proton and neutron there is a vast number of energy levels that can be studied, and of course the appearance of energy levels is the key indication of the existence of an underlying structure. There is absolutely no indication of excited states of electrons, muons and quarks. These are not excited states which decay into each other, but are simply systems with different quantum numbers.

Casimir: When you say that most questions of yesteryear have been solved, it seems to me that you have still many adjustable parameters which you cannot explain.

That signals an incomplete theory. Why the fine-structure constant is $\frac{1}{137}$ is an unsolved problem. I am surprised how well the theory works despite the fact that it is incomplete. Just think of how many of the predicted particles, both in the 1960s and 1970s have turned out actually to exist.

Glashow: I fully agree with you. The existence of so many adjustable parameters, of which the fine-structure constant is just one, is terrible. It is one of the primary goals of the ultimate theory to resolve this ancient problem. Also I agree with you that the succesful predictions of new hadrons in the 1960s, the charmed quark and the W and Z can be viewed as surprising.

Rabi: Often new discoveries are made by closer examination or an increase in accuracy of measurements of already known results. This is not possible any more. Often I listen to talks in experimental high-energy physics and I ask the speaker, "have you published this?". He answers, "of course I have", but really he has not. Only conclusions are published, not experimental data that can be subjected to criticism. Moreover the experiments are too expensive to be repeated. The experimental physicists are reduced to technicians testing some theoretical predictions. I am afraid that what we see today sometimes may be artifacts of the theory.

Glashow: Professor Rabi, you are absolutely right. The only solution is that you stop attending all these symposia and get back to your laboratory.

The Lesson of Quantum Theory, edited by J. de Boer, E. Dal and O. Ulfbeck

Observability of Quarks

James D. Björkén

Fermi National Accelerator Laboratory
Batavia, Illinois, USA

Contents

Abstract

Even if stable hadrons with fractional charge do not exist, most of the criteria of observability used for ordinary elementary particles apply in principle to quarks as well. This is especially true in a simplified world containing only hadrons made of top quarks and gluons. In the real world containing light quarks, essential complications do occur, but most of the conclusions survive.

1. Introduction

It is an honor and privilege to be here to participate in this centenary of Niels Bohr's birth. I am not at all of his generation. I glimpsed him only once in the lunchroom, where a friend pointed him out to me during my first visit to Copenhagen as a fresh postdoc. So his personal influence on me is indirect—mainly through the style and atmosphere of the institute which he created, which to this day so splendidly and directly perpetuates his influence on science and his way of doing science.

The topic I have chosen to discuss—quarks—is not of Niels Bohr's generation. Nevertheless, the issue of how we observe them and how the interior machinery of that observation process works is very much of his generation. And the topic, as it turns out, is even very Scandinavian. The deeper side is studied here in Copenhagen, especially by Holger Bech Nielsen and his colleagues. The more pragmatic side which has its heritage in Bohr's work on propagation of particles through matter, can be found across the Øresund in Lund. So in choosing this topic for a talk here, I risk uttering mere banalities. So be it.

The problem addressed in this talk originated almost as soon as the quark hypothesis was enunciated: if all hadrons are made of fractionally charged constituents, why do we not eventually reach an energy scale of collisions where the constituents are liberated, thereby yielding at least one stable, isolated hadron of net fractional charge? (This goes under the name of the confinement problem.) And given the empirical absence of fractionally charged objects in bulk matter that this is *not* the case, what meaning is there in ascribing reality to these constituents within hadrons? In particular, *how* do the quarks confine themselves even in the most violent of collisions?

Nowadays, the problem is believed to be resolved in the context of the theory of the strong force, quantum chromodynamics, or QCD. This theory did not emerge until a decade after the emergence of the quark, and it was at least another half-decade before it was generally accepted. While to this day QCD is not universally accepted, it is not my purpose here to entertain any doubts about it, but rather to assume that QCD is true. Likewise, I will not try to look at the question from very much of a historical perspective, but go directly to the modern viewpoint, expressed in as simple terms as I can muster.

2. Quantum chromodynamics without light quarks

An immediate nonrelativistic answer to the confinement problem is found in the simple harmonic oscillator. If the quarks in hadrons were bound together by harmonic oscillator-like forces, then they never would be "ionized". In order to separate them by a macroscopic amount, one could, with enough energy, accomplish this: they could be placed into a macroscopic orbit. The problem with quarks lies in reconciling this old-fashioned viewpoint with relativistic quantum mechanics. Surprisingly, QCD seems to allow this to happen, at least in a simplified, albeit artificial limit.

Let us start with a review of the essential features of QCD as a theory of the strong force. The most remarkable is the renunciation of the Yukawa picture of meson exchange as the essence of the strong force. Indeed the essence of the QCD strong force is best seen if *all* known mesons—and their quark constituents—are disregarded. This leaves only the unknown—or at least not very well known—tricolored top quarks and the gluon carriers of the QCD force as the remaining degrees of freedom. In this limit, the natural range of the strong force emerges in full clarity as being determined by the QCD confinement scale-parameter Λ. This parameter, with dimensions of mass, is by chance believed to have about the same value as the pion mass, even when pions are removed from the theory.

Thus the confinement distance $\hbar/\Lambda c$ is of order 10^{-13} cm. For distances small compared to this, the QCD force is approximately inverse-square and not too strong; its "fine structure constant" is small compared to unity, and there are many analogies to quantum electrodynamics (QED). But at large distances it is believed—and there are good reasons to believe so—that the force becomes constant and the flux lines becomes concentrated in a tube of roughly the size $\hbar/\Lambda c$.

This top-quark limit has a splendid simplicity, largely devoid of all the complications of relativistic quantum fields—in particular multiparticle production and pair creation. Why is this? The flux tube has dimensions large compared to the Compton wavelength of the top quark, known to be less than 10^{-15} cm. Hence the color field contained in the flux tube is too feeble to pair-create the top quarks; this mechanism is indeed exponentially suppressed. And emission of gluons or quark–antiquark pairs by short-distance mechanisms, while occasionally present, is suppressed because of the smallness of the QCD fine-structure constant governing these processes.

The net result of this is that the effective Hamiltonian controlling the dynamics of a $t\bar{t}$ meson or ttt baryon is no worse than a relativistic potential-model. And the harmonic-oscillator analogy therefore still holds, the only differences being relativistic kinetic energies for the quarks and between them a potential energy which depends linearly, not quadratically, on their separation.

How now do we observe the quark? An easy way to try is to illuminate, say, a $t\bar{t}$ meson with a weakly interacting probe, such as a photon or a lepton. The amplitude A_n for finding the system in quantum state $|n\rangle$ is essentially

$$A_n \sim \langle n | e^{iq \cdot x} | 0 \rangle. \tag{1}$$

If the excitation energy is large (implying large momentum q transferred to the quark), the states $|n\rangle$ can be approximated by their semiclassical WKB formulae. In the region of overlap with the initial state $|0\rangle$, these are essentially plane waves. The important excitation energies will then be established as

$$E_q \sim \left(\sqrt{q^2 + m^2} - m \right) + \text{binding corrections}, \tag{2}$$

in accordance with classical kinematics and the Bohr correspondence principle. Hence, if a coarse-grained energy average is admitted, a wave packet

$$\psi_q(x) = e^{iq \cdot x} \psi_0(x) \tag{3}$$

is created by the collision, which propagates classically toward the turning point.

The picture could hardly be simpler. The probability for the collision to take place is given by a perturbative calculation. Coherence between, say, the contributions of interaction of the probe with t and $\bar{t}$ will for large q clearly be negligible, and the impulse approximation and semi-classical picture of the subsequent motion may be justified.

Need we observe the struck quark? What happens to it? It is easiest to first view the evolution in the center-of-mass reference frame. In that frame the t and $\bar{t}$ quarks simply oscillate back and forth between their classical turning points. The string tension, or energy per unit length of the QCD string, is about 1 GeV/fermi, i.e. about one proton mass per proton diameter. Were the t and t̄ quarks to be given relativistic momenta by the collision, for example 20 TeV, they would have not inconsiderable oscillation amplitudes. For 20 TeV the maximum separation is of order 0.4 Å, almost atomic dimensions. Eventually the oscillations will be damped. Two mechanisms come to mind but there may be more.

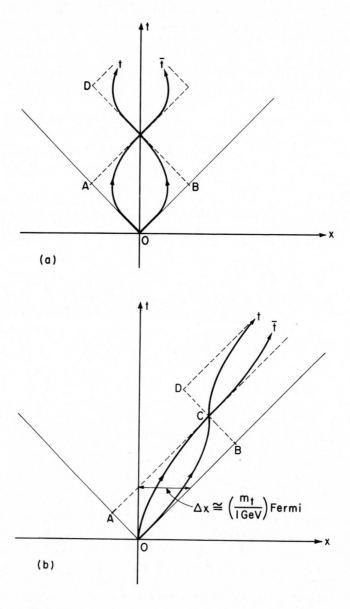

Fig. 1. (a) Space-time picture of t–t̄ motion in the center-of-mass frame; (b) the same in the laboratory frame. (tt̄ initially at rest; momentum q imparted to the t quark only).

The first is the emission of gluonia or glueballs. These are globs of pure gluons—perhaps better characterized as bits of closed flux-tube—which have a mass and size of the order of the confinement scale. (The typical mass estimates are a little larger than the proton mass.) Explicit computation is most easily done in the frame where the glueball in question is emitted at the turning point. But there is poor overlap between the wavefunctions of the initial and final t quarks (the level density is too high). This appears to imply a low probability per oscillation cycle for glueball emission.

A second mechanism is the emission of hard, "perturbative" gluons at birth (analogous to internal bremsstrahlung in QED) and at every half-period when the t and t̄ pass by each other. This mechanism appears to be the most important. There may be other damping mechanisms that I have not found. But it is a near certainty that the oscillations will be highly underdamped.

There is an additional subtlety which occurs when our process of Compton scattering from a tt̄ meson is viewed not in the center-of-mass frame but in the laboratory frame. The t quark struck by a photon recoils with a momentum (and energy) q, large compared with the rest mass m. This quark indeed moves a long distance, with a constant momentum loss of 1 GeV/fermi, while being decelerated by the string. However, at the other end of the string the antiquark is being accelerated. Soon it is traveling at essentially the speed of light behind the decelerating t quark. And some straightforward relativistic kinematics shows that, *no matter how large the initial momentum q*, the t̄ antiquark and t quark never separate by a distance greater than a fixed amount proportional to the t quark mass, essentially,

$$\Delta x \leq \left(\frac{m_t}{1 \text{ GeV}} \right) \times (1 \text{ fermi}). \tag{4}$$

When the leading t quark finally is decelerated to rest, the antiquark passes it up and the roles reverse; this commences the second half of the oscillation period. All this is shown in the space-time diagrams in fig. 1.

Thus it is *not* possible in the laboratory frame to observe the t quark in isolation, no matter how high the energy. But, fortunately, we may take recourse to the more practical colliding-beam processes $e^+e^- \rightarrow t\bar{t}$, or gluon + gluon $\rightarrow t\bar{t}$ in hadron colliders. These do provide concrete, in-principle, ways of producing macroscopically isolated quarks.

So if one had enough energy and the will, this oscillating top quark could, in a world devoid of light quarks, be observed in just as real terms as any other elementary particle.

3. Effects of the ordinary light quarks

The real world contains much more than top quarks, and these create fundamental complications. An essential change is that the stable string of the previous section cannot exist. It becomes unstable, due to pair creation of the light up and down quarks and antiquarks by the strong color fields in the flux tubes. There is now a

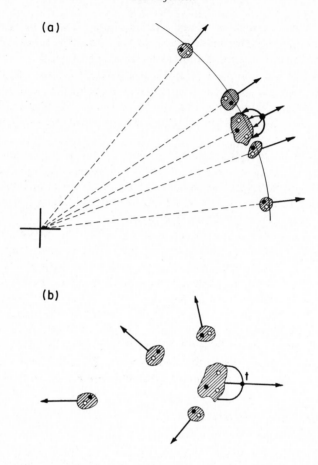

Fig. 2. Sketch of a t quark and its associated "fireball" cloud of quarks, antiquarks, and gluons during its evolution into a hadron jet; (a) laboratory frame, and (b) rest frame of "fireball".

good match between the size of the QCD flux tube and the light-quark Compton wavelength. The energy stored in the string can be used to create these $q\bar{q}$ pairs, and the string breaks into many pieces on a natural time scale $\leq 10^{-23}$ s. Since the string pieces contain quarks as well as glue, they are simply ordinary hadrons.

The lifetime of a piece of string can be estimated from experimental observed lifetimes of highly excited $c\bar{c}$ or $b\bar{b}$ meson states. These observations give a width per unit length of order tens of MeV/fermi of string length. This time scale is barely long enough to maintain viability of the concept of flux tube in the presence of the light-quark instability mechanism.

Let us now again look at our $t\bar{t}$ system—again literally with a highly inelastic γ-ray. Now the struck t quark, which we assume to be relativistic, does not grow a long string. Instead the incipient string invariably fragments into mesons. As the t recedes from the spectator, what emerges is a system as shown in fig. 2. The t quark again loses energy at a rate of order 1 GeV/fermi of transit. But this energy is no longer stored in the lengthening string, but is liberated into its decay products. The decay products emitted at late times will be of higher momenta and found adjacent to the excited system containing the top quark. It is tempting to think of that system, with dimensions of the order of one fermi, as a "fireball", emitting mesons as it cools off, and finally becoming a top meson or baryon. But this is a little over-simplified. A large fraction of the momentum imparted to the top quark in the original collision remains with the final hadron containing that quark. But the rest mass of the "fireball" in the early stages of the collision process is much larger than the top quark mass, while in the final stages it is very close to the top-quark mass. Thus the "fireball" mass *decreases* during the evolution of the event much more than its momentum does. Hence the Lorentz $\gamma \simeq p^*/m^* \simeq (1 - v^2/c^2)^{-1/2}$ (and therefore velocity) of the "fireball" *increases*—the "fireball" is accelerated. It is something like a rocket. In the "fireball" rest frame one sees the top quark at the front edge behind which there is a gluon "wake" which creates the quark–antiquark pairs. These quarks and antiquarks materialize into mesons which are emitted out the *back* of the fireball (fig. 2b).

The time-scale for this process is again set by the top-quark energy loss of ~ 1 GeV/fermi, just as for the elastic string. But because the process is dissipative, it terminates at a time of the same order as, but somewhat less than, what was needed to reach the turning point in the simplified "elastic" situation. This is still a large time at high energies. It again scales linearly with energy, as must be the case from basic special relativity: rapidly moving clocks slow down, so the laboratory time to get the job done grows accordingly. Thus the 200 GeV jets found at the CERN Sp$\bar{\text{p}}$s collider already evolve over a distance scale of up to 10^{-11} cm.

The picture we sketched implies that the quantum numbers of the source of the jet (t quark in our example) are linked to the portion of the jet carrying most of the momentum. Hence the most energetic hadrons of the jet will carry these quantum numbers. This is found to hold experimentally. On average the bottom meson carries more than 70% of the total b-jet momentum, and a charm meson about 50% of the momentum of a c-jet. For light quarks, the net fractional charge of the quark is, on average, found in the leading particles. Evidently this can hold only statistically. But this has been checked both in e^+e^- annihilation and hadron–hadron collisions.

4. Summary of ways to observe quarks

This is not the place to recite a long compilation of evidence for the quarks, but is meant only to underline the fact that we "see" them in ways not very different from the way we "see" electrons or protons. The methods include spectroscopy, inelastic scattering, and secondary interactions. We discuss these in turn very briefly.

4.1. Spectroscopy

The pattern of energy levels of a bound system tells us about its structure. This goes back, of course, to Niels Bohr himself. And the long history of spectroscopy which is relevant for the quark structure of hadrons goes back at least twenty-five years. The spectroscopy of baryons provided especially beautiful evidence for quarks even though, to this day, it is not obvious why a nonrelativistic quark model should work so well. More recently, the spectroscopy of $c\bar{c}$ (ψ) and $b\bar{b}$ (Υ) systems, which looks so similar to positronium spectroscopy, is equally decisive in convincing us that these states are built from fractionally charged quarks.

4.2. Inelastic scattering

The presence of electrons in matter can be inferred from the kinematics of the Compton effect. Many similar examples exist, not all of which use photons. Inelastic scattering of electrons from nuclei directly exhibit the presence of individual nucleons and determine their internal motion within the nucleus. For quarks in hadrons, lepton rather than photon scattering has also played the leading role. While lepton scattering from a $t\bar{t}$ system can be viewed as we have in the previous sections, there were grave obstacles in doing so for ordinary hadrons. They are not as reliably describable in terms of potential models. The light-quark pointlike constituents can be expected to move at relativistic velocities, spoiling the impulse approximation. In addition, virtual and real pair creation can be expected to be important as well.

The resolution of these difficulties came in exploiting the relativistic nature of the problem. When the hadron to be probed moves ultra-relativistically, its internal clocks slow down, and the external lepton probe, moving in the opposite direction, sees essentially a static distribution of constituents during the period of collision. The picture therefore reverts to something very similar to the familiar examples from atoms, nuclei, and even excitations in condensed matter. The initial internal motion of the constituent is slow compared to its motion after being struck by the probe. Free-particle kinematics can be used to estimate the collision probability, and the distribution of the scattered probe-particles again measures the initial velocity distribution of the constituent.

4.3. Secondary interactions

The previous methods observe the quark as it is bound within the hadron, just as the analogous methods observe the electron as it is within the atom, or the nucleon within a nucleus. To many, the essence of a real observation of a particle "in isolation" would be to follow and observe its subsequent motion. This implies—especially here in Copenhagen—additional interactions with a medium through which the particle propagates. For example, the ionization loss of a charged particle provides a mechanism by which its path can be followed and indeed defined, in the sense of quantum measurement theory. In the simplified case of section 2, QCD with only top quarks, this could be done straightforwardly given high enough excitation energy. In the general case which includes light quarks, the top quark is invariably immersed in its "fireball" of glue and $q\bar{q}$ pairs. It again would propagate a distance comparable to the "elastic" case and clearly leaves behind a track as well, but one somewhat harder to interpret.

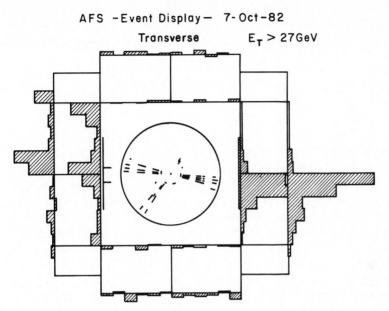

Fig. 3. View along the beam direction of a two-jet final state, most probably a quark–antiquark elastic scattering via gluon exchange, as seen at the CERN Intersecting Storage Rings by the Axial-Field Spectrometer experiment. (The date of the event is singularly appropriate).

Nevertheless, there is a practical way of probing the structure of such a newly formed quark system. It consists of highly inelastic lepton scattering from *nuclei*. If the energy scale in the collision exceeds hundreds of GeV—something attainable especially well in upcoming muon-scattering experiments at Fermilab—the quark system will traverse a considerable amount of nuclear matter before becoming independent hadrons, and its interior structure can thereby be probed. The main effects on the quark motion are anticipated to be multiple scattering and bremsstrahlung through the *strong* force. This gives, for large atomic numbers, a characteristic broadening of the angular distribution of the most energetic hadrons, as well as an attenuation (because of the gluon bremsstrahlung in nuclear matter) in the number of energetic hadrons.

In addition to these means, the quark and gluon jets seen in e^+e^- and $p\bar{p}$ collisions are in some sense the residue of the track of the quark fireball as it propagated through the vacuum. A nice example of this, kindly provided to me by Knud Hansen, is shown in fig. 3.

Carlo Rubbia pointed out to me that the ultimate high-energy physics experiment would be to somehow find the magnetic monopoles and antimonopoles anticipated in grand-unified theories and annihilate them. These monopoles might have a mass of at least 10^{15} proton masses. Were they to annihilate they would liberate quarks of comparable energies. The characteristic distance the quark would travel before full "hadronization" occurred would approach one meter. In this case macroscopic means could be used to follow and (again in the sense of measurement theory) define the course of the quark and its wake of gluons, strings, and $q\bar{q}$ pairs. It may even be that the mean ionization density of the "fireball", which fluctuates in charge as it emits the hadrons comprising its jet, corresponds (when averaged over many events) to a fractionally charged object.

5. Reflections and conclusions

When I look at the preceding arguments, they seem so self-evident that it is hard to recognize a problem at all. Was there ever a problem? The answer, I think, is yes. It existed in acute form before the development of QCD, and was divided into two parts: *why* quarks should be confined within hadrons, and then *how* they did not get out in high-energy collisions. While vague ideas about strings were available, there was little in the way of a relativistic theoretical structure within which such ideas of confinement could be developed. *How* quarks did not get out, and the importance of large distance scales in this process, could be—and was—attacked in the interim, even without appreciation of the color degree of freedom and QCD. The simple example of the top quark bound to elastic strings came with the full comprehension of QCD as the theory of the strong force. In addition the ψ and Υ spectroscopy provided much-needed stimulation from experiment.

What most distinguishes the observability of quarks from the observability of other particles is the technical complication of the light quarks. This makes the traversal of an energetic quark through vacuum a dissipative process, something like (but not identical to) ionization loss of a charged particle in matter. Instead of a mean energy loss of 2 MeV/gm cm^{-2}, we have a value of order 1 GeV/fermi. And the presence of light quarks also implies that a quark will not be found in isolation, but will inevitably be accompanied by a polarization cloud of quarks, antiquarks, and glue, which screens - in *any* frame of reference - its color field and fractional charge at distances beyond the confinement scale of 10^{-13} cm.

To me this additional complication is more technical than truly fundamental, but others may well disagree. The quark has been observed, even in the absence of quark tracks, and there need be little if any mystery associated with that. The real mystery lies in the nature of the medium through which the quark propagates—that is, the nature of the vacuum itself. It has by now taken on much dynamical character of its own, very much like the ground states of the solid-state analogues. The question of the observability of the vacuum itself has become the big problem. I wonder what Niels Bohr would say about that.

Discussion, session chairman H. Bethe

Schrieffer: First a remark. There is a very beautiful example of the near-observability of fractional charge in condensed-matter physics. We have the advantage of being able to pull apart the fractional charge to very large distances in quasi one-dimensional conductors. But this is only in very few conductors. What I did not mention in my talk yesterday was that in most cases we have confinement forces between what is the analogue of solid-state quarks, if you like. These confinement potentials are in fact linear as soon as you get outside the form factor of the excitations themselves. When you pull them far enough apart they break into what you might call color singlets, which have quite weak forces between them. This can be seen beautifully in photo-excitation experiments. There is a remarkable coincidence between the two fields, even if the origin is probably quite different.

Then a question: Do you have any idea as to how one can account for fractional charge starting out with integer charged fields in a relativistic context?

Björkén: There is an old and beautiful idea by Han and Nambu, which is now obsolete. The model has three triplets of quarks with integer charges. However, it simply does not fit well with present days' standard model.

Rubbia: Concerning the experimental verification of the top quark I would like to bring an update on this: There is a handful of events with two jets and an electron and a neutrino (or a muon and a neutrino) in the collider results. A possible interpretation for these events is a decay of a W particle into a top-quark and a beauty-quark. If this is correct then the top-quark mass is in the vicinity of 40 GeV.

The Lesson of Quantum Theory, edited by J. de Boer, E. Dal and O. Ulfbeck
© Elsevier Science Publishers B.V., 1986

The Transformation of Elementary Particle Physics into Many-Body Physics

Léon Van Hove

CERN
Geneva, Switzerland

Contents

1. A tribute to Niels Bohr

It is a great honour and a deep pleasure to contribute to this symposium which recalls in today's perspective the momentous scientific achievements of Niels Bohr, the extraordinary radiance and kindness of his personality, and the unique atmosphere of scientific excellence and human warmth which he created at the Copenhagen Institute of Theoretical Physics. I was privileged to enjoy twice the hospitality of the Institute, briefly in 1947 and for a longer period in the following year. The formative value of these stays at Copenhagen was enormous, for me as for so many other young physicists.

An experience which was striking for the newcomers in Copenhagen was the broad spectrum of nationalities present at the Institute on Blegdamsvej and the astounding ease with which everybody was taken up and integrated in the international group. It is therefore not surprising that after World War II Niels Bohr was among the strongest proponents of a collaborative European effort in experimental physics, and that he supported the initiatives which led to the creation of CERN. Although he did not share the prevailing opinion that CERN should start immediately with the construction of a large synchrotron, Niels Bohr put himself and his Institute at the disposal of the provisional CERN organization, set up by an intergovernmental conference at Geneva in February 1952. He accepted the leadership of the Theoretical Study Group which was based in his Institute for several years, and participated with the Council and the other officials of the provisional CERN in the crucial deliberations leading to the establishment of the definitive organization in 1954. From then onward, he served on the Scientific Policy Committee until his death in 1962. As a CERN physicist, it is my privilege at this symposium to pay tribute to Niels Bohr for his contributions to the foundation of CERN and for the devotion with which he put his wisdom, his experience and his influence to the service of this first great venture in European scientific collaboration.

2. Three ways in which particle physics transformed into many-body physics

When the organizers of the Niels Bohr Centenary Symposium asked me to speak on what they called the "transmutation" of elementary particle physics into many-body physics, I first was surprised by the suggestion that the branch of physics which is commonly regarded to be the most fundamental one, seemed somehow to be degradated to the rank of an additional chapter in the everwidening physics of complex systems.

I soon realized, of course, that the relevance of the subject does not lie in the question of fundamentality; the search for basic constituents of matter and basic laws of interaction remains the central task of particle physics. What has become apparent, however, is that in recent times most problems encountered experimentally and theoretically in particle physics have characteristics typical of many-body physics, with many features which are of great interest in their own right. As I shall now explain, this occurs in three directions.

2.1. Multiple production of particles

To search for constituents of matter means to probe matter over short distances, which by the fundamental quantum relation $\lambda p = h$ (λ = wavelength, p = momentum) implies that high-energy probes must be used. But one learned very early from cosmic-ray experiments that most high-energy collisions produce large multiplicities of newly created particles. This in itself makes high-energy physics a new chapter of many-body physics, in fact the first chapter of relativistic many-body physics since most particles created have energies large compared to their masses times c^2.

Heisenberg pointed out in 1939 [1] that multiple particle production could be understood in principle in the "meson theories of nuclear forces" (we would now say "strong interaction theories") as a consequence of the non-linear nature of the interactions. This line of work was pursued by Heisenberg and his collaborators during and after the war [2] and was taken up also by Oppenheimer and collaborators [3] in 1948. In the early fifties, it was largely replaced by Fermi's statistical model [4] and Landau's hydrodynamical model [5], which made the linkage with many-body physics particularly obvious. It is remarkable that the hydrodynamical model reappeared in force recently (with improved assumptions on the initial conditions of the hydrodynamical expansion) in the field of relativistic heavy-ion collisions, a recent joint venture of high-energy physics and nuclear physics.

The sixties and early seventies saw two main lines of parallel development in what is now called multiparticle dynamics; one based on specific dynamical models (multiperipheral, Regge, dual and early string models) and the other one based on purely phenomenological approaches (uncorrelated jet model, longitudinal phase space analysis, limiting fragmentation, Feynman scaling, short-range order, Koba–Nielsen–Olesen scaling). The picture was radically changed around 1973 by two breakthroughs, the experimental discovery of collisions producing high transverse momenta (high p_T) at the CERN Intersecting Storage Rings (ISR) and the emergence of Quantum Chromodynamics (QCD) as the new field theory of strong interactions, two very successful advances which have dominated the scene of strong interaction physics in the last ten years.

The many-body problem of multiple particle production in high-energy collisions has thereby come to stand in a completely new light, quite different from what is commonly encountered in other branches of physics. The reason is that the basic fields of QCD have quanta, the *quarks* and *gluons*, which are not appearing as free particles. They carry a new set of (non-commuting) charges called *colour charges*, and all observable hadrons (this is the generic name given to the strongly interacting particles) are supposed to be composites of the above "partons" with vanishing colour charges. This is the principle of *confinement* which is required to make QCD compatible with the facts. For the moment it is an assumption, but lattice calculations suggest that it is likely to be a consequence of the QCD equations. We shall illustrate in section 3 how multiple particle production is now understood in the framework of QCD.

2.2. Non-trivial structure of the vacuum

Another development of modern particle physics connecting it with many-body physics concerns the properties of the vacuum state in the field theories describing the various particles and interactions. In practically all field theories now used, one is led to assume *spontaneous symmetry breaking*, in the sense that the vacuum state is supposed to be one of a set of degenerate groundstates, each of which violates one or several symmetries of the Lagrangean. This type of situation is, of course, entirely familiar in condensed-matter physics, e.g. in crystals, ferromagnetic spin systems, superconductivity, etc.

In particle physics, one of its simplest versions is the so-called *Higgs mechanism* [6] for spontaneous breaking of gauge symmetries, which postulates the existence of one or more scalar fields Φ and of a potential $V(\Phi)$ which reaches its minimum $V_{\min}$ at non-vanishing values of Φ. The true vacuum is then one of the states for which the expectation value(s) $\langle\Phi(x)\rangle$ are constant and obey $V(\langle\Phi(x)\rangle) = V_{\min}$. These non-vanishing *scalar condensates* are non-perturbative in the sense that the perturbative vacuum has $\langle\Phi\rangle = 0$. Perturbative methods can nevertheless be applied by expanding in $\delta\Phi = \Phi - \langle\Phi\rangle$. The Higgs mechanism has the remarkable property of giving masses to gauge bosons. It is postulated to apply in the *electroweak theory* of Glashow, Weinberg and Salam which successfully unifies the electromagnetic and weak interactions. Although the predicted weak gauge bosons W, Z^0 were discovered in 1983 in experiments at CERN, there is still no experimental verification of the Higgs mechanism.

Also quantum chromodynamics (QCD) is assumed to have a nontrivial vacuum structure, but it is of a more subtle type than the Higgs mechanism, in the sense that it should follow from the QCD Lagrangean without addition of *ad hoc* fields and interaction terms. The non-trivial vacuum structure is manifesting itself in this case through *two-field condensates*, i.e. non-vanishing vacuum expectation values of type $\langle\bar{q}q\rangle$ and $\langle g^2 G_{\mu\nu}G^{\mu\nu}\rangle$, where q stands for the quark field operators ($q = u, d, s, \ldots$, for the various quark flavours), $G_{\mu\nu}$ for the gauge-field tensor and g for the coupling constant. These condensates are supposed to be of *non-perturbative* nature, which means that their finite numerical values are supposed to remain after the divergent expressions obtained for the expectation values in perturbation theory are eliminated by renormalization.

The quark condensate $\langle\bar{q}q\rangle$ is a manifestation of spontaneous breaking of chiral symmetry. That the latter symmetry is approximately valid and would hold in the limit of zero pion-mass was recognized in the fifties, and Nambu proposed in 1960 [7] that it should be regarded as spontaneously broken. This proposal was inspired by Nambu's fundamental paper elucidating the question of gauge invariance in the Bardeen–Cooper–Schrieffer theory of superconductivity [8]. With the advent of QCD in 1973 it was readily understood that the quark condensate was the appropriate "order parameter" for spontaneously broken chiral symmetry (although chiral symmetry may also be broken in a more trivial fashion by non-vanishing mass terms for the quarks).

The story is different for the QCD gauge-field condensate $\langle g^2 G_{\mu\nu}G^{\mu\nu}\rangle$, also called gluon condensate because the gauge-field quanta are called gluons. The first indication that the QCD vacuum state must be abnormal with respect to the gauge-fields came in 1975 with Polyakov's discovery of non-perturbative "instanton" solutions of the classical gauge-field equations [9]. These instantons are localized both in space and in imaginary time, and they have an energy lower than the perturbative vacuum. As one possible structure the true groundstate of QCD can contain a gas of such instantons in Euclidean space-time (ordinary space and imaginary time). It would give a vacuum correlation $\langle g^2 G_{\mu\nu}(x)\, G^{\mu\nu}(x')\rangle$ which would vary rapidly in x and x', a situation referred to as "twinkling vacuum" in a recent review by Shuryak [10]. A second indication for an abnormal QCD vacuum was found by Savvidi who pointed out in 1977 that the energy of the perturbative vacuum gets lowered by the introduction of an external, constant Gauge-field of magnetic type [11]. If this is the main mechanism of instability of the perturbative vacuum, the true vacuum might have a smooth correlation $\langle g^2 G_{\mu\nu}(x)\, G^{\mu\nu}(x')\rangle$ and would be "homogeneous" in Shuryak's terminology [10].

In both cases one wonders about the translation and Lorentz invariance of the vacuum state, a basic property automatically verified in perturbation theory but easily violated by condensates. For example, is $\langle g^2 G_{\mu\nu}(x)\, G^{\mu\nu}(x')\rangle$ invariant and therefore depending only on the invariant space-time distance $(x - x')^2$, or could the vacuum have a non-invariant micro-structure which is invisible at presently accessible scales? This question is not answered (the opposite question of apparent Lorentz invariance for theories with non-invariant Langrangeans has been address-ed by various authors, see e.g. ref. [12], where earlier work is quoted). Another problem connected with QCD instantons is that they induce a violation of *CP* and *T* invariance which must be assumed *ad hoc* to be extremely small [13].

A very interesting development initiated in 1979 by Shifman, Vainshtein and Zkharov [14] has permitted to relate the quark and gluon condensates to observable quantities in a rather direct way. We shall briefly describe their method in section 4.

2.3. Field theory at positive temperature

The third domain where particle physics meets many-body physics concerns the behaviour of modern field theories at positive temperatures, or in other words the statistical mechanics of relativistic fields and their associated particles. It began to attract considerable attention when it was realized that the modern field theories of particles and interactions have interesting links with cosmology, more precisely with the evolution of the early universe which is believed to have been filled with matter and radiation at very high temperature. In 1972 Kirzhnits and Linde [15] pointed out that the electroweak theory predicts a thermodynamic phase transition at sufficiently high temperatures, because in the Higgs mechanism the expectation value $\langle \Phi \rangle$ of the scalar field is then no longer fixed by the minimum of $V(\Phi)$ and will in general vanish, corresponding to restoration of the symmetry. In the actual theory the transition temperature cannot be estimated reliably, but is expected to lie in the range $T \sim 50\text{--}500$ GeV (by T we mean the temperature multiplied by the Boltzmann constant, $T = 1$ GeV corresponds to 1.16×10^{13} K). In the high-temper-ature phase (which in the standard Big Bang model of the expanding universe prevailed at times $\leq 10^{-12}$ s) all gauge bosons are predicted to be massless and additional scalar bosons appear instead of the longitudinally polarized states of the massive gauge bosons. Cosmological effects of the electroweak phase transition are generally believed to be unimportant.

Also QCD and strongly-interacting matter are predicted to behave quite differ-ently at high temperatures than in laboratory situations studied so far, correspond-ing mostly to $T \simeq 0$ MeV or, in exceptional cases, to perhaps $T < 10$ MeV in quasi-thermal excitations of heavy nuclei (compound-nucleus states). The con-densates $\langle g^2 G_{\mu\nu} G^{\mu\nu} \rangle$ and $\langle \bar{q}q \rangle$ are expected to "melt away" at high T, with the vanishing of $\langle \bar{q}q \rangle$ restoring chiral symmetry (except for possible quark-mass effects). In addition, a more spectacular transition is expected to occur, usually called the *deconfinement phase transition* first considered by Collins and Perry in 1975 [16].

Under normal conditions, hadronic matter is composed of well separated hadrons, each hadron being a composite of QCD partons, quarks, antiquarks and gluons (see the end of section 2.1), with a typical size of order 1 fermi $= 10^{-13}$ cm. At temperatures $T \sim 100$ MeV, blackbody radiation contains a substantial fraction of hadrons, especially pions which are the lightest hadrons (mass ~ 140 MeV/c^2). As T increases further, the density of these hadrons grows rapidly, and when it reaches about 1 fermi^{-3} it is expected that the hadrons somehow coalesce, allowing the partons to form a so-called *quark–gluon plasma* in which they circulate rather freely over distances $\gg 1$ fermi. This is the *deconfinement phase transition* which is clearly predicted by lattice QCD calculations and is generally expected to occur at a temperature in the range $T \sim 150$–250 MeV. The corresponding time in the early expansion of the universe is of order 10^{-6}–10^{-5} s. We shall return in section 5 to the problems related to possible effects of field-theoretical phase transitions in the early universe.

In addition to the temperature, there is a second variable which, for large values, is expected to have a drastic effect on hadronic matter. It is the baryon-number density which in QCD is one third of the net quark density (density of quarks minus density of antiquarks), or the corresponding chemical potential μ. Even at low T it is expected that hadronic matter will turn into a quark–gluon plasma when μ becomes sufficiently large, and this presumably happens in the collapse of heavy stars [16]. It is conceivable that a cold quark–gluon plasma ($T \sim 10$ MeV) could exist in the core of neutron stars. There are also more extreme speculations on abnormal, stable or metastable states of cold nuclear matter containing a large number of strange quarks [17].

Contrary to the electroweak phase transition mentioned earlier, the QCD deconfinement phase transition will perhaps become accessible to laboratory experiments in the near future. The hope exists that small, short-lived droplets of quark–gluon plasma could be formed in heavy-ion collisions at relativistic centre-of-mass energies, and much discussion is going on concerning the possibility to detect their occurrence and thereby determine some of their properties [18].

3. Multiple production of particles in quantum chromodynamics

As mentioned already, the confinement property of QCD has a profound effect on the particle-production process in high-energy collisions, making it a many-body problem of a totally novel type. In this section we attempt to describe in pictorial terms the essentials of this process as it is understood at present in the framework of QCD, concentrating on features which are common to the main theoretical models currently used to describe the data. Figure 1A represents the incident state of a meson–nucleon collision. The meson contains a quark labelled 1 and an antiquark (open circle), the nucleon contains three quarks labelled 2, 3, 4 (these five partons are presumably surrounded by clouds of quark–antiquark pairs and gluons). The shaded areas represent the volumes of the two incoming hadrons. They are supposed to be occupied by colour fields (the gauge fields of QCD) and have diameters of order 1 fermi. We have drawn them as if they were spherical, neglecting Lorentz contraction. The arrows indicate the motion.

Figure 1B depicts what is conjectured to be the situation a few fermi/c after the beginning of the collision (1 fermi/$c = 1/3 \times 10^{-23}$ s) in the large majority of cases, the so-called *soft collisions* which produce only particles with low transverse momenta (low p_T). Presumably through parton-exchange processes the incident hadrons have picked up colour charges as they traversed each other, and a

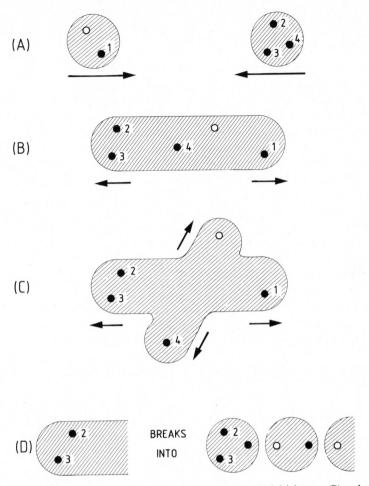

Fig. 1. Time development o a high-energy meson–nucleon collision. (A) initial state; (B) early state after soft collision; (C) early state after hard collision; (D) hadronization (see text for details).

colour-field region has formed between them (shaded area). It stretches out more and more but is believed not to expand significantly in the transverse direction, so that it forms a so-called *flux tube* of width ~ 1 fermi. By the Gauss theorem it contains a longitudinal colour electric field of constant flux as the opposite colour charges at the ends remain constant. The corresponding tension can be estimated to be ~ 1–2 GeV/fermi. The data suggest that one parton of each incident hadron gets "held back" and is located rather centrally in the tube whereas the others (quarks 1, 2 and 3 in fig. 1B) are located near the ends. As the tube stretches it breaks up, and this is believed to take place as illustrated in fig. 1D, by creation of additional quark–antiquark pairs in the colour field and regrouping of all partons into colourless hadrons, which then form the final state of the collision. This last phase of the collision is called *hadronization*.

At all stages the processes involved are characterized by energy–momentum transfers < 0.1–1 GeV/c, for which perturbative QCD calculations are not valid. We have so far no reliable way to calculate these processes. The present state of theory is that various phenomenological models have been developed in close contact with the abundant experimental material. Two of them, the Lund model [19] and the dual-parton model [20] are now quite elaborate and play an important role in experimental work. They have shown considerable predictive power at various stages, but in other respects have been and continue to be guided in their development by unforeseen experimental facts. While fitting in the qualitative picture sketched above, these models differ considerably in the assumed detailed specification of the early collision phase (fig. 1B) and the hadronization phase (fig. 1D).

As mentioned in section 2.1, we know since 1973 that in a fraction of high-energy hadron collisions, particles of high p_T are produced. These so-called *hard collisions* have been studied extensively, mainly at the CERN Intersecting Storage Rings and the Proton-Antiproton Collider. Although very rare, they form a clearly distinct class of collisions characterized by the occurrence of the high-p_T particles in two lateral jets on opposite sides of the longitudinal (i.e. incident) direction. Their existence and properties were predicted by QCD and brilliantly confirmed by experiment. In our pictorial description they are examplified by fig. 1C. Their characteristic is the occurrence of a hard collision (i.e. a collision with energy–momentum transfer $\gg 1$ GeV/c) between one parton of one incident hadron and one parton of the other (quark 4 and the antiquark in fig. 1C). The other partons fly through. The former two partons create the lateral jets and the other partons create the two longitudinal jets of fig. 1C, all four jets then breaking up by hadronization as depicted in fig. 1D. The new element is now that the hard collision of the two first partons can be calculated reliably from QCD because perturbative methods apply at large momentum transfers and indeed these calculations are successful in accounting for the high-p_T data.

Similarly, perturbative QCD calculations are very successful in accounting for hard collisions involving leptons (deep-inelastic scattering of electrons, muons and neutrinos on nucleons; high-energy electron–positron annihilation into hadrons). In all cases, nevertheless, the non-perturbative aspects connected with hadronization (fig. 1D) must be included by means of phenomenological models of the type mentioned above. Vice-versa, these models are being developed to the point where they are able to deal with the various types of collisions, soft and hard.

In principle, the non-perturbative properties of QCD should all be calculable numerically by lattice field-theory methods, but for scattering and particle-production processes this prospect is a long way off. The situation is very much better for hadron spectroscopy, vacuum condensates and QCD at positive temperature.

4. Quantum chromodynamics vacuum condensates

In this section we sketch how one can estimate the values of the quark and gluon vacuum condensates of QCD, $\langle \bar{q}q \rangle$ and $\langle g^2 G_{\mu\nu} G^{\mu\nu} \rangle$ (see section 2.2). The general idea can be described as follows. Consider the Feynman diagrams which are

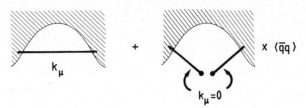

Fig. 2. Contribution of the quark condensate to a Feynman diagram (see text).

relevant for the perturbative calculation of a physical quantity, and suppose this quantity to be such that the integrands of the Feynman integrals pertaining to the diagrams vary little when the four-momentum k_μ of any line varies over an interval $|k_\mu| \sim 0.1$ GeV. This happens when large momentum-transfers or large masses are involved. The effect of the condensates can then be incorporated in the calculation by adding for each line a "condensate contribution", as illustrated in fig. 2 for a quark line. In the figure the diagram on the left represents the perturbative contribution and the one on the right the additional term. In the latter, the cut line of the diagram is given by $k_\mu = 0$ and the multiplicative factor is the condensate $\langle \bar{q}(x) \, q(x') \rangle$ at $x \simeq x'$. This is justified under the assumption that the diagram contribution is constant for k_μ varying around $k_\mu = 0$ over the range of the condensate in momentum space. The latter range is believed to be $\ll 1$ GeV. The contribution of the gluon condensate is added similarly by cutting gluon lines, the multiplicative factor being $\langle g^2 G_{\mu\nu}(x) \, G^{\mu\nu}(x') \rangle$ at $x \simeq x'$ up to a numerical factor.

The theoretical basis for this method, called the operator product expansion, goes back to a fundamental paper of K. Wilson in 1969 [21]. Then years later, Shifman, Vainshtein and Zkharov [14] showed how it could be used to determine the vacuum condensates by the study of correlation functions between currents. One considers sum rules for correlation functions in cases where large momentum-transfers are involved, which occurs for currents containing heavy quarks. The operator-product expansion expresses the correlation function in terms of vacuum condensates as explained above (fig. 2). Using the sum rule one can estimate the value of the correlation from experimentally known resonances and one can guess it for higher-mass contributions, so that the values of the condensates can be extracted. Many studies of this type have been carried out in recent years. Individual papers often give rather precise values for the vacuum condensates, but they show considerable variations from paper to paper. We quote below more conservative estimates which we took from Shuryak's review [10]: $\langle \bar{q}q \rangle \sim -(0.010\text{–}0.015)$ GeV^3 for u and d quarks, and $\langle g^2 G_{\mu\nu} G^{\mu\nu} \rangle \sim 0.1\text{–}0.8$ GeV^4.
The condensate $\langle \bar{q}q \rangle$ for s quarks (strange quarks) is believed to be comparable to its value for u and d.

5. Field-theoretical phase transitions and the early universe

In the last few years the possible role of field-theoretical phase transitions in the early expansion of the universe has given rise to an outburst of studies which, although very speculative, are undoubtedly of great interest and have strongly increased the contacts between astrophysicists and particle physicists. We shall briefly review the two main lines of work. References and additional information can for example be found in the surveys of Abbott [22] and Brandenberger [23].

5.1. The hadronic phase transition

Most probably the main effect of the transition from quark-gluon plasma to hadron gas in the early universe has been to stop or strongly slow down the cooling for a few microseconds when the transition temperature $T_c \sim 200$ MeV was reached, i.e. at an age of the universe which was itself a few microseconds. During this time interval, the energy needed to drive the continued expansion ($dE = -p \, dV$, $E =$ energy in volume V, $p =$ presure) was provided by the apparently large latent heat of the deconfinement transition, whereby both temperature and pressure remained constant or almost constant. This process modified the time evolution of the expansion, but it probably had no after-effects in the present universe if it took place adiabatically (no entropy production).

More interesting possibilities occur if one assumes that the transition was irreversible despite the fact that its time scale ($\sim 10^{-6}$ s) was very long compared to hadronic times ($> 10^{-23}$ s). One can speculate that the transition showed strong supercooling followed by violent release of the latent heat. As pointed out by Witten [24] such violent processes with a microsecond time scale could produce gravitational waves which for present observers, by the red shift due to the expansion, would lead to a characteristic time of the order of one year. In favourable cases this gravitational remnant of the hadronic phase transition could be detectable through its effect on pulsars.

Another type of effect, not requiring such violent processes, could be the appearance of an inhomogeneous distribution of nucleons in space as a result of the transition. During the latter, growing regions of space were occupied by hadron gas and shrinking regions by quark–gluon plasma. Due to the high mass of the nucleon (~ 940 MeV/c^2) compared to the transition temperature (~ 200 MeV), the net quark number (i.e. the small relative surplus of quarks over antiquarks which at that time was of order 10^{-9}) was probably concentrated in the plasma regions more than in the gas regions, giving spatial inhomogeneities in the nucleon distribution when all plasma had disappeared. Most probably these inhomogeneities were smeared out long before nuclei began to be synthetized, in which case they had no after-effects. It is conceivable, however, that they were stable enough to survive until the epoch of nucleosynthesis and to affect the cosmic abundance of light elements [25]. Witten has also speculated that they could have caused the production of nuggets of "strange quark matter", if such matter would be stable [24]. De Rújula has reviewed how such nuggets could be searched for [26].

Further progress on these problems will profit from the results of future experiments on relativistic ion collisions if these experiments reveal manifestations of the hadronic phase transition and provide facts concerning its characteristics.

5.2. The possibilities of very early phase transitions

Great efforts have been made since the early seventies to go beyond the electroweak theory and to find more comprehensive field theories unifying the electroweak interactions with QCD (the so-called grand unification programme) and also with general relativity (supergravity programme). Despite numerous attempts, these endeavours have not been successful so far, and many theorists are now turning to another possible line of unification, the so-called superstring theories which in pre-QCD days grew out of the dual theories of strong interactions. It is much too early to evaluate the prospects of this new approach [27].

Theoretical developments of unified field theories can potentially make crucial contributions to our understanding of some very puzzling cosmological problems by offering possible solutions in terms of the behaviour of matter at temperatures in the range 10^{10}–10^{19} GeV (10^{19} GeV corresponds to the Planck temperature where gravity can no longer be treated classically). Thus, one of the earliest predictions of grand unified theories [28] was that the baryon number B would not be strictly conserved, giving rise to the possibility that the net baryon number now observed in the universe would have originated during the expansion when the temperature was somewhere in the range mentioned above, the earlier state of the universe being symmetric between matter and antimatter ($B = 0$). This would imply that the present value of the net baryon number relative to the photon number in the cosmic background radiation could be calculated from particle physics and general relativity, a possibility which would have seemed totally out of reach twenty years ago.

Another idea which attracted an enormous interest in the last five years is that a field-theoretical phase transition at a temperature T_0 in the above range could have had a deep effect on the early expansion of the universe. The simplest example is a period of "inflation", i.e. of exponential expansion [29], as would happen if for a period the energy–momentum tensor were dominated by a vacuum energy-density of order T_0^4 (in units where $h = c = 1$). The great interest of this possibility is that such a period of abnormal expansion would permit causal contact between various regions of the universe to have extended very far at early times, providing an elegant explanation for the high degree of isotropy observed in the cosmic background radiation. It would also explain why the present universe is very flat, any initial curvature having been stretched out by the early expansion.

It is true that it has so far not been possible to put forward a convincing theoretical scheme realizing the above aims. The constraints imposed by theoretical coherence and compatibility with known properties of particles and cosmology turn out to be very severe [23]. Also the hope to observe B violation experimentally in the form of proton decay has so far not been fulfilled. But success may still be forthcoming and the speculative work done so far has greatly widened our outlook on early cosmology.

6. Concluding remarks

We have tried in this lecture to illustrate the domains of particle physics where the theoretical problems and methods have much in common with many-body and condensed-matter physics. The multitude of diverse physical systems accessible to experimentation in condensed-matter physics, and the numerous concepts developed for their theoretical understanding provide a rich store of ideas and analogies to the particle physicist. This can help him to overcome the great handicap that in his own discipline the experimental facts are very hard to come by and are often extremely incomplete. On the other hand, particle physics brought us such truly fundamental advances as non-Abelian gauge theories, electroweak unification with the heavy weak bosons, and quantum chromodynamics with the confinement principle for the field quanta. As our understanding of these novel schemes deepens, possibly with further progress toward unification, we can expect that they will slowly have an impact on the rest of physics, just as the concepts and techniques of Abelian field theories have gradually invaded most of condensed-matter physics.

In closing, I would like to thank the organizers of the Bohr Centenary Symposium for having recreated the stimulating and delightful atmosphere which so many of us fondly remember from our stays in Copenhagen in Bohr's days.

References

[1] W. Heisenberg, Z. Phys. 113 (1939) 61.
[2] W. Heisenberg, in: Vorträge über Kosmische Strahlung (Springer, Berlin, 1943) pp. 57 and 115; Z. Phys. 126 (1949) 569; 133 (1952) 65.
[3] H.W. Lewis, J.R. Oppenheimer and S.A. Wouthuysen, Phys. Rev. 73 (1948) 127.
[4] E. Fermi, Progr. Theor. Phys. 5 (1950) 570.
[5] L.D. Landau, Izv. Akad. Nauk SSSR Ser. Fiz. 17 (1953) 51; Collected Papers of L.D. Landau (Gordon and Breach, New York, 1965).
[6] F. Englert and R. Brout, Phys. Rev. Lett. 13 (1964) 321; P.W. Higgs, Phys. Lett. 12 (1964) 132; 13 (1964) 508.
[7] Y. Nambu, Phys. Rev. Lett. 4 (1960) 380.
[8] Y. Nambu, Phys. Rev. 117 (1960) 648.
[9] A.M. Polyakov, Phys. Lett. 59B (1975) 82; A.A. Belavin, A.M. Polyakov, A.A. Schwartz and Yu.S. Tyupkin, Phys. Lett. 59B (1975) 85.
[10] E.V. Shuryak, Theory and phenomenology of the QCD vacuum, Preprints 83-157, 83-158, 83-159, 83-164, 84-20, 84-23, 84-24 (Institute of Nuclear Physics, Novosibirsk, 1983–1984).
[11] G.K. Savvidi, Phys. Lett. 71B (1977) 133.
[12] S. Chadha and H.B. Nielsen, Nucl. Phys. B217 (1983) 125 and references cited therein.
[13] V. Baluni, Phys. Rev. D19 (1979) 2227 and references cited therein.
[14] M.A. Shifman, A.I. Vainshtein and V.I. Zakharov, Nucl. Phys. B147 (1979) 385, 448, 519.
[15] D.A. Kirzhnits, JETP Lett. 15 (1972) 529; D.A. Kirzhnits and A.D. Linde, Phys. Lett. 42B (1972) 471.
[16] J.C. Collins and M.J. Perry, Phys. Rev. Lett. 34 (1975) 1353.
[17] G. Baym and S.A. Chin, Nucl. Phys. A262 (1976) 527; G. Chapline and M. Nauenberg, Nature 264 (1976) 23.
[18] See for example Quark Matter 84, Lecture Notes in Physics 221, Helsinki Conf. Proc., ed. K. Kajantie (Springer, Berlin, 1985).
[19] B. Andersson, G. Gustafson, G. Ingelman and T. Sjöstrand, Phys. Rep. 97 (1983) 33.
[20] A. Capella and J. Tran Thanh Van, Z. Phys. C23 (1984) 165 and references cited therein; A.B. Kaidalov and K.A. Ter Martirosyan, Phys. Lett. 117B (1982) 247.
[21] K.G. Wilson, Phys. Rev. 179 (1969) 1499.
[22] L.F. Abbott in ref. [18] p. 106.
[23] R. Brandenberg, Rev. Mod. Phys. 57 (1985) 1.

[24] E. Witten, Phys. Rev. D30 (1984) 272.
[25] J.H. Applegate and C.J. Hogan, Phys. Rev. D31 (1985) 3037.
    S.A. Bonometto, P.A. Marchetti and S. Matarrese, Phys. Lett. 157B (1985) 216.
[26] A. De Rújula, Nucl. Phys. A434 (1985) 605c.
[27] S. Weinberg and, for a less optimistic view, S.L. Glashow, chapters in this volume.
[28] H. Georgi and S.L. Glashow, Phys. Rev. Lett. 32 (1974) 438.
[29] D. Kazanas, Astrophys. J. 241 (1980) L59;
    A.H. Guth, Phys. Rev. D23 (1981) 347;
    A. Linde, Phys. Lett. 108B (1982) 389;
    A. Albrecht and P. Steinhardt, Phys. Rev. Lett. 48 (1982) 1220.

Discussion, session chairman H. Bethe

Cook: Is the implication of the EMC effect that it is more appropriate to describe the heavy nuclei as made out of quarks than made out of nucleons?

Van Hove: Some would like to put it that way, but there is no consensus. People have taken simple views, like saying that there are pions in the nucleus that give you additional quark–antiquark pairs. Other people have said: The building blocks are closely packed nucleons, and their quark content can therefore be different from what it is in free nucleons. Prof. Bleuler and his collaborators have shown that even low-energy nuclear phenomena can be understood in terms of the color properties of the quarks.

Bang: A comment: There are still simpler ways to explain the EMC effect than the ones you mentioned. Taking the energy–momentum distribution of the nucleons properly into account can explain the effect. I refer to the work of Akulinichev and Vagradov.

Nielsen: Do you say that there is experimental evidence that the strings, which are pulled out when quarks are moving away from each other, are pulled out hot, but don't get their entropy from this?

Van Hove: Yes. The experimental evidence is obtained by applying the Lund model to the multiplicity distribution measured at the highest energies of the $p\bar{p}$ collider. It is observed that the Lund model does not give the necessary multiplicity and entropy creation in the important central region.

The Lesson of Quantum Theory, edited by J. de Boer, E. Dal and O. Ulfbeck
© Elsevier Science Publishers B.V., 1986

Physics in Terms of Difference Equations

T.D. Lee

Columbia University
New York, USA

Contents

1. Introduction

The invention of quantum mechanics by Niels Bohr, and others, brought a profound change in our thinking. As a result of quantum mechanics, all dynamics is now described in terms of observables, and each observable is expressed as an operator or a matrix. Its eigenvalues denote the possible outcome of any measurements on the observable, with a probability amplitude given by the appropriate component of the state vector. These ideas were truly revolutionary at the time of their creation. By now they are accepted universally and taught to students everywhere. Mankind will be forever grateful to Niels Bohr, the originator and the developer of quantum mechanics.

Nearly thirty years ago Pauli came to Columbia University to give a seminar on his joint work with Heisenberg. Many people were in the audience, including Niels Bohr. There were a lot of questions, discussions and criticisms, some not so friendly. In the end Pauli said, in an uncharacteristic way: "Perhaps my idea is crazy." Whereupon Bohr stood up, walked a few steps towards Pauli, looked down, pointed a finger at Pauli and said, "The problem is not whether your idea is crazy, but whether it is crazy enough."

It is in this spirit that I wish to analyze further some of the basic concepts of quantum mechanics. Let me first raise two questions:

(i) In a measurement of, say, the electric field $E(r, t)$ is it not true that the precise values of r and t, like E, can only be determined after the measurement?

(ii) Is the concept of local field theory applicable to very small distances?

1.1. Space and time as dynamical variables

To take up the first question, consider, say, an experimentalist who proposes to the CERN program committee to measure the electric field in an e^+e^- collision at LEP in 1989. The proposal is approved. The experimentalist and his several hundred collaborators then set up their gigantic instruments. Before the measurement, they have some expectation of the amplitude of the electric field E that they are going to measure. They also know the approximate location and the time that the collision will take place. The precise value of E, the exact location of the reaction and the precise time of the collision can only be determined after his measurement. In other words, in terms of observation, we do treat space–time on the same footing as the electric field. Therefore, in terms of conceptual thinking, should we not also treat space–time on the same footing as the electric field? Wouldn't it be more in the spirit of Niels Bohr to regard space and time both as *dynamical variables*, the same as the electric field?

In the conventional description of local quantum field theory, we view the fields as the dynamical variables, represented by operators, but space r and time t are only parameters. Even in Einstein's general relativity, although the metric is a dynamical variable, the continuous four-dimensional space–time that the metric is embedded in is not. In this paper I shall explore the alternative view, treating space and time as dynamical variables, playing a similar role as the electromagnetic field, the gluon field, etc. As we shall see, this new description has a connection with my second question.

1.2. Locality at the Planck length

Let me present an argument to show that the concept of local field theory is possibly inapplicable to distances of the order of the Planck length l_P. In fig. 1 we consider two points A and B, separated by a spatial distance Δx and a time difference Δt with ($\hbar = c = 1$)

$$\Delta t < \Delta x < l_P \sim 10^{-33} \text{ cm},$$

in which the first inequality ensures that A and B are outside each other's light cone. Local field theory then states that these two experiments can be done independently of each other, no matter how close the points. Yet, just based on the uncertainty principle, we expect a fluctuation of energy ΔE caused by these two measurements,

$$\Delta E \sim \frac{1}{\Delta t} > \frac{1}{l_P}.$$

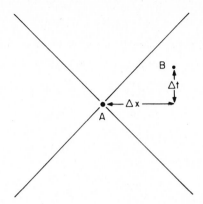

Fig. 1. Space–time points A and B outside each other's light cone.

The gravitational field associated with such a ΔE is very strong at small distances. Indeed, its Schwarzschild radius (i.e. black hole radius) is

$$R \sim G\Delta E > \frac{G}{l_{\mathrm{P}}},$$

where $G = l_{\mathrm{P}}^2$ is Newton's gravitational constant. Thus we find

$$R > \Delta x!$$

It seems quite unreasonable that these two measurements A and B could be viewed as independent. Consequently, the concept of locality very likely breaks down at such distances. Likewise, the usual linear superposition principle of quantum mechanics may also be questioned. If locality is not satisfied at the Planck length, then the correct physical theory must be non-local in character. The fact that the Planck length is small is beside the point. We would like this non-local fundamental theory to retain all the good features of the usual continuum theory: Lorentz invariance, Poincaré invariance, non-Abelian gauge symmetries, unitarity and the general coordinate invariance of general relativity. In addition it should not contain divergence difficulties, so that quantization of gravity can be carried out.

There are perhaps two different directions one may follow: one is to add degrees of freedom to the usual local field theory. The other is to subtract degrees of freedom. The former is followed by those working on string theories, and the latter will be the subject of this chapter.

2. Time as a dynamical variable

In this new theory I shall treat time as a dynamical variable [1]. This will lead to a dynamics which is formulated in terms of difference equations, instead of the usual differential equations. We will first review briefly the classical theory of this new mechanics, called discrete mechanics, and then go over to the quantum theory.

2.1. Classical mechanics

Take the simplest example of a non-dimensional nonrelativistic particle of unit mass moving in a potential $V(x)$. In the usual continuum mechanics the action is

$$A(x(t)) = \int_0^T \left[\tfrac{1}{2}\dot{x}^2 - V(x)\right] \, dt, \tag{2.1}$$

where $x(t)$ can be any smooth function of time t. Keeping the initial and final positions fixed, say x_0 and x_f, at $t = 0$ and T, we determine the orbit of the particle by the stationary condition

$$\frac{\delta A}{\delta(x(t))} = 0, \tag{2.2}$$

which leads to Newton's equation

$$\ddot{x} = -\frac{dV}{dx}. \tag{2.3}$$

In the above, x is the dynamical variable and t is merely a parameter. Next, we shall see how this customary approach may be modified in the discrete version.

Let the initial and final positions of the particle be the same:

$$x_0 \quad \text{at } t = 0 \quad \text{and} \quad x_f \quad \text{at } t = T. \tag{2.4}$$

In the discrete mechanics we restrict the usual smooth path $x(t)$ to a "discrete path" $x_D(t)$, which is continuous but piecewise linear, characterized by N vertices (as shown in fig. 2). In fig. 2a we have the usual smooth path $x(t)$ of a nonrelativistic particle in classical mechanics. Moving along $x(t)$ from $t = 0$ to $T > 0$, the time t increases monotonically; this property is retained under the constraint restricting

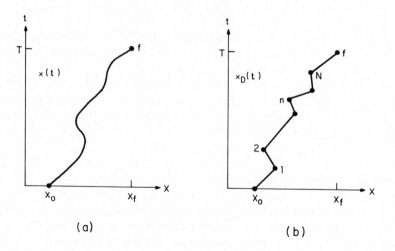

(a)   (b)

Fig. 2. Usual smooth path (a) and discrete path (b).

$x(t)$ to $x_\mathrm{D}(t)$. Thus, as in fig. 2b, we may label the N vertices of $x_\mathrm{D}(t)$ consecutively as $n = 1, 2, \ldots, N$, each of which carries a space–time position x_n and t_n with

$$0 < t_1 < t_2 < t_3 < \cdots < t_N < T. \tag{2.5}$$

The nearest-neighboring vertices are linked by straight lines, forming the discrete path $x_\mathrm{D}(t)$, which also appears as a one-dimensional lattice with n as lattice sites. In fig. 2b, a variation of the space–time positions of these vertices changes the discrete path $x_\mathrm{D}(t)$. However, a mere exchange of any two vertices clearly defines the same $x_\mathrm{D}(t)$. This is because only the discrete path with unlabeled vertices has a physical meaning. There is no "individual" identity of any of the vertices. Thus, the time-ordered sequence (2.5) is not an additional restriction, but one that arises naturally when we pass from the usual nonrelativistic path $x(t)$ to the discrete $x_\mathrm{D}(t)$.

In the following, we shall keep the site-density

$$\frac{N}{T} \equiv \frac{1}{l} \tag{2.6}$$

fixed, and regard l as a fundamental constant of the theory. The action integral (2.1) evaluated on such a discrete path $x_\mathrm{D}(t)$ is

$$A_\mathrm{D} = A\big(x_\mathrm{D}(t)\big) = \sum_n \left[\frac{1}{2} \frac{(x_n - x_{n-1})^2}{t_n - t_{n-1}} - (t_n - t_{n-1}) \, \overline{V}(n) \right], \tag{2.7}$$

where

$$\overline{V}(n) = \frac{1}{x_n - x_{n-1}} \int_{x_{n-1}}^{x_n} V(x) \, \mathrm{d}x \tag{2.8}$$

is the average of $V(x)$ along the straight line between x_{n-1} and x_n.

Because the path $x_\mathrm{D}(t)$ is completely specified by its vertices $n(x_n, t_n)$, a variation in $x_\mathrm{D}(t)$ is equivalent to a variation in all the positions of its vertices

$$\mathrm{d}\big[x_\mathrm{D}(t)\big] = \prod_n [\mathrm{d}x_n][\mathrm{d}t_n]. \tag{2.9}$$

Correspondingly, the dynamical eq. (2.2) becomes the difference equations

$$\frac{\partial A_\mathrm{D}}{\partial x_n} = 0, \tag{2.10a}$$

and

$$\frac{\partial A_\mathrm{D}}{\partial t_n} = 0. \tag{2.10b}$$

We see that in this new mechanics the roles of x_n and t_n are quite similar. Both appear as *dynamical variables*. For each x_n or t_n we have one difference equation, (2.10a) or (2.10b). The former gives Newton's law on the lattice and the latter gives the conservation of energy

$$E_n \equiv \frac{1}{2}\left(\frac{x_n - x_{n-1}}{t_n - t_{n-1}}\right)^2 + \overline{V}(n) = E_{n+1}. \qquad (2.11)$$

In the usual continuum mechanics, conservation of energy is a consequence of Newton's equation. Here, the two eqs. (2.10a) and (2.10b) are independent. Altogether there are $2N$ such equations, matching in number the $2N$ unknowns x_n and t_n in the problem. Because the action A_D is stationary under a variation in x_n and in t_n for all n, the discrete theory retains the translational invariance of both space and time, and that leaves the conservation laws of energy and momentum intact. *

For a free particle $V(x) = 0$, eqs. (2.10a) and (2.10b) become degenerate; both give

$$v_n = \frac{x_n - x_{n-1}}{t_n - t_{n-1}} = \text{constant}.$$

The corresponding trajectory is a straight line, the same as the continuum case.

When $V(x) = gx$ with g a constant, the solution of eqs. (2.10a) and (2.10b) can be readily found. We find in this case the spacing between successive t_n to be independent of n:

$$t_n - t_{n-1} = \epsilon = \text{constant}.$$

Correspondingly, $t_n = t_0 + n\epsilon$ and

$$x_n = x_0 + n v_1 \epsilon - \tfrac{1}{2} n(n-1) g\epsilon^2,$$

where v_1 is the initial velocity $(x_1 - x_0)/(t_1 - t_0)$.

When $l \to 0$, the site density $\to \infty$ and the discrete path $x_D(t)$ can assume the form of any smooth path $x(t)$; consequently the discrete mechanics approaches the usual continuum mechanics. Introduce

$$\tau \equiv nl,$$

which varies from 0 to T, as n runs from 0 to N. Consider the quantities

$$x_n = x(\tau) \quad \text{and} \quad t_n = t(\tau).$$

From eqs. (2.10a) and (2.10b), it can be shown that in the limit $l \to 0$, but keeping T

---

* Here, conservation of momentum means that the change of particle momentum is equal to the "impulse" generated by the potential.

fixed (hence, $N \to \infty$),

$$\frac{d^2x}{dt^2} = -\frac{dV}{dx}, \tag{2.12a}$$

and

$$\left(\frac{dt}{d\tau}\right)^3 \left(\frac{dV}{dx}\right)^2 = \text{constant}; \tag{2.12b}$$

the former is Newton's equation, and the latter gives the asymptotic distribution of t_n versus n. The constant in eq. (2.12b) is determined by the boundary condition (2.4), so that when τ varies from 0 to T, t also changes from 0 to T. In the usual continuum mechanics, only eq. (2.12a) is retained. Therefore, even in this limit, the discrete mechanics contains more information than the usual continuum mechanics. From eq. (2.12b), we see that, except for $V(x) = gx$, the spacing $t_n - t_{n-1}$ is not a constant.

It is of interest to examine the distribution $t(\tau)$ near the point $V'(x) \equiv dV/dx = 0$, which occurs at, say, $x = \bar{x}$. Let the particle trajectory in the continuum limit be $x = x(t)$. When $x = \bar{x}$, we have $V'(\bar{x}) = 0$ and, for the solution under consideration, $t = \bar{t}$ so that $\bar{x} = x(\bar{t})$. In the neighborhood x near $\bar{x}$, we may write, with $V''(x) \equiv d^2V/dx^2$ and $\dot{x} \equiv dx/dt$,

$$V'(x) \simeq (x - \bar{x}) V''(\bar{x})$$

$$= (t - \bar{t})\dot{x}(\bar{t}) V''(\bar{x}).$$

Substituting this expression into eq. (2.12b) we find

$$(t - \bar{t}) \propto (\tau - \bar{\tau})^{3/5}.$$

Hence, as $\tau \to \bar{\tau}$ (correspondingly $n \to \bar{\tau}/l$), although $dt/d\tau \to \infty$ one sees that $t \to \bar{t}$ and remains finite. Information such as this is lost if one concentrates only on Newton's equation (2.12a).

In the following, we are interested in $l \neq 0$, in which case the discrete mechanics is fundamentally different from the continuum theory.

2.2. Nonrelativistic quantum mechanics

When we go over from classical to quantum mechanics, in the usual continuum theory the particle can take on any smooth path $x(t)$; each path carries an amplitude e^{iA} where $A = A(x(t))$ is the same action integral (2.1). In Feynman's path integration formalism, the matrix element of e^{-iHT} in the usual continuum quantum mechanics is given by

$$\langle x_f | e^{-iHT} | x_0 \rangle = \int e^{iA(x(t))} \, d[x(t)], \tag{2.13a}$$

in which all paths $x(t)$ have the same end-points (2.4) and

$$H = -\frac{1}{2}\frac{\partial^2}{\partial x^2} + V(x). \tag{2.13b}$$

Sometimes it is more convenient to consider the analytic continuation of T to $-iT$. The operator e^{-iHT} becomes then e^{-HT}, and its matrix element is given by

$$\langle x_f | e^{-HT} | x_0 \rangle = \int e^{-\mathscr{A}(x(t))} \, d[x(t)], \tag{2.14}$$

where

$$\mathscr{A}(x(t)) = \int_0^T \left[\tfrac{1}{2}\dot{x}^2 + V(x)\right] dt. \tag{2.15}$$

In the corresponding discrete theory, we again restrict the particle to move only along the discrete path $x_D(t)$. By using eqs. (2.7) and (2.9), we see that the right-hand side of eq. (2.13a) becomes

$$\int e^{iA_D} \prod_n [dx_n][dt_n]. \tag{2.16}$$

Likewise, eqs. (2.14) and (2.15) become

$$\langle x_f | G_N(T) | x_0 \rangle \equiv \int e^{-\mathscr{A}_D} \prod_{n=1}^{N} [dx_n][dt_n], \tag{2.17}$$

where

$$\mathscr{A}_D = \mathscr{A}(x_D(t)). \tag{2.18}$$

When the vertices $n = 1, 2, \ldots,$ are arranged in a time-ordered sequence (2.5), by using eqs. (2.15) and (2.18) we see that the discrete action $\mathscr{A}_D$ is given by

$$\mathscr{A}_D = \sum_{n=1}^{N+1} \left[\frac{(x_n - x_{n-1})^2}{2(t_n - t_{n-1})} + (t_n - t_{n-1})\, \bar{V}(n)\right], \tag{2.19}$$

with $x_{N+1} = x_f$ and $t_{N+1} = T$, as shown in figure 2b.

In the integration over $\prod_n[dt_n]$, whenever t_i appears larger than, say, t_{i+1}, we should re-link the vertices so that the newly linked ones are in a time-ordered sequence. Alternatively, we may re-label them so that eq. (2.5) remains valid; such a relabeling of vertices clearly does not change the discrete path $x_D(t)$. (As explained before, this follows from the usual nonrelativistic continuum mechanics in which the path $x(t)$ is a single-valued function of t.)

In the quantum version of the discrete mechanics it is more convenient to regard the constraint (2.6) as a condition on the *average* site-density. This can be most easily arranged by considering an ensemble sum over N:

$$\mathscr{G}(T, l) \equiv \sum_{N=0}^{\infty} \frac{1}{N!} \left(\frac{1}{l}\right)^N G_N(T),
\tag{2.20}$$

where $G_N(T)$ refers to the matrix defined by eq. (2.17). One may readily verify that this Green function satisfies

$$\frac{\partial}{\partial(1/l)} \mathscr{G}(T, l) = \int_0^T \mathscr{G}(\tau, l)\, \mathscr{G}(T - \tau, l)\, d\tau,
\tag{2.21}$$

from which it follows that for large T and neglecting $e^{-T/l}$ the operator $\mathscr{G}(T, l)$ becomes

$$\mathscr{G}(T, l) \sim e^{-\mathscr{H}T},
\tag{2.22}$$

where $\mathscr{H}$ is Hermitian. When $l \to 0$, $\mathscr{H}$ reduces to the continuum Hamiltonian H, given by eq. (2.13b). The analytic continuation of $\mathscr{G}(T, l)$ from T to iT leads at large T to the unitary operator $e^{-i\mathscr{H}T}$, which is the S-matrix of the theory. Therefore, the unitarity of the S-matrix is maintained in the new mechanics [2], at least when $\mathscr{H}$ is O(1).

3. Relativistic quantum field theory

As an example, let $\phi(x)$ be a scalar field in the usual continuum theory with x denoting the space–time coordinates. In the path integration formulation the operator e^{-HT} is given by, similar to eq. (2.14),

$$e^{-HT} = \int e^{-\mathscr{A}} [d\,\phi(x)],
\tag{3.1}$$

where H is the Hamiltonian operator, $\mathscr{A}$ the usual continuum action in the Euclidean space and T the total "time" interval. (Here, as in eqs. (2.14) and (2.15), "time" refers to the Euclidean time.) Because in the usual continuum theory the space–time coordinates x are parameters and only $\phi(x)$ are dynamical variables, the functional integration in eq. (3.1) is over $[d\,\phi(x)]$, not over $[dx]$.

In the discrete version, we impose a constraint on the (average) number N of experiments that can be performed within any given space–time volume Ω, with $N/\Omega \equiv l^{-4}$ = fundamental constant. Each measurement determines the field $\phi(i)$ as well as the space–time position $x(i)$ with $i = 1, 2, \cdots, N$. The i will be referred to as lattice sites, as illustrated by fig. 3a.

As we shall see, the Green function (3.1) will be replaced by

$$\int e^{-\mathscr{A}_D} [d\,x(i)] [d\,\phi(i)].
\tag{3.2}$$

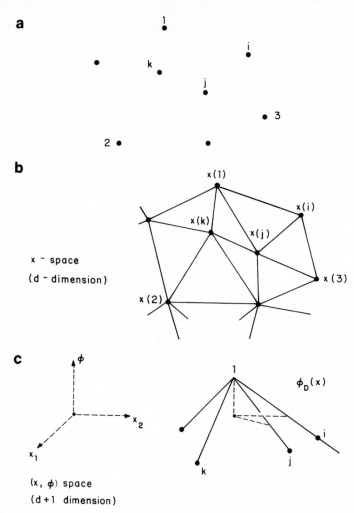

Fig. 3. (a) Lattice sites in space–time. (b) Lattice sites coupled to neighboring sites. (c) The "discrete" function $\phi_D(x)$ in $x-\phi$ space.

Because $\phi(i)$ and $x(i)$ are all dynamical variables, in the discrete theory we integrate over [d $\phi(i)$] as well as over [d $x(i)$]. The latter integration makes it obvious that rotational and translational symmetries can be maintained in the discrete theory.

To simulate the local character of the usual continuum theory, each site in the discrete theory is coupled only to its neighboring sites, as illustrated in fig. 3b. The whole volume is then divided into triangles if the dimension of $x(i)$ is $d = 2$, tetrahedra if $d = 3$, four-simplices when $d = 4$, etc. An example of such a simplicial lattice when $d = 2$ appears in fig. 3b.

We give the algorithm [3] of linking an arbitrary distribution of sites into a simplicial lattice for any dimension d. Select any group of $d + 1$ sites; consider the hypersphere (in the d-dimensional Euclidean space) whose surface passes through these $d+1$ sites. If the interior of the sphere is empty of sites, link these sites to form a d-simplex; otherwise, do nothing. Proceed to another group of $d+1$ sites, and repeat the same steps. The d-simplices thus formed never intersect each other, and the sum total of their volumes fills the entire space.

Each site i carries, in addition to its space–time coordinates $x(i)$, also a $\phi(i)$. Viewed in the x–ϕ space, the lattice forms a d-dimensional surface represented by $\phi_D(x)$, called the "discrete" function; it is continuous but piece-wise flat within each d-simplex as illustrated in fig. 3c.

The discrete action $\mathscr{A}_D$ in eq. (3.2) can be readily evaluated by using the usual continuum action $\mathscr{A}(\phi(x))$, but restricting $\phi(x)$ to the discrete function:

$$\mathscr{A}_D \equiv \mathscr{A}(\phi_D(x)). \tag{3.3}$$

For example, if

$$\mathscr{A}(\phi(x)) = \int \left[\tfrac{1}{2}(\nabla\phi)^2 + V(\phi)\right] dx, \tag{3.4}$$

where dx is the d-dimensional volume element in the x-space, then setting $\phi(x)$ to be the discrete function $\phi_D(x)$, we find

$$\mathscr{A}_D = \mathscr{A}(\phi_D(x)) = \tfrac{1}{2}\sum_{l_{ij}}\lambda_{ij}\left([\phi(i) - \phi(j)]^2\right) + \sum_i \omega_i V(\phi(i)), \tag{3.5}$$

where the first sum is over all links l_{ij} and the second over all sites i, ω_i is the volume of the Voronoi cell that is dual to the site i, and [4]

$$\lambda_{ij} = -\frac{1}{d^2}\sum \frac{1}{V(ij)}\tau(i)\cdot\tau(j), \tag{3.6}$$

in which the sum extends over all d-simplices $V(ij)$ that share the link l_{ij}. In $V(ij)$, each vertex, say k, faces a $(d-1)$-dimensional simplex $\tau(k)$. In eq. (3.6), $V(ij)$ denotes also the volume of the d-simplex and $\tau(i)$ is the outward normal vector of $\tau(i)$ times its $(d-1)$-dimensional volume, as illustrated in figure 4. As in the previous section, mathematically the discrete theory can be regarded as a special case of the usual continuum theory: one in which $\phi(x)$ is restricted to those continuous but piece-wise flat functions $\phi_D(x)$ with a fixed average density of vertices (i.e. lattice sites). Because the site density is an invariant, rotational and translational invariances can both be preserved in the discrete theory.

Since the discrete surface, described by $\phi_D(x)$, is characterized by the positions $\phi(i)$ and $x(i)$ of its vertices, a variation over the functional space $\phi(x)$ in the usual continuum theory becomes

$$[d\,\phi_D(x)] = \prod_i [d\,\phi(i)][d\,x(i)]. \tag{3.7}$$

T.D. Lee

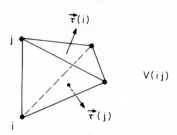

Fig. 4. Simplex $V(ij)$ and the associated outward normals $\tau(i)$.

Correspondingly, eq. (3.1) becomes eq. (3.2). As $x(i)$ changes, the linking algorithm keeps track of how these vertices should be linked, so that the discrete action $\mathscr{A}_D$ is extensive; i.e. $\mathscr{A}_D$ is proportional to the overall space–time volume Ω when Ω is large. Thereby, the unitarity of the S-matrix can be established, as before.

In the usual continuum theory, the equation of motion is given by the partial differential equation

$$\frac{\delta\mathscr{A}(\phi(x))}{\delta\phi(x)} = 0. \tag{3.8}$$

Here in the discrete version it is replaced by the difference equations

$$\frac{\partial\mathscr{A}_D}{\partial\phi(i)} = 0 \quad \text{and} \quad \frac{\partial\mathscr{A}_D}{\partial x(i)} = 0; \tag{3.9}$$

the former is the field equation on the lattice and the latter expresses the conservation law of the energy–momentum tensor.

In the integrand of eq. (3.2), the locations of x can be arbitrary. Hence, the discrete action $\mathscr{A}_D$ is identical to that of a random lattice [5].

4. Gauge theory

We review briefly the random lattice results on Abelian (QED) and non-Abelian (QCD) gauge theories.

The lattice gauge theory was introduced by K. Wilson. In the strong-coupling limit (square of coupling constant $g^2 \to \infty$), any lattice gauge theory gives confinement. This holds for both QED and QCD, and for arbitrary space dimension d. The realistic case corresponds, however, to the weak coupling. Thus, a key question is whether the transition from strong to weak coupling is smooth or not. If smooth, then the confinement property of the strong coupling can be carried over to weak coupling, otherwise not. When $\beta = 1/g^2$ changes from 0 (strong coupling) to ∞ (weak coupling), we would like the transition to be smooth for the non-Abelian case, but not smooth for the Abelian, so that the confinement holds for QCD, but not for QED. In a hypercubic lattice, there appears to be a phase transition in β for the

U(1) gauge, consistent with the fact that QED is not confined. However, numerical results in SU(2) and SU(3) indicate that the transition from $\beta = 0$ to $\beta = \infty$ is also far from smooth for a hypercubic lattice. While there is probably no bona fide phase transition in the non-Abelian case, the change from cubic ($g^2 = \infty$) symmetry to spherical ($g^2 = 0$) symmetry is sufficiently hazardous that it is difficult to infer, from the strong-coupling result, that confinement would remain valid in the weak coupling.

On the other hand, for the random lattice, its strong-coupling limit behaves like a relativistic string theory, with full rotational symmetry: the string thickness t is related to the string tension T by

$$t^2 = \frac{1}{2\pi T} \ln a, \tag{4.1}$$

where a is the area enclosed by the string. Furthermore, the mass of the glueball m_J

a

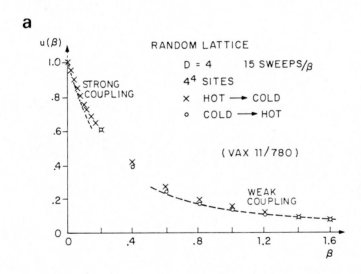

b

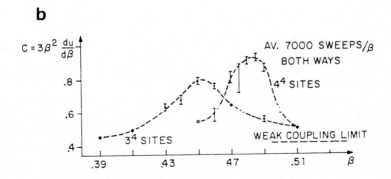

Fig. 5. (a) Average plaquette energy u and (b) specific heat C in $U(1)$ theory, as functions of $\beta = g^{-2}$.

a

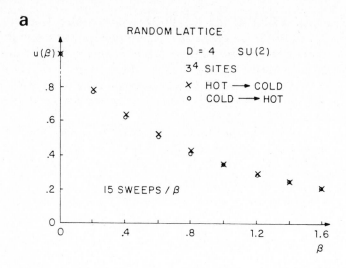

b

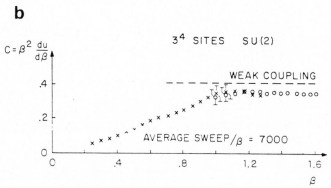

Fig. 6. (a) Average plaquette energy u and (b) specific heat C in SU(2) theory, as functions of $\beta = g^{-2}$.

for large angular momentum J varies as

$$m_J \propto \sqrt{J}, \tag{4.2}$$

exhibiting the typical Regge behavior of a rotating relativistic string. Both eqs. (4.1) and (4.2) are valid in the strong-coupling limit.

Numerical programs for a random lattice gauge theory were set up by Friedberg and Ren at Columbia; the computations were carried out by Ren [6]. In fig. 5 we give the average plaquette energy u and specific heat C vs. $\beta = 1/g^2$ for the U(1) theory.

The corresponding plots for an SU(2) theory are given in fig. 6. We see that the specific heat has a peak in the U(1) theory, but not in the SU(2) theory. For U(1), the peak becomes steeper when the number of lattice sites increases, suggesting that

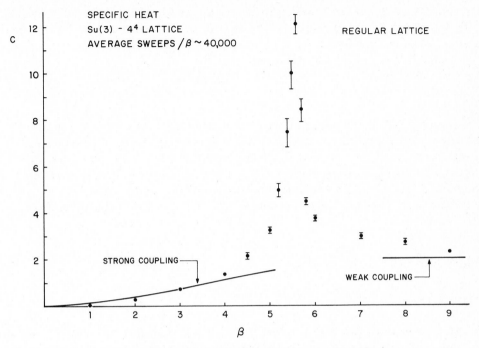

Fig. 7. Specific heat for SU(3) gauge theory on a regular lattice.

there is a phase transition. On the other hand, the specific heat curve for SU(2) has no peak, indicating that the passage from strong to weak coupling is a smooth one. Consequently, while both theories are confined in the strong coupling limit, the weak coupling limit is consistent with deconfinement in the U(1) theory (QED), but with confinement in a non-Abelian gauge theory (QCD).

In contrast, we give in fig. 7 the numerical calculation by N.H. Christ and A.E. Terrano [7] for the SU(3) gauge theory on a regular lattice. As we can see, there is a sharp peak in the specific heat, suggesting that the transition from strong to weak coupling in a regular lattice is by no means smooth, unlike that in a random lattice.

5. Lattice gravity

The usual Einstein action in general relativity is

$$A(S) = \int_S \sqrt{|g|} \, \mathscr{R} \, \mathrm{d}x, \tag{5.1}$$

where S is a d-dimensional smooth continuous surface, $|g|$ is the absolute value of the determinant of the matrix of the metric tensor $g_{\mu\nu}$ on S, $\mathscr{R}$ is the scalar curvature and $\mathrm{d}x$ is the d-dimensional volume element in the space–time coordinate x.

For lattice gravity, we consider first a (random) lattice $\mathring{L}$ in a flat d-dimensional Euclidean space R_d. Label each site by $i = 1, 2, \ldots$. For every linked pair of sites i and j there is a link-length $\mathring{l}_{ij}$.

Consider now an arbitrary variation

$$\mathring{l}_{ij} \to l_{ij}. \tag{5.2}$$

Correspondingly, each d-simplex, say $\mathring{\tau}$ in $\mathring{L}$, becomes a new d-simplex τ with the same vertices, but different link-lengths. These new link-lengths l_{ij} are assumed to satisfy all simplicial inequalities, so that each d-simplex τ, by itself, can still be realized in a flat d-dimensional space R_d. In general the entire new lattice cannot fit into R_d. This then defines [8] a d-dimensional non-flat lattice surface L.

Sometimes, it is convenient to embed L in a flat space R_N. This is possible if

$$N = d + n$$

is sufficiently large; in that case

$$l_{ij}^2 = \left[r(i) - r(j) \right]^2 \quad \text{in } R_N, \tag{5.3}$$

with $r(i)$ the Cartesian N-dimensional position vector of the ith site in R_N. Since, as we shall see, we shall deal only with the intrinsic geometric properties of the lattice surface, this embedding is merely a convenience.

Next we wish to evaluate Einstein's action (5.1) when S is restricted to the lattice surface L. At first sight, it might appear difficult because the metric g_{ij} would change discontinuously from simplex to simplex, the Christoffel symbol would then acquire δ-functions, and the scalar curvature δ'-functions. Since Einstein's action is nonlinear in g_{ij}, one might expect the resulting expression to be totally unmanageable. It turns out that this is not the case.

It can be shown that the Einstein formula (5.1) evaluated on any d-dimensional lattice space L gives the discrete action [9,10]

$$A(L) \equiv \int_L \sqrt{|g|} \, \mathcal{R} \, dx \tag{5.4}$$

$$= 2 \sum_s s \epsilon_s, \tag{5.5}$$

where dx is the d-dimensional volume element, s is the volume of the $(d-2)$ simplex, ϵ_s is Regge's deficit angle around s and the sum extends over all s in the lattice. (See ref. [11] for the definition of ϵ_s.) The right-hand side of eq. (5.5) is *precisely* the formula of Regge calculus [11].

In Regge's original approach, he considered the discrete action as an approximation to Einstein's continuum action. Here we are reversing the role and regarding the discrete action $A(L)$ as more fundamental. It is therefore satisfying to realize that Regge's action is identical to Einstein's action, but evaluated on L.

The quintessance of Einstein's theory of general relativity lies in its invariance under a general coordinate transformation

$$x \to x' \tag{5.6}$$

that leaves ds^2 unchanged. Since the action for the lattice space L is the discrete action

$$A_D \equiv A(L) = \int_L \sqrt{|g|} \, \mathcal{R} \, dx = 2 \sum_s s\epsilon_s, \tag{5.7}$$

the discrete theory clearly remains invariant under the coordinate transformation (5.6). Thus, the entire apparatus of coordinate invariance in the usual continuum theory automatically applies to the lattice theory as well. In addition, as we shall see, the lattice theory enjoys still another total new class of symmetries which does not exist in the usual continuum theory. Aesthetically, this adds greatly to the appeal of lattice gravity. For physical applications, when the link-length l is small, our general formula (5.7) ensures that all known tests of general relativity are automatically satisfied. Furthermore, by keeping l nonzero, we see that the lattice action A_D per volume possesses only a finite degree of freedom. The normal difficulty of ultra-violet divergence that one encounters in quantum gravity disappears in the lattice theory. All these suggest that the lattice theory with a nonzero l may be more fundamental. The usual continuum theory is quite possibly only an approximation.

To amplify the aforementioned symmetry properties, let us consider any lattice L. From eq. (5.7), we see that the discrete action A_D, through its right-hand side, is a function of the link-lengths l_{ij},

$$A_D = A_D(l_{ij}). \tag{5.8}$$

We may also characterize the lattice by other means of parametrization. We assume all the lattice sites i to lie on a d-dimensional smooth enveloping surface S, with $z_\mu(i)$ as the coordinates of the site i on S, where $\mu = 1, 2, \ldots, d$. Embed both S and L in a flat space R_N, which is always possible provided that N is sufficiently large. Because of eq. (5.3), l_{ij} can also be determined by giving S and $z_\mu(i)$. Hence we can also express A_D as a function of the enveloping surface S and the site positions on S,

$$A_D = A_D(S, z_\mu(i)). \tag{5.9}$$

Thus, we can have new symmetry transformations,
    (i) fix $z_\mu(i)$, but vary $S \to S'$,
    (ii) fix S, but vary $z_\mu(i) \to z'_\mu(i)$.
These symmetries are exact if l_{ij} are unchanged; they can be approximate even if l_{ij} does change, provided that, e.g. the link-lengths are sufficiently small and $\sqrt{|g|} \, \mathcal{R} \, dx$ remains the same on the enveloping surface, in which case $A_D \simeq A(S)$ of eq. (5.1).

In the usual continuum theory, the physical space–time points and the underlying four-dimensional manifold are the same. Here, they are distinct; the former is related to measurements, while the latter is purely a mathematical artifice (like the choice of gauge in the usual continuum theory of a spin 1, or 2, field).

6. Concluding remarks

For more than three centuries we have been influenced by the precept that fundamental laws of physics should be expressed in terms of differential equations. Difference equations are always regarded as approximations. Here, we try to explore the opposite: Difference equations are more fundamental, and differential equations are regarded as approximations.

As we have shown, such a difference equation formulation leads to the discrete mechanics which can also be viewed as the mathematical limit of the usual continuum mechanics, but with a fixed density of lattice sites. Because this is an invariant constraint, the discrete theory shares the same symmetries of the usual continuum theory. In this way, we have succeeded in the creation of theories with finite degrees of freedom, but which retain the good properties of the usual continuum theory. We suggest that this discrete formulation might be more fundamental.

In this new mechanics, space and time are treated as dynamical variables, on the same footing as electromagnetic fields, gluon fields, etc. In this sense, I hope the thinking of Niels Bohr can be extended further to bring our theoretical concepts even closer to actual experimental observation.

References

[1] T.D. Lee, Phys. Lett. 122B (1982) 217; in: Shelter Island II, eds. R. Jackiw, N. Khuri, S. Weinberg and E. Witten (MIT Press, Cambridge, 1985), p. 38–64.
[2] R. Friedberg and T.D. Lee, Nucl. Phys. B225 [FS9] (1983) 1.
[3] N.H. Christ, R. Friedberg and T.D. Lee, Nucl. Phys. B202 (1982) 89.
[4] R. Friedberg and T.D. Lee, unpublished.
[5] N.H. Christ, R. Friedberg and T.D. Lee, Nucl. Phys. B210 [FS6] (1982) 310.
[6] H.C. Ren, Field Theory on a Random Lattice, Dissertation (Columbia University, New York, 1984).
[7] N.H. Christ and A.E. Terrano, Nucl. Instrum. Meth. 222 (1984) 534.
[8] For physical application to general relativity, in order to maintain the quasi-local character of the discrete action, we must link only neighboring sites. Thus, when the new link-lengths l_{ij} are too large the sites have to be re-linked. Details will be given elsewhere.
[9] R. Sorkin, Phys. Rev. D12 (1975) 385.
[10] R. Friedberg and T.D. Lee, Nucl. Phys. B242 (1984) 145.
[11] T. Regge, Nuovo Cimento 19 (1961) 558.

Discussion, session chairman A. Salam

Regge: You are right in saying that the approximation used in the 1961 paper was designed to get away from coordinate systems. However, I would be surprised if this system represents reality since it does not cure the diseases of the conventional formulation of quantum gravity. For example, there still is the problem that the action does not have a lower bound. So further work appears to be necessary.

Lee: Well, I hope that this is the beginning of the work and not the end.

Rubia: To those of us in experimental science, can you say what we should be looking for?

Lee: What I could tell you at this time is not yet mature. If we introduce a fundamental length which does not violate Poincaré invariance, it is reasonable that it should be somewhere between 10^{-16} cm and the Planck length, 10^{-33} cm. The worst situation we can imagine is that it is near the latter. Now, consider a system with a strong gravitational field, namely a black hole. The size of the black hole will decrease due to Hawking radiation. In conventional theory, it will ultimately vanish, but not in the theory which I have described. To answer your question, we can ask what the probability is of a proton of 10^{-13} cm dimension to shrink suddenly to, say, 10^{-33} cm and to become a tiny (lattice) black hole of a smaller mass plus soft radiation. A crude estimate gives the probability as the ratio of these two volumes, about $(10^{-33}/10^{-13})^3 = 10^{-60}$. The natural time scale is 10^{-23} s, determined by the larger dimension 10^{-13} cm. This gives a proton lifetime of about 10^{30} years, which is experimentally accessible.

Casimir: I remember vaguely that Heisenberg played with the idea of a discrete space–time around 1930. Perhaps Weizsäcker knows more about this.

Weizsäcker: It is true that in the thirties Heisenberg thought about the idea of a fundamental length, but he did not develop it, and his last theory did not contain it. But let me ask whether there is any connection between your theory and the ideas of David Finkelstein, published in 1968.

Lee: Finkelstein did have similar thoughts, but his work did not offer a concrete proposition. I should say that what I have discussed is not meant to be the ultimate theory (as I said in response to Professor Regge's remark), but only a new way of thinking and a method of getting rid of divergences while maintaining continuum symmetries.

Kohn: A continuum trajectory, in your first example, can be approximated by discrete points in many different ways. So isn't there an ill condition in your equations, leading to indeterminacy?

Lee: No.

The Lesson of Quantum Theory, edited by J. de Boer, E. Dal and O. Ulfbeck
© Elsevier Science Publishers B.V., 1986

Some Lessons of Renormalization Theory

Arthur S. Wightman

Princeton University,
Princeton, New Jersey, USA

Contents

Introduction

In their fundamental paper on quantum mechanics, Born, Heisenberg and Jordan (1926) gave the first quantum mechanical treatment of a system of an infinite number of degrees of freedom: the vibrating string. They encountered the first divergence difficulty of the quantum theory of fields: the zero-point energy of the ground state. They dropped it, thereby performing the first infinite subtraction in the history of renormalization theory. In the course of the sixty years that have since elapsed, the quantum theory of systems of an infinite number of degrees of freedom has been extended greatly in scope and depth and renormalization theory along with it.

By now the term renormalization has a variety of associations both mathematical and physical. On the one hand, renormalization in one broad sense has often come to include any procedure by which infinite or ambiguous expressions in quantum field theory are replaced by well defined mathematical objects. A more precise definition would here distinguish *renormalization* and *regularization*, the latter being any rule that produces finite answers while the former is reserved for the special case in which the rule gives answers associated with a self-consistent field theory. On the other hand, renormalization is often used as a catch word for a family of methods of analyzing the significant parameters labeling the states of a theory and of their relations to the parameters actually appearing in the Hamiltonian. In the following we shall have the occasion to use renormalization in both these senses.

We are far from having a complete renormalization theory for the quantum theory of fields. The main point of the present chapter is to examine some of the significant developments in the history of renormalization theory for the light they can throw on our present unsolved problems. The text is arranged in the form of seven lessons with commentary.

1. Zero-point energy

Described more explicitly, what Born, Heisenberg and Jordan did was to write a formal expansion of the real valued function u, which specifies the transverse displacement of an unquantized string with ends fixed at 0 and l:

$$u(x, t) = \sum_{k=1}^{\infty} q_k(t) \sqrt{\frac{2}{l}} \sin \frac{\pi k}{l} x, \tag{1.1}$$

where

$$q_k(t) = \int_0^l dx \sqrt{\frac{2}{l}} \sin \frac{\pi k}{l} x \, u(x, t).$$

If the tension in the string is τ and the mass per unit length is ρ, the equation of motion is

$$\rho \frac{\partial^2 u}{\partial t^2} = \tau \frac{\partial^2 u}{\partial x^2},$$

so insertion of expression (1.1) gives

$$\ddot{q}_k(t) + \omega_k^2 q_k(t) = 0,$$

with $\omega_k = \sqrt{\frac{\tau}{\rho}} \frac{\pi k}{l}$. The total energy is

$$H = \frac{1}{2} \int_0^l dx \left\{ \rho \left(\frac{\partial u}{\partial t} \right)^2 + \tau \left(\frac{\partial u}{\partial x} \right)^2 \right\}$$

$$= \frac{1}{2} \sum_{k=1}^{\infty} \left[\rho (\dot{q}_k(t))^2 + \left(\frac{k\pi}{l} \right)^2 \tau q_k(t)^2 \right]$$

$$= \frac{1}{2} \sum_{k=1}^{\infty} \left[P_k^2 + \omega_k^2 Q_k^2 \right],$$

if we define $P_k = \rho \dot{q}_k$, $Q_k = \rho q_k$. The quantization procedure was (and is today) to quantize each oscillator independently, replacing P_k by the operator $P_k^{\mathrm{op}} = -i\hbar(\partial/\partial Q_k)$ and Q_k by the multiplication operator Q_k^{op}. Then

$$\tfrac{1}{2}\left[\left(P_k^{\mathrm{op}}\right)^2 + \omega_k^2\left(Q_k^{\mathrm{op}}\right)^2\right] = \hbar\omega_k\left[N_k + \tfrac{1}{2}\right],$$

where N_k is the number operator for the kth oscillator; it has the eigenvalues 0, 1, 2,... . The infinite subtraction of Born, Heisenberg and Jordan referred to above is the omission of the sum $\tfrac{1}{2}\sum_{k=1}^{\infty}\hbar\omega_k$ from H.

Of course, if this sum were finite its omission would have no effect whatever on the time evolution of observables defined by H:

$$q(0) \to q(t) = e^{iHt}\, q(0)\, e^{-iHt},$$

since such a constant cancels out. Nevertheless, this infinite subtraction raises a question. Does the omission of the zero-point energy mean that the zero-point vibrations are without physical significance? This question has a very unambiguous negative answer in a number of other contexts. For example, the low-lying vibrational states of a molecule can be approximately described in terms of a quantum-mechanical harmonic oscillator and measurements of the ground-state energy relative to the energy of the dissociated atoms can be compared to the minimum potential energy; the result is an ambiguous support for the reality of the zero-point energy.

Born, Heisenberg and Jordan's discussion of the problem was directed toward a resolution of longstanding difficulties in the theory of energy fluctuations in black body radiation. In his famous report on quanta at the 1911 Solvay Conference, Einstein (1911) had displayed a calculation of the energy fluctuations of electromagnetic radiation in equilibrium with atoms, modeled as a set of oscillators. He showed that there were two contributions to the mean square energy fluctuation, only one of which would be present according to classical wave theory. Born, Heisenberg and Jordan found agreement with Einstein's result.

It was some years later that Heisenberg (1931) showed that the three-man derivation is incorrect; the energy fluctuation calculated is in fact infinite. Heisenberg convinced himself that the infinity in question had nothing to do with the zero-point energy by calculating the analogous (infinite) quantity for a free Schrödinger particle for which there is no zero-point energy. He attributed the infinity of the fluctuation to the fact that the fluctuation was calculated for a sharply defined region. If the energy in a region R of space averaged over a time interval $\{t, t+T\}$,

$$\frac{1}{T}\int_t^{t+T}\mathrm{d}t \int_R \mathrm{d}^3x\, T_{00}(\mathbf{x}, \tau),$$

is replaced by

$$\int\int f(x)\, \mathrm{d}^4x\, T_{00}(x),$$

where f is a smooth function, equal to 1 inside R during the time interval $(t, t + T)$, and dropping rapidly to zero outside, then the fluctuation turns out to be finite, provided that the zero-point contribution to $T_{00}(x)$ is dropped. This fact that, in general, quantized fields like $T_{00}(x)$ only make sense as operators on states when they are smeared with sufficiently smooth functions f, was a key idea in the paper of N. Bohr and L. Rosenfeld (1933) on the measurability of the electromagnetic field. For the electromagnetic field, it was often possible to permit f to be discontinuous, say 1 inside a region and zero outside. However, Heisenberg (1934) showed that the smeared charge and current of a Dirac spin $\frac{1}{2}$ field has infinite fluctuations in the vacuum, unless function f has bounded first derivatives. That the energy density and charge density require somewhat smoother test functions is, roughly speaking, a consequence of the fact that they are quadratic in the annihilation and creation operators for photons, whereas the electromagnetic fields are linear. (All these calculations are for free fields, the coupling constant having been set equal to zero.)

Nowadays, following the general mathematical ideas developed by L. Schwartz (1952) in his great treatise of 1951–1952, one regards quantized fields as generalized functions i.e. linear functionals $T(f)$ of a test function f; in Schwartz's terminology $T_{00}(x)$ is an operator-valued distribution. So far, this mathematical concept has proved itself adequate for the purposes of quantum field theory.

While all this discussion of fields as operator-valued distributions is indispensible for understanding field fluctuations, it does not address directly the significance of the zero-point energy, except that it provides the rule: drop the terms in T_{00} which make its expectation value in the vacuum different from zero. This rule still provoked some uneasiness and efforts were made by Rosenfeld and Solomon (1931) and by Pauli (1933) in his "Handbuch" article to write the energy density as a function of fields in such a fashion that the zero-point contribution automatically drops out. For example for the electromagnetic field

$$T_{00}(x) = \tfrac{1}{2}\left[E^2(x) + B^2(x) \right] + i\left[E(x),\ \frac{1}{\sqrt{-\Delta}}\ \nabla x B \right],$$

where $E(x)$ and $B(x)$ are the electric and magnetic field strengths, respectively. (See Pauli's article p. 256.) Exactly what the significance of such an expression is was not clear at the time, apart from the fact that it leads to an energy with no contributions from zero-point vibration.

It was in 1950, long after, that G.C. Wick (1950) introduced the operation that we now call Wick ordering, which permits us to write this expression as

$$T_{00}(x) = \tfrac{1}{2}\left[:E^2: (x) + :B^2: (x) \right], \tag{1.2}$$

with the Wick ordering $: :$ defined by

$$:E^2(x): (x) = \lim_{x \to y} \left[E(x) \cdot E(y) - \left(\Phi_0,\ E(x) \cdot E(y)\Phi_0 \right) \right]. \tag{1.3}$$

Here Φ_0 is the vacuum state. The effectiveness of this formula can be described in words as follows: The quantity $E(x) \cdot E(y)$ has a singularity at $x = y$; if the singularity is cancelled in the vacuum expectation value of the quantity by the subtraction of $(\Phi_0, E(x) \cdot E(y)\Phi_0)$, it is cancelled in all physically significant matrix elements.

In retrospect, the occurrence of such singularities in the vacuum expectation value of the product of two fields should not have been a great surprise. The vacuum expectation values are wave-equation solutions intimately related to fundamental solutions and such fundamental solutions are well known to have singularities on the light cone. These singularities had already been calculated explicitly by Jordan and Pauli in 1928. With the wisdom of hindsight, one can say that what should have been surprising is that cancelling singularities in the vacuum expectation value should suffice to cancel all singularities. That fact was the beginning of the general subject of operator-product expansions, of which more will be said later.

That the above procedure for defining the energy density does not eliminate all the physical consequences of electromagnetic zero point vibrations is clear from the Casimir effect. As Casimir (1948) pointed out, the introduction of a pair of metal plates into the vacuum of the electromagnetic field alters the zero-point vibrations of the field and thereby produces an attraction between the plates. The answer would be infinite if calculated naively, but when the zero-point contribution for the vacuum without plates is subtracted, the remainder is finite, depends on the distance between the plates, and agrees with experiment. This result gives a precise meaning to the statement that after the infinite contribution of the zero-point vibration to the energy has been dropped, there remain physically significant finite contributions which should be regarded as consequences of zero-point vibrations.

The question of the proper definition of the energy–momentum density rose again in the 1970s in the context of quantum field theory on curved space–time. The problem there is that no unique vacuum state exists, in general, so that a more refined definition of Wick ordering is necessary. It has been shown that for a class of states Ψ, and a free field which for simplicity will be assumed scalar, the expectation value of the "point-split" energy momentum tensor $T_{\mu\nu}(x, y)$ has singularities of the form $A(x, y)/\mathrm{d}(x, y)$, $B(x, y) \ln \mathrm{d}(x, y)$, where A and B are smooth functions depending in the limit $x \to y$ only on the curvature of the space–time manifold and its derivatives. Here $\mathrm{d}(x, y)$ is the geodesic distance between x and y. Cancellation of these terms provides the required generalization of the cancellations implicit in the definition (1.3) of Wick ordering. This procedure also provides a first step toward a semi-classical theory of gravitation coupled to a quantized field. In such a theory, the expectation value of the energy–momentum tensor is treated as the source in the classical Einstein equations

$$R_{\mu\nu} - \tfrac{1}{2}g_{\mu\nu}R = \lim_{x \to y} \left[\left(\Psi, \, T_{\mu\nu}(x, \, y)\Psi \right) - \mathrm{sing} \right]. \tag{1.4}$$

It is instructive to regard this equation together with the operator equation for the scalar field as describing a self-consistent analogue of the Casimir effect. Here, for each metric $g_{\mu\nu}$, one can solve the operator equation for the scalar field and, given Ψ, compute the right-hand side of eq. (1.4). Then the self-consistent $g_{\mu\nu}$ is one for which the left- and right-hand sides coincide.

2. The off-diagonal density matrix

The definition of Wick ordering in eq. (1.3) as

$$\lim_{x \to y} \left[E(x) \cdot E(y) - (\Phi_0, \, E(x)) \cdot E(y) \Phi_0) \right]$$

is often referred to as "point splitting". In fact, point splitting was first used in another context, the problem of vacuum polarization. In his report to the Solvay Conference of 1933, P.A.M. Dirac (1934a) computed the polarization of the vacuum by a given external electric field according to his hole theory of electrons and positrons and found that, in lowest approximation in an expansion in the electric charge e, an external field produces a polarization charge density ρ, proportional to the external charge density, ρ_{ext}, with an infinite proportionality constant as well as a polarization charge density proportional to $\Delta \rho_{\text{ext}}$:

$$\rho(x) = - \frac{e^2}{\hbar c} \left(\frac{2}{3\pi} \right) \left[\ln \left(\frac{2P}{mc} \right) - \frac{5}{6} \right] \rho_{\text{ext}}(x) - \frac{1}{15\pi} \left(\frac{e^2}{\hbar c} \right) \left(\frac{\hbar}{mc} \right)^2 \Delta \rho_{\text{ext}}(x).$$

$$(2.1)$$

Here P is a cutoff which should be taken as infinity but which Dirac took as $137 \, mc$ for purposes of discussion. He speculated on the possibility that this vacuum polarization might lead to deviations from the Klein–Nishina formula for the Compton effect and the Rutherford formula for Coulomb scattering.

To derive this formula, Dirac introduced a quantity $\langle x, \, t \, | \, R \, | \, x't' \rangle$ which may be described as the subtracted off-diagonal density matrix. This terminology is partly justified by the fact that, in the absence of an electromagnetic field and before the subtraction, $\langle x, \, t \, | \, R \, | \, x', \, t' \rangle$ reduces to

$$\sum_{\text{occupied}} \psi_n(x, \, t) \, \overline{\psi_n(x', \, t')}, \tag{2.2}$$

where the ψ_n are the eigenfunctions of the Dirac equation in a large box chosen for convenience. The summation is over occupied states. According to the original ideas of Dirac's hole theory of the positron, in the unperturbed vacuum all negative energy states are occupied, but there is a subtraction prescription which causes their contributions to charge, current, and energy density to vanish. A state with a finite number of free positive-energy electrons and positrons is then to be described by a density matrix with holes in the negative-energy sea describing the positrons—the subtraction prescription leads to positive-energy contributions from the holes appropriate to positrons.

In the presence of an external electromagnetic field there are ambiguities in this procedure, because one cannot in general separate unambiguously into positive and negative energy states. Dirac (1934b) found a way to resolve these ambiguities by displaying the characteristic singularities of $\langle xt \, | \, R \, | \, x't' \rangle$ on the light cone $(x - x')^2 - c^2(t - t')^2 = 0$. The cancellation of the infinities to which these singularities give rise in the charge density etc. then provided the required definition of subtraction. Applied to the vacuum polarization eq. (2.1), it would mean that the first (infinite) term should be dropped—that is, charge renormalization. Dirac's definition of these subtractions had the reputation of being deep and dark. At least that was the way it appeared to me when I heard him lecture on it in the spring of 1947.

Dirac's work on $\langle x | R | x' \rangle$ was further developed, clarified and generalized by Heisenberg (1934) who introduced the operator

$$R(x, x') = \psi(x) \, \psi(x')^*, \tag{2.3}$$

which for a free Dirac field ψ has an expectation value in the vacuum equal to eq. (2.2). In the presence of an external field incapable of creating pairs, the expectation value of $R(x, x')$ in the vacuum reduces to the $\langle x | R | x' \rangle$ considered by Dirac. However, Heisenberg was using the charge-symmetric formalism which treats positrons and electrons on an equal footing, so what was a mysterious subtraction of a perturbed negative-energy sea in the hole theory became simply the removal of a singularity on the diagonal $x = x'$ of the operator $R(x, x') = 0$.

Nowadays, for conceptual clarity, Heisenberg's discussion of the quantized Dirac field in an external electromagnetic field would be divided into two parts. In the first, one would solve the Dirac equation for the quantized field ψ in a given external field. It is a linear equation and for a reasonable class of external fields gives rise to no divergences whatsoever. The second part would be to define the electric current and the energy–momentum tensor in terms of the operator R, e.g. the current

$$j^\mu(x) = -e \lim_{x' \to x} \mathrm{Tr}\big[(R(x, x') - \text{subtraction terms})\gamma^0 \gamma^\mu\big], \tag{2.4}$$

where the trace is over the Dirac spinor indices. This arrangement of the argument makes it clear that the choice of the subtraction terms is a matter of definition. If one wants $j^\mu(x)$ to exist, and to be conserved $\partial^\mu j_\mu(x) = 0$, that imposes a constraint on the subtraction terms, a constraint Heisenberg showed could be satisfied. What he did not notice was that with the same subtraction terms the axial current defined by

$$j^{5\mu}(x) = \lim_{x' \to x} \mathrm{Tr}\big[(R(x, x') - \text{subtraction terms})\gamma^0 \gamma^5 \gamma^\mu\big] \tag{2.5}$$

has an anomaly in the sense that

$$\partial^\mu j_\mu^5(x) = 2m \, \mathrm{Tr}\big[(R(x, x') - \text{subtraction})\gamma^5\big] + \frac{e^2}{12\pi} \boldsymbol{E}(x) \cdot \boldsymbol{B}(x), \tag{2.6}$$

where $\boldsymbol{E}$ and $\boldsymbol{B}$ are the external electric and magnetic fields. In the unquantized theory only the first term would appear on the right-hand side of eq. (2.6). This anomaly was discovered in 1968 by Adler (1969), although in the context of π^0 decay into two photons it had been known since the work of Steinberger (1949) and Schwinger (1951).

Heisenberg also considered the problem of defining $R(x, x')$ in the full quantum electrodynamics in which both the Dirac spinor field and the electromagnetic field are quantized. He noted that the subtraction terms for $R(x, x')$ can then be determined order by order in perturbation theory but that high-order terms are afflicted with divergences similar to that occurring in the electron self-energy and therefore beyond the control of the then existing theory. Progress on that front had to await the development of perturbative renormalization theory.

3. The dimension of coupling constants

It was Heisenberg (1936) who first attempted to distinguish theories according to the dimension of their coupling constants. He compared quantum electrodynamics with Fermi's theory of nuclear β decay and noted that while a cross-section for the production of n pairs calculated in lowest-order perturbation theory in quantum electrodynamics (QED) is proportional to α^{2n} where α is the fine-structure constant $\alpha = e^2/\hbar c$, in Fermi's theory it is proportional to $[G_{\text{Fermi}}/\hbar c)k^2]^{2n}$ where k is some typical wave number associated with the reaction. The conclusion is that when k becomes as large as $(\sqrt{G_{\text{Fermi}}/\hbar c})^{-1}$ so the expansion parameter is near 1, perturbation theory should break down and multiple production processes should become frequent. On the other hand, in QED the probability of multiple processes gets smaller as n increases whatever the value of the energy. Heisenberg hoped that this fundamental distinction between the Fermi interaction and the electromagnetic interaction could account for the phenomenon of bursts in cosmic rays. He regarded these arguments and the experimental evidence on bursts as evidence that there is a fundamental length in nature which has to be incorporated into the description of elementary particles. In fact, as Heisenberg noted, the critical value

$$\hbar c k \sim \frac{\hbar c}{\sqrt{G_{\text{Fermi}}/\hbar c}} \sim 600 \text{ GeV}$$

is too high. At first, he tried to appeal to the Konopinski–Uhlenbeck theory of nuclear β-decay to lower this critical value, but later, after the success of the Yukawa theory of mesons, and the development of the cascade theory of electron–photon showers, he abandoned the idea but nevertheless pursued the notion of a fundamental length. He also argued (Heisenberg 1938a) that the Yukawa theory of mesons, because its dimensionless coupling constant $g^2/\hbar c$ is so much larger than the fine-structure constant, should give rise to multiple production of mesons.

In a paper in the Planck Festschrift (Heisenberg 1938b), he gave semi-philosophical arguments for the existence of a fundamental length. He noted that $\hbar$ and c are universal constants in the sense that they correspond to restrictions on physical laws applicable under all circumstances: the condition $0 < c < \infty$ is associated with the requirement that all laws be consistent with the special theory of relativity; the condition $0 < \hbar < \infty$ is associated with the requirement that all physical laws be consistent with quantum mechanics. He proposed that there be a fundamental length λ of similarly universal character, and suggested that if it was to be associated with multiple meson production, it would be reasonable if $\lambda \sim e^2/mc^2 = 2.8 \times 10^{-13}$ cm. As he noted, the existence of λ would bring a certain tidiness to elementary particle theory: every physical quantity could be given as a dimensionless number times appropriate powers of $\hbar$, c and λ.

Heisenberg despaired of being able to incorporate a fundamental length into conventional field theory. So he was led to develop elementary particle theories based on the S-matrix (Heisenberg 1943). In a sequel (Heisenberg 1944) he gave examples of S-matrix theories with a fundamental length and multiple meson production.

Nevertheless, the distinction between dimensional and non-dimensional coupling constants persisted, and later, in 1952 it was given a deep interpretation in terms of perturbative renormalization theory by Sakata, Umezawa and Kamefuchi (1952). They pointed out that there are three categories of theories according to the dimension of their coupling constants

$$g = [M]^n \qquad \text{super-renormalizable,}$$
$$g = [M]^0 \qquad \text{renormalizable,}$$
$$g = [M]^{-n} \qquad \text{non-renormalizable.}$$

Here $n > 0$. I will discuss the meaning of this distinction in the next section.

4. Perturbative renormalization theory

The field theory of the 1930s was largely based on formal perturbation theory so that solutions, if they could be constructed at all, were expected to be formal power series in the coupling constant. The expansions typically had well defined first terms, but all higher orders were infinite or ambiguous — the so-called divergences. There was no systematic theory giving control of the divergences, until a theory was created in the 1940s by Tomonaga (1948), Feynman (1949), Dyson (1949), Schwinger (1958) and others.

The now familiar methods involve the expansion of the quantity in question, typically an S-matrix element or Green function, into terms corresponding to Feynman diagrams, their regularization to obtain well-defined expressions and finally their renormalization by the isolation of a finite part. The divergent part which has to be separated is located by the method of power counting. That method attributes to an integral of a rational function P/Q, where P and Q are polynomials,

$$\int \frac{P(k)}{Q(k)} \, d^n k$$

a degree of divergence $D = \deg P - \deg Q + n$. If $D \geq 0$ the integral, in general, turns out to be divergent, while if $D < 0$ it is convergent. The subtle part of the analysis arises when one deals with the divergence and convergence of subintegrals and shows how to separate contributions which would arise from counter terms in the Lagrangean. If there is at most a finite number of divergent graphs occurring in a theory, the theory is said to be super renormalizable; if an infinite number of divergent graphs occur but they all can be cancelled by a finite number of counter terms, the theory is renormalizable, if the theory has an infinite number of divergent graphs and requires an infinite number of counter terms it is non-renormalizable.

These definitions give a precise meaning to the result of Sakata, Umezawa and Kamefuchi (1952) quoted above: theories with coupling constants which have a dimension of a positive power of mass are super-renormalizable, the ones with dimensionless coupling constants are renormalizable, and the ones with negative power of mass are non-renormalizable. For example:

Super-renormalizable: $P(\phi)_2$, ϕ_3^4, Y_2, $\mathrm{QED}_{\leq 3}$,

Renormalizable: QED_4, ϕ_4^4,

Non-renormalizable: $\mathrm{Fermi}_{\geq 3}$, $\phi_{\geq 5}^4$.

The success of perturbative renormalization theory was twofold. First, in QED of electrons and muons it yielded theoretical predictions which have been verified to an accuracy greater than that in any other measurement made by mankind. Second, it revolutionized the prevailing attitude toward quantum field theory as a language for describing Nature—perhaps quantum field theory is more consistent than the founding fathers Heisenberg and Pauli believed in 1929.

However, even in perturbative renormalization theory there were loose ends. The problem of overlapping divergences remained somewhat elusive. It was not until Bogoliubov and Shirkov invented the R-operation in 1955 (Bogoliubov and Shirkov 1959), and Hepp (1966), completing the work of Bogoliubov and Parasiuk (1957), gave a proof that the R-operation really works, that perturbative renormalization theory was put on a sound mathematical footing. [Here, I risk offending my friend Abdus Salam; I have never understood whether his papers of 1951 (Salam 1951a, b) really solved the problem of overlapping divergences in QED or not. Maybe I should study them again.] This was followed by Zimmermann's formula of 1969 (Zimmermann 1969) which gave an explicit expression for the R-operation. The resulting BPHZ renormalization scheme is now standard. There were others, analytic renormalization developed by Speer in 1967 (Speer 1968) and dimensional renormalization with numerous parents shortly thereafter. All were shown to be equivalent up to finite renormalizations. I will say more about dimensional renormalization because it has played an important role in recent developments.

The simplest treatment of dimensional renormalization takes as its starting point the so-called α-parametrization of the contribution of a Feynman diagram. It is an integral over a bounded domain in an α-space of which the number of dimensions is determined by the number of lines in the Feynman diagram. The dimension of space–time appears only in the power to which a polynomial in the denominator is raised. The integral turns out to be a meromorphic function of the dimension with poles only on the real axis at the dimensions for which the integral is divergent. Some typical diagrams and their poles for $\lambda \phi_d^4$ theory are shown in fig. 1.

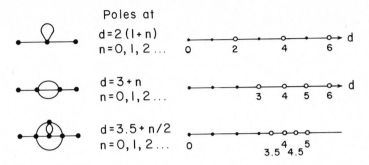

Fig. 1. Some typical diagrams and their dimensional poles for the two-point function in $\lambda\phi_d^4$.

When the poles for all the diagrams contributing to the two-point function are plotted on the same line, one finds that all are positive and ≥ 2. Below $d = 4$ the poles are isolated but have 4 as a limit point, while above $d = 4$, all rational numbers give poles. Thus, viewed from the point of view of dimensional renormalization, the distinction between super-renormalizable ($d < 4$), renormalizable ($d = 4$), and non-renormalizable ($d > 4$) is vividly displayed for $\lambda\phi_d^4$ theory in this distribution of poles.

I would now like to make a digression to describe an analogy between this distribution of dimensional poles and a situation occurring in Poincaré's thesis (1882). This analogy is probably at best poetic, but it may possibly give us some solace in our struggle to understand quantum field theory.

Poincaré's thesis treated, among other things, the normal-form problem for an ordinary differential equation at an equilibrium point. It is formulated as follows. Let x be a real n-dimensional vector and the differential equation be

$$\frac{dx}{dt} = F(x),$$

with $F(0) = 0$ and $F(x) = Ax + O(x^2)$, where A is some real linear transformation. We ask: when does there exist a change of coordinates near the origin $x = h(y)$ so that the equation for y is

$$\frac{dy}{dt} = Ay,$$

i.e. so that the differential equation can be reduced by change of coordinates to its linear part? To state Poincaré's result we use a definition: let the eigenvalues of A be $\lambda_1 \ldots \lambda_n$, then A is *resonant* if for some non-negative integers $m_1 \ldots m_n$ and s with $1 \leq s \leq n$ and $\sum_{j=1}^n m_j \geq 2$

$$\lambda_s = \sum_{j=1}^n m_j \lambda_j.$$

A belongs to the *Poincaré region* if the origin does not belong to the convex hull of the set of complex numbers $\{\lambda_1, \ldots, \lambda_n\}$.

Theorem (Poincaré) *If F is a formal power series, then h exists as a formal power series if, and only if, A is not resonant.*

If F is analytic in a neighborhood of the origin the series for h converges and defines an analytic change of coordinates if A belongs to the Poincaré region.

In the terminology of V.I. Arnold (1980), whose book I recommend for an account of these matters, the exterior of the Poincaré region is the Siegel region. The normal-form problem for A in the Siegel region remained open for more than half a century until the work of Siegel (1952). The qualitative difference between the two regions is that resonant A are isolated in the Poincaré region and dense in the Siegel region. Below follows the result of Siegel which accounts for his name appearing in the terminology.

Theorem (Siegel) *Let the hypotheses of the preceding theorem be satisfied except that A lies in the Siegel region. Suppose further that for all positive integers $m_1 \ldots m_n$ and s such that $\Sigma m_j = |m| \geq 2$ and $1 \leq s \leq j$ there are positive constants c and v such that*

$$\left| \lambda_s - \sum_{j=1}^{n} m_j \lambda_j \right| \geq \frac{c}{|m|^v} ,$$

then the series for h converges and defines an analytic change of coordinates.

My analogy juxtaposes the Siegel region for A in the normal-form problem and the range of dimensions, $d \geq 4$, for which $\lambda \phi_d^4$ theory is non-renormalizable. Let us hope that it does not take as long for us to understand non-renormalizable field theory as the six decades from Poincaré to Siegel.

5. Summing the perturbation series; renormalons and instantons

The output of perturbative renormalization theory is a set of well-defined formal power series for the S matrix elements and Green function of a field theory. A natural question is whether the series converge, if they converge whether the resulting sums define solutions of the theory, and if they do not converge whether some summability method can be used to recover functions which are solutions.

It was argued by Hurst (1952), Thirring (1953) and by Petermann (1953) that the renormalized perturbation series for the two-point function in $\lambda \phi_4^4$ theory is divergent. Their work still left some uncertainties about the possible effect of renormalization in causing cancellations and therefore making convergence possible. (Renormalization changes the sign of some terms which would be positive in the unrenormalized expansion.) It was pointed out by Jaffe (1965) that in $P(\phi)_2$ theory no such difficulty occurs; he gave a complete proof of the divergence of the perturbation series for an arbitrary connected Green function. For $\lambda \phi_3^4$ the difficulty does arise, but de Calan and Rivasseau (1982) showed that by rearrangement of the terms they could obtain a divergent series of positive terms. Although it is widely believed that the renormalized perturbation series of most field theories diverge, there is still no complete proof in most cases. For example, for spinor QED_4, Dyson (1952) already gave an ingenious and suggestive argument supporting divergence, but there is no complete proof to this day. It is worth noting that there is a counter example: The Euclidean Green functions of the massive sine-Gordon theory in two-dimensional space–time are analytic in the mass for sufficiently small masses, as was shown by Fröhlich in 1975 (Fröhlich 1976).

If the perturbation series for a Green function is divergent, it may still be asymptotic to a solution of the theory. In the mid 1970s as a consequence of the deep results of constructive field theory, the existence of solutions was established for $P(\phi)_2$, $\lambda\phi_3^4$, Y_2 and a number of other super-renormalizable theories. The question whether the perturbation series was asymptotic was then open to direct attack. Dimock (1974) answered it affirmatively for the $\lambda P(\phi)_2$ theory. He proved that the Schwinger functions and Green functions of $\lambda P(\phi)_2$ are infinitely differentiable in λ for $0 \leq \lambda < \infty$ and derivatives from the right of all orders exist at $\lambda = 0$. If the Schwinger function in question is denoted $S(\lambda)$, Taylor's theorem with remainder then gives

$$\left| S(\lambda) - \sum_{n=0}^{N} c_n \lambda^n \right| = 0(\lambda^{N+1}) \quad \text{near } \lambda = 0, \tag{5.1}$$

where

$$c_n = \frac{1}{n!} \left. \frac{d^n S(\lambda)}{d\lambda^n} \right|_{\lambda = 0+}. \tag{5.2}$$

The estimate (5.1) is by definition what is meant by the formal power series $\sum_{n=0}^{n} c_n \lambda^n$ being asymptotic to $S(\lambda)$: $f(\lambda \sim \sum_{n=0}^{\infty} c_n \lambda^n$. Here, at last, was proof that the traditional expansions in terms of the contributions from Feynman diagrams, when renormalized, give an asymptotic series for the solution of the theory.

The knowledge of an asymptotic series for a function does not permit one to determine a function uniquely, but with further information it can happen that not only is the function uniquely determined but that a more or less explicit method exists for recovering it from the coefficients of the asymptotic series. Such methods are called summability methods. I will describe briefly two such methods and then state some of the results obtained by applying them in field theory.

The Padé method gives for each formal power series a table of rational functions labeled by a pair of positive integers

$$f^{[M,N]}(\lambda) = \frac{P^{[M]}(\lambda)}{Q^{[N]}(\lambda)}, \tag{5.3}$$

where P is a polynomial of degree M, and Q a polynomial of degree N. P and Q are so determined that $f^{[M,N]}$ has a Taylor series agreeing with the given formal series up to terms of degree $M + N + 1$, which is the number of free parameters in the ratio. A sufficient condition that f be recoverable from diagonal sequence of Padé approximants is given by the following theorem of Stieltjes.

Theorem *Let f be a function of the form*

$$f(\lambda) = \int_0^\infty \frac{d\rho(t)}{1 + \lambda t},$$

where ρ is a positive measure with finite moments of all orders

$$\int_0^\infty t^n \, d \, \rho(t) < \infty, \quad n = 0, 1, 2, \ldots .$$

Then $f(\lambda)$ has an asymptotic expansion

$$f(\lambda) \approx \sum_{n=0}^\infty c_n \lambda^n, \quad with \quad c_n = (-1)^n \int_0^\infty t^n \, d\rho(t),$$

and

$$f(\lambda) = \lim_{N \to \infty} f^{[N, N+j]}(\lambda),$$

for $j = 0, 1, 2, \ldots$, and the convergence is uniform in any compact set of the complex plane cut along the negative real axis.

In the late sixties the Padé method was tried on a test problem, the eigenvalues of the quartic anharmonic oscillator, a system which can be regarded as $\lambda\phi_1^4$, a theory of a Hermitean scalar field in space–time of one dimension. The Hamiltonian is taken as

$$H = \tfrac{1}{2}(\pi^2 + m^2\phi^2) + \lambda\phi^4, \quad \pi = -i\frac{d}{d\phi}, \tag{5.4}$$

with m, $\lambda > 0$. Regarded as acting in the space of square integrable functions on the real line, H has a pure discrete spectrum bounded below. By a scaling argument one has for the ground state energy

$$E_0(m^2, \lambda) = mE_0\left(1, \frac{\lambda}{m^3}\right). \tag{5.5}$$

Bender and Wu in 1968 showed that the perturbation series for $E_0(1, \lambda/m^3)$ in the parameter (λ/m^3) is divergent (Bender and Wu 1969). Loeffel et al. (1969) showed that the eigenvalues are functions satisfying the hypotheses of Stieltjes theorem. Thus, the Padé method works for the eigenvalues of the quartic anharmonic oscillator with $m^2 > 0$. Unfortunately, it turns out that, if the quartic anharmonicity $\lambda\phi^4$ is replaced by $\lambda\phi^{2n}$ with $n > 3$, one has to modify the Padé method as a function of n so one cannot use the above theorem to prove the convergence of the diagonal Padé approximants. Fortunately, as was shown by Graffi, Grecchi and Simon (1970), there is a more powerful summability method, the Borel summability, which does cover these cases. To this Borel summability I will now turn.

If f has the asymptotic expansion $f \sim \sum_{n=0}^{\infty} c_n \lambda^n$, its Borel transform f_B is defined as the asymptotic series

$$f_B \sim \sum_{n=0}^{\infty} \frac{c_n}{n!} \lambda^n,$$

f_B may converge even if the series for f diverges. Then under conditions first obtained by Nevanlinna in 1919, one can recover f. [See also the paper by A. Sokal (1980).]

Theorem *If f is analytic in the interior of the circle $C_R : \mathrm{Re}(1/z) = 1/R$ and its derivatives satisfy*

$$\left| \frac{\mathrm{d}^n f}{\mathrm{d}\lambda^n}(\lambda) \right| \le A \sigma^n (n!)^2$$

in the closure of the region then the Borel transform, f_B, of the Taylor series of f at the origin

$$f_B(t) = \sum_{n=0}^{\infty} \frac{t^n}{(n!)^2} \left(\frac{\mathrm{d}^n f}{\mathrm{d}x^n} \right) \bigg|_{\lambda=0},$$

is analytic in the half strip

$$\mathrm{d}(t, \boldsymbol{R}^+) \le \frac{1}{\sigma},$$

and satisfies the inequality

$$|f_B(t)| \le C \, \mathrm{e}^{t/R}, \quad 0 \le t < \infty.$$

Furthermore, f can be recovered from f_B by the inverse Borel transform

$$f(\lambda) = \frac{1}{\lambda} \int_0^{\infty} \mathrm{e}^{-t/\lambda} f_B(t) \, \mathrm{d}t, \tag{5.6}$$

valid for λ in the interior of C_R.

One of the pleasant consequences of the constructive field theory treatment of the $\lambda P(\phi)_2$ and $\lambda \phi_3^4$ field theory is the proof that, for degree $P = 4$, the perturbation series for the Euclidean Green functions are Borel summable. This was shown by Eckmann, Magnen and Sénéor (1975) for $P(\phi)_2$ and by Magnen and Sénéor

(1977) for $\lambda\phi_3^4$. One could speculate that Borel summability might provide a general method for obtaining solutions of quantum field theories from their perturbation series. However, this illusion did not last long. It was soon realized that things are likely to be much more complicated. Two of the additional complications are indicated by the catchwords *instantons* and *renormalons*. Both give rise to singularities in the Borel transform of perturbation series and these singularities sometimes appear on the positive real axis, making the Borel inversion (5.6) inapplicable as it stands. Thus to understand the physical meaning of these singularities one must understand the physical mechanisms for producing instantons and renormalons. Of course, assuming that one can understand their occurrence, and prove that no other singularities occur, one still has the problem of recovering the non-perturbative solutions from the Borel transform in the presence of such singularities. Such questions are at or beyond the limits of our present understanding of the perturbation series, so what follows is an impressionistic account of recent developments with an emphasis which reflects my personal taste.

Instantons first appeared in semi-classical approximations of functional–integral solutions of quantum field theories. They are solutions of the classical field equations which give stationary points of the action appearing in the quantum mechanical expressions. They played a significant role in answering the following basic question: how can a field theory of bosons imply the existence of fermions. It was Skyrme (1961) who studied what is now known as the sine-Gordon equation and attempted to construct a fermion theory which would contain it. To say that Skyrme's work was not widely understood at the time is surely an understatement. A second contribution concerning this question, but with an entirely different flavor, was the work of Haag and Kastler (1964) on algebras of observables. In their general framework, the Hilbert space of states decomposes into sectors in each of which there is a representation of the algebra of operators generated by local observables. The representations are unitary inequivalent in different sectors, but the corresponding states look like the vacuum state in the limit in which all observations take place far from one another and at approximately the same time. If one is given any one of the representations all the others are uniquely determined. If this general theory is applied to the sine-Gordon equation with a vacuum sector determined by the sine-Gordon vacuum, it turns out that there is an infinity of other sectors labelled by the (non-vanishing) fermion number of Skyrme's fermions. However, Haag and Kastler did not apply their general theory to Lagrangean field theory models and several years passed before Streater and Wilde (1970) worked out the details for the special case in which the sine-Gordon equation reduces to the massless free wave equation. Even then it took some time before it was realized by Streater and Dell'Antonio that the fermion theory in question is the massless Thirring model. Finally, Coleman (1975) gave the connection between the general sine-Gordon model and the massive Thirring model.

The second important role played by instantons involves tunneling phenomena. When the potential occurring in a classical action has several degenerate minima, quantum mechanical tunneling between them has an important effect on the structure of low-lying states. The non-degeneracy of the ground state depends on the existence of instantons connecting the degenerate minima. This non-degeneracy is a non-perturbative effect; the splitting of the ground and first excited state as a function of the coupling constant has a zero-perturbative expansion. Thus, here the presence of instantons invalidates Borel summability. The simplest case in which this phenomenon can be studied is the anharmonic oscillator, the $\lambda\phi_1^4$ theory whose Hamiltonian is displayed in eq. (5.4). By the same scaling argument that produced eq. (5.5), we have

$$E_n(m^2, \lambda) = \lambda^{1/3}E_n(m^2\lambda^{-2/3}, 1).$$

For a field with $m^2 > 0$, the case discussed in eq. (5.5), the right-hand side is analytic in λ for λ on the three sheeted Riemann surface of the cube root, except at a set of singularities which have zero as a limit point and lie inside two horn-shaped regions on the second and third sheets. When $m^2 = 0$, these singularities all disappear into the origin only to reappear on the first sheet when $m^2 < 0$, but where? Parisi (1977) argued that there would be two horns tangent to the real axis containing the singularities. Crutchfield (1979) argued that they would be tangent to the imaginary axis and on circles tangent at the origin. He based his statement on a semi-classical approximation. If Parisi is right the violation of the hypotheses of analyticity in Nevanlinna's theorem on Borel summability is blatant. On the other hand, Crutchfield has to appeal to a failure in the remainder estimate to account for the failure of Borel summability. To my knowledge, there has as yet been no rigorous discussion, which is a bit of a scandal, especially since Zinn-Justin (1982) has conjectured an intriguing and numerically very successful asymptotic expansion involving logarithms as well as powers of the coupling constant.

In his Erice lectures of 1977, 't Hooft (1979) analyzed the singularities of the two-point function of various theories using renormalization group methods to relate singularities in the coupling constant to singularities in momentum–space. He concluded that for asymptotically free theories there is a horn of singularities in the right-half plane like that described by Parisi (1977).

Instantons are intimately related to the asymptotics of the perturbation series. Such a connection was suggested in 1968 by Bender and Wu (1969) for the anharmonic oscillator and generalized to a finite number of dimensions, but it was Lipatov (1977) who gave a general steepest-descent method usable in quantum field theory. Brezin, Le Guillou and Zinn-Justin (1977) perfected the method and applied it to a wide variety of examples. The result is that the nth coefficient of a

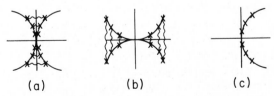

(a)　　　　(b)　　　　(c)

Fig. 2. The singularities in λ of $E_n(m^2, \lambda)$ for (a) $m^2 > 0$, according to Bender and Wu (1969) and Simon (1970); (b) $m^2 < 0$, according to Parisi (1977); (c) $m^2 < 0$, according to Crutchfield (1979).

perturbation series is typically of the form

$$(-a)^n n^b n! A\left[1 + 0\left(\frac{1}{n}\right)\right],$$

where a turns out to be the reciprocal of the radius of convergence of the Taylor series for the Borel transform at the origin. Since a can also be expressed in terms of the instanton solution it is natural to call the corresponding singularity an instanton singularity. The Lipatov method for ϕ_2^4 was put on a rigorous footing in work by Breen (1983). Magnen and Rivasseau (1986) recently showed that the Lipatov method correctly gives the radius of convergence of the Taylor series for the Borel transform in the $\lambda\phi_3^4$ model.

Now let me turn to renormalons. They turned up in various connections in the work of Gross and Neveu (1974), of Lautrup (1977) and of 't Hooft (1977). For the massive $\lambda\phi_d^4$ model, they are absent for $d = 1, 2, 3$ and first appear for $d = 4$, when renormalization produces logarithmic corrections to the momentum dependence of the propagator. An $(n + k)$th order Feynman diagram containing a chain of n bubbles then has an amplitude proportional to $n!$ If all $(n + k)$th order Feynman diagrams were of this magnitude, the $(n + k)$th series coefficient for the Borel transform would be of order $Ac^{n+k}\lambda^{n+k}(n + k)!$ instead of $Ac^{n+k}\lambda^{n+k}$ and the series for the Borel transform would diverge for all non-vanishing values of the coupling constant. Thus, the result of de Calan and Rivasseau (1981), that the total nth order contribution to the series for the Borel transform is bounded by $Ac^n\lambda^n$, is decidedly non-trivial. It gives a toehold for a possible non-perturbative approach by analytic continuation to the definition of renormalon singularities. So far no one seems to have succeeded along this line.

From each renormalon singularity trails a cut of the Borel transform. It was Parisi (1979a) who first attempted to calculate the jump of the Borel transform across these cuts and recognized that they are related to matrix elements of composite operators of arbitrarily high dimension. Parisi's conjectures were established to all orders in a $1/N$ expansion for the $\lambda(\phi^2)_4^2$ model by Bergère and David (1986a, b). In 1979, Parisi (1979b) gave an analogous discussion of the infrared singularities in massless theories, showing that multilocal counter-terms can cancel the divergences. It had already been known since the work of Symanzik (1973) that these infrared singularities require counter terms which are nonanalytic in the coupling constant. Parisi offered evidence that the resulting finite perturbation expansion has a Borel transform with characteristic singularities which it is natural to call infrared renormalons. In 1985, Bergère and David (1986a, b) were able to verify Parisi's conjectures to all orders in a $1/N$ expansion.

This information on ultraviolet and infrared renormalons together with the knowledge of instantons represents qualitatively new information about the nature of the solutions of renormalizable and super-renormalizable field theories which has not been given a complete nonperturbative treatment in the sense of constructive quantum field theory. If past experience is any guide, a full understanding will require a nonperturbative construction of solutions without reference to the Borel transform of the perturbation series. Then a posteriori, as was the case for $P(\phi)_2$

and $\lambda\phi_3^4$, one may verify that the solution is uniquely determined from its perturbation expansion. Current work using rigorous renormalization group methods by Feldman et al. (1986), Gawedzki and Kupiainen (1985), Balaban (1984, 1985), King (1986) and by Gallavotti and Nicolo (1985) give constructions of the Gross–Neveu model in two- and three-dimensional space–time (the first non-trivial solutions of renormalizable and nonrenormalizable theories, respectively), as well as the first steps toward a constructive field theory treatment of gauge field theories in four dimensions. These developments are taking place rapidly, at least by the standards of speed of constructive field theory, so it is perhaps not too optimistic to hope that in a few years we may finally understand in what sense the perturbation series of a quantum field theory determine solutions.

6. Renormalizable versus non-renormalizable–non-perturbatively

It is characteristic of the axiomatic field theory that was invented in the early 1950s that the fields are operator-valued distributions whose definition involves a choice of test functions. From the beginning, the customary assumption was that the test functions are infinitely differentiable functions of fast decrease at infinity. More precisely, the test functions are assumed to be in Schwartz's space $\mathscr{S}(\mathbf{R}^4)$. The space $\mathscr{S}(\mathbf{R}^n)$ consists of all infinitely differentiable functions, f, on $\mathbf{R}^n$ for which the following semi-norms are finite:

$$\| f \|_{k,l} = \sup_{x \in \mathbf{R}^n} \sup_{|\alpha| \leq k} \sup_{|\beta| \leq l} |x^\alpha D^\beta f(x)|, \tag{6.0}$$

where

$$x^\alpha = x_1^{\alpha_1} x_2^{\alpha_2} \dots x_n^{\alpha_n}, \quad D^\beta = \left(\frac{\partial}{\partial x_1}\right)^{\beta_1} \dots \left(\frac{\partial}{\partial x_n}\right)^{\beta_n},$$

$$|\alpha| = \sum_{j=1}^n \alpha_j, \quad |\beta| = \sum_{j=1}^n \beta_j.$$

Then the quantized fields are tempered operator-valued distributions, by assumption, i.e. the matrix elements, $(\Phi, \phi(f)\Psi)$, regarded as linear functionals in f are elements of Schwartz's space $\mathscr{S}'(\mathbf{R}^n)$.

Although this assumption of temperedness was supported by evidence of perturbation theory, it was clear that alternatives are possible, and might be necessary for the construction of some theories. For example, if the matrix elements $(\Phi, \phi(x)\Psi)$ of the field increase too rapidly as $x \to \infty$ one could restrict the test functions to have compact support. With an appropriate definition of convergence, these functions constitute Schwartz's space $\mathscr{D}(\mathbf{R}^n)$. The corresponding continuous linear functionals form the space $\mathscr{D}'(\mathbf{R}^n)$ of distributions, and one would be led to assume $(\Phi, \phi(f)\Psi) \in \mathscr{D}'(\mathbf{R}^n)$.

Replacement of $\mathscr{S}'(\boldsymbol{R}^n)$ by $\mathscr{D}'(\boldsymbol{R}^n)$ enables one to treat fields which are singular at $x = \infty$. To treat fields which are more singular than distributions in neighborhoods of a finite point, requires quite a different set of test functions. The standard example arises if one attempts to define the exponential of a free scalar field. [This example was already studied in the 1950s; see Wightman (1981b) for a brief review with references.] Formally,

$$:\exp g\phi: (x) = \sum_{h=0}^{\infty} \frac{g^n}{n!} :\phi^n: (x) \tag{6.1}$$

and a ready calculation shows

$$\left(\Phi_0, :\exp g\phi: (x) :\exp g\phi: (y)\Phi_0\right) = \exp\left[g^2(\Phi_0, \phi(x)\,\phi(y)\Phi_0)\right]. \tag{6.2}$$

Whatever the choice of operator convergence in eq. (6.1) it is reasonable to expect eq. (6.2) to hold. Now in space–time of four dimensions, the worst singularity of $F(x - y) = (\Phi_0, \phi(x)\,\phi(y)\Phi_0)$ is const. $[(x - y)^2]^{-1}$. Furthermore, a distribution in $\mathscr{S}'(\boldsymbol{R}^4)$ or $\mathscr{D}'(\boldsymbol{R}^4)$ can have singularities which are at worst bounded by a power, so whatever one does to define :exp $g\phi$: it cannot be a distribution in $\mathscr{S}'(\boldsymbol{R}^4)$ or $\mathscr{D}'(\boldsymbol{R}^4)$. [A rigorous argument to back up this somewhat loose statement goes as follows. If the two point function is in $\mathscr{D}'(\boldsymbol{R}^n)$ and satisfies the positivity condition

$$\int \bar{f}(x)\,F(x - y)\,f(y)\,\mathrm{d}^4 x\,\mathrm{d}^4 y \geq 0 \quad \text{for } f \in \mathscr{D}(\boldsymbol{R}^4),$$

then F must be tempered and the Fourier transform must be of a positive measure on momentum space. By the spectral condition this measure must vanish outside the positive cone $E^2 - p^2 \geq 0$, $E \geq 0$ and therefore the two-point function has an analytic continuation $F(y - x + i\eta)$, where η lies in the future light cone. This analytic function is Lorentz invariant and so a function of $(y - x + i\eta)^2$. It can grow no faster than const. $[((y - x) + i\eta)^2]^{-k}$ for some k, as $\eta \to 0$, unlike eq. (6.2).]

There is a wide variety of other classes of generalized functions which are candidates for the treatment of such an example. I mention first hyperfunctions in the sense of Sato, because Nagamachi and Mugibayasi have developed a general theory of hyperfunction fields and proved that the exponential function :exp $g\phi$: belongs to that class. (For the general theory see their papers of 1976; for the exponential function their paper of 1986.) I also mention the Jaffe (1967) classes which have the conceptual advantage that their test functions include many functions of compact support. Fields evaluated on test function of compact support should commute (or anti-commute) when the supports are space-like separated. Since hyperfunctions use analytic test functions, for them there are no test functions of compact support. It is necessary to introduce the notion of the carrier of a hyperfunction to formulate local commutativity for hyperfunction fields.

Most of the preceding distinctions and some others were discussed at length, but in a somewhat different terminology in a paper by Schroer (1964), who argued that field theories that are non-renormalizable in the sense of perturbation theory should require generalized functions worse than distributions. This would give a non-perturbative definition of non-renormalizability.

There is evidence that it would be wiser to make the distinction without trying to force the definition. There may well be theories with tempered fields which are non-renormalizable in the sense of perturbation theory. Gawedzki and Kupiainen (1985) have announced an existence theorem for solutions of (Gross–Neveu)$_3$, just such a theory. On the other hand in 1984 Gallavotti and Nicolo (1985) have shown that an infinite number of distinct renormalization counter-terms is necessary to extend the range of the coupling constant over the interval, $4\pi \le \beta^2 < 8\pi$ for the sine-Gordon equation. The situation for $\beta^2 > 8\pi$ is not yet clear. Schroer and Truong (1977) conjectured that there would be non-renormalizable solutions in the sense of solutions which are not distributions.

Furthermore, there are other distinctions within the class of perturbatively non-renormalizable theories. The Green functions of a theory may fail to be C^∞ in the coupling constant because all sufficiently high derivatives blow up. On the other hand, for theories with instantons there are contributions to the Green functions which are C^∞ but have all zero derivatives. All these distinctions appear in a major reexamination of the non-renormalizable field theories undertaken by Symanzik (1975) and by Parisi (1975). On the basis of this exploratory work, there is now major effort in constructive field theory to understand the structure of non-renormalizable and renormalizable theories. The work of Felder and Gallavotti (1985) deserves mentioning in this connection.

The problem is particularly acute with $\lambda\phi_4^4$. There are no-go theorems of Fröhlich (1982) and of Aizenman and Graham (1983) which assert that the traditional ferromagnetic lattice approximation converges to a trivial solution, if the fields have anomalous dimension as everyone expects. The puzzle is: if the only non-perturbative solution is trivial, how can it have, as it does, a non-trivial renormalized perturbation series? Gallavotti in his review of 1985 concludes that there could be other approximation procedures which would converge to a non-trivial solution. The evidence he displays does not say much about the necessity of solutions which would be generalized functions more singular than distributions. Rigorous methods have not yet penetrated deeply enough so that one can see the effects of the renormalon singularities described above. As was mentioned there, the results of Parisi (1979a) and Bergère and David (1986a, b) indicate that to control the solution of $\lambda\phi_4^4$ one will have to control the powers ϕ^{2n} for all integers $n \ge 2$, and would seem to say that the solution of $\lambda\phi_4^4$ would have more to do with the non-renormalizable theories $\lambda\phi_{4+\epsilon}^4$ than the super-renormalizable theories $\lambda\phi_{4-\epsilon}^4$. This is a point of view that I have been urging for some time, the cogency of which has not been clear, for lack of rigorous results on non-renormalizable theories, (Wightman 1977, 1979, 1981a, b).

From a physical point of view, the striking feature of the recent work on non-renormalizable theories is that the solutions have only a finite number of arbitrary parameters. This is contrary to their description in terms of the perturbation series, but not unreasonable; an illegitimate expansion of a function can easily give rise to an infinite number of parameters to be "renormalized".

7. Coupling constant bounds; the ultraviolet phase transition and beyond

One of the discoveries of constructive quantum field theory in the 1970s was the existence of coupling constant bounds in $\lambda\phi_\nu^4$ theories. Originally found by Glimm and Jaffe (1975) in a form valid for all space–time dimensions, ν, these bounds were improved to be optimal for $\nu = 0, 1$ by Newman (1981). Newman considered the dimensionless scale-invariant coupling constant

$$g = -m^\nu \frac{\bar{u}_4}{(\bar{u}_2)^2},$$

where m is the mass gap between the vacuum Ψ_0, and the first excited state, $\bar{u}_4$ is the integrated four-point Ursell function

$$\bar{u}_4 = \int \int \int d^\nu x_2 \, d^\nu x_3 \, d^\nu x_4 \, u_4(x_1, x_2, x_3, x_4),$$

where $u_4(x_1, x_2, x_3, x_4)$ is the analytic continuation to imaginary time of

$$(\Psi_0, \phi(x_1) \, \phi(x_2) \, \phi(x_3) \, \phi(x_4)\Psi_0)$$
$$- (\Psi_0, \phi(x_1) \, \phi(x_2)\Psi_0)(\Psi_0, \phi(x_3) \, \phi(x_4)\Psi_0)$$
$$- (\Psi_0, \phi(x_1) \, \phi(x_3)\Psi_0)(\Psi_0, \phi(x_2) \, \phi(x_4)\Psi_0)$$
$$- (\psi_0, \phi(x_1) \, \phi(x_4)\Psi_0)(\Psi_0, \phi(x_2) \, \phi(x_3)\Psi_0)$$

and $\bar{u}_2$ is the integrated two-point Ursell function

$$\bar{u}_2 = \int d^\nu x_2 \, u_2(x_1, x_2),$$

where $u_2(x_1, x_2)$ is the analytic continuation to imaginary time of $(\Psi_0, \phi(x_1) \, \phi(x_2)\Psi_0)$. He proved

$$\nu = 0, \qquad 0 \le g \le 2,$$

$$\nu = 1, \qquad 0 \le g \le 6,$$

and showed that the upper bound is reached only for the Ising model in which $[\phi(x)]^2$ is equal to a constant. Analogous statements for $\nu \ge 4$ alluded to above can be interpreted as upper bounds of zero analogous to those above. It is a striking feature of Newman's results that he has to make no assumptions about his Hamiltonian except that is has a mass gap $m > 0$.

An obvious question suggested by these bounds: is there any solution for g above or below the bounds? This question was investigated in 1985 by Baker and Wightman (1986). They showed that a one-parameter family of solutions of the equations of motion for $\nu = 0$, found by Sokal (1982) gave theories violating the bounds above and below. However, an analogous procedure for $\nu = 1$, gave no satisfactory example. Baker and Wightman attributed this to the fact that the form of the equations of motion they assumed is not appropriate for a solution beyond the bounds, when $\nu = 1$.

It is natural to seek guidance on the above question from exactly soluble models. Here the work of Ruijsenaars (1983) on the Federbush model deserves mentioning. Recall that the Federbush model is a parity violating Fermi interaction between two massive fermion fields $\psi^{(1)}$ and $\psi^{(2)}$ of the form $\lambda\, j_\mu^{(1)}\epsilon^{\mu\nu}j_\nu^{(2)}$. Ruijsenaars constructed operator solutions for λ in the restricted range $|\lambda| \leq 2\pi$ and conjectured that the theory makes sense as a non-renormalizable field theory for $|\lambda| > 2\pi$.

For $-\lambda\phi_4^4$, i.e. $\lambda\phi_4^4$ with the "wrong" sign for the coupling constant, it was pointed out by Symanzik (1973a, b) that asymptotic freedom holds, so one might hope for an easier existence proof for solutions. First 't Hooft (1982) and then Rivasseau (1984) gave a complete construction for planar $-\lambda\phi_4^4$, and then Feldman et al. (1986) constructed the Green functions for $-\lambda\phi_4^4$ without the planar restriction. Presumably, neither of these solutions gives rise to an honest quantum dynamics. The powerful rigorous renormalization group methods used for the latter and the ingenious special methods for the former do not prove reflection positivity, which is necessary to give a positive probability interpretation to a field theory formulated in the Euclidean world.

I draw the conclusion from all this that one should take the possibility of a phase beyond the ultraviolet phase transition seriously and should not exclude a priori the possibility of non-renormalizable solutions or solutions in indefinite metric.

In these seven little lessons I hope to have offered evidence that there is still a lot of life left in quantum field theory.

References

Adler, S., 1969, Phys. Rev. **177**, 2426.

Aizenman, M., and R. Graham, 1983, Nucl. Phys. **B225**, 261.

Arnold, V.I., 1980, Chapitres Supplémentaires de la Théorie des Équations Différentielles Ordinaires (Mir, Moscow).

Baker, G., and A. Wightman, 1986, Trying to violate the coupling constant bound, to appear in the Umezawa Festschrift.

Balaban, T., 1984, Commun. Math. Phys. **96**, 223.

Balaban, T., 1985, Commun. Math. Phys. **99**, 75.

Bender, C., and T.T. Wu, 1969, Phys. Rev. **184**, 1231.

Bell, J., and R. Jackiw, 1969, Nuovo Cimento **60A**, 47.

Bergère, M., and F. David, 1986a, Nonanalyticity of the Perturbative Expansion of Super-renormalizable Massless Field Theories, Ann. Phys., to appear.

Bergère, M. and F. David, 1986b, Ambiguities of renormalized ϕ_4^4 field theory and the singularities of its Borel transform, to appear.

Bogoliubov, N., and O. Parasiuk, 1957, Acta Math. **97**, 227.

Bogoliubov, N., and D. Shirkov, 1959, Introduction to the Theory of Quantized Fields (Interscience, New York).

Bohr, N., and L. Rosenfeld, 1933, Mat.-Fys. Medd. Dan. Vidensk. Selsk. **XII-8**, 1.

Born, M., W. Heisenberg and P. Jordan, 1926, Z. Phys. **35**, 557.

Breen, S., 1983, Commun. Math. Phys. **92**, 179.

Brezin, E., 1977, Phys. Rev. **D15**, 1544, 1558.

Brezin, E., J. Le Gillou and J. Zinn-Justin, 1977, Phys. Rev. **D15**, 1544, 1558.

Brydges, D., J. Fröhlich, T. Spencer, C. de Calan and V. Rivasseau, 1981, Commun. Math. Phys. **82**, 69.

Casimir, H.B.G., 1948, Proc. K. Ned. Akad. Wet. Amsterdam **5**, 793.

Coleman, S., 1975, Phys. Rev. **D11**, 2088.

Crutchfield II, W.Y., 1979, Phys. Rev. **D19**, 2370.

de Calan, C., and V. Rivasseau, 1981, Commun. Math. Phys. **82**, 69.

de Calan, C., and V. Rivasseau, 1982, Commun. Math. Phys. **83**, 77.

Dimock, J., 1974, Commun. Math. Phys. 35, 347;

Dimock, J., 1976, J. Funct. Anal. **21**, 340.

Dirac, P.A.M., 1934a, Institut Solvay Septième Conseil de Physique, Bruxelles, 22–29 Octobre 1933 (Gauthier-Villars, Paris) pp. 203–219.

Dirac, P.A.M., 1934b, Proc. Comb. Phil. Soc. **30**, 150.

Dyson, F.J., 1949, Phys. Rev. **75**, 1736.

Dyson, F.J., 1952, Phys. Rev. **85**, 631.

Eckmann, J.P., J. Magnen and R. Sénéor, 1975, Commun. Math. Phys. **39**, 251.

Einstein, A., 1911, Institut Solvay, 1911 Conseil de Physique (Gauthier-Villars, Paris) pp. 407–415.

Felder, G., and G. Gallavotti, 1986, Perturbation Theory and Non-Renormalizable Scalar Fields, to appear.

Feldman, J., J. Magnen, R. Sénéor and V. Rivasseau, 1986, A Renormalizable Field Theory: The Massive Gross–Neveu Model in Two Dimensions, to appear.

Feynman, R., 1949, Phys. Rev. **76**, 769.

Fröhlich, J., 1976, in: Renormalization Theory, ed. G. Velo, A. Wightman (Reidel, Dordrecht) pp. 371–414.

Fröhlich, J., 1982, Nucl. Phys. **B200**, 281.

Gallavotti, G., 1985, Rev. Mod. Phys. **57**, 471.

Gallavotti, G., and F. Nicolo, 1985, Commun. Math. Phys. **100**, 545.

Gawedzki, K., and A. Kupiainen, 1985, Phys. Rev. Lett. **54**, 2192.

Glimm, J., and A. Jaffe, 1975, Ann. Inst. H. Poincaré, **XXII**, 97.

Graffi, S., V. Grecchi and B. Simon, 1970, Phys. Lett. **32B**, 631.

Gross, D., and A. Neveu, 1974, Phys. Rev. **D10**, 3235.

Haag, R., and D. Kastler, 1964, J. Math. Phys. **5**, 747.

Heisenberg, W., 1931, Sitzber. Sächs. Akad. Leipzig **83**, 3.

Heisenberg, W., 1934, Z. Phys. **90**, 209.

Heisenberg, W., 1936, Z. Phys. **101**, 533.

Heisenberg, W., 1938a, Z. Phys. **110**, 251.

Heisenberg, W., 1938b, Ann. Phys. **32**, 20.

Heisenberg, W., 1943, Z. Phys. **120**, 513, 673.

Heisenberg, W., 1944, Z. Phys. **123**, 93.

Hepp, K., 1966, Commun. Math. Phys. **2**, 301.

Hurst, C.A., 1952, Proc. Camb. Phil. Soc. **48**, 625.

Jaffe, A., 1965, Commun. Math. Phys. **1**, 127.

Jaffe, A., 1967, Phys. Rev. **158**, 1454.

Jordan, P., and W. Pauli, 1928, Z. Phys. **47**, 151.

King, C., 1986, Commun. Math. Phys., **103**, 323.

Lautrup, B., 1977, Phys. Lett. **69B**, 109.

Lipatov, L.N., 1977, Sov. Phys. JETP **45**, 216.

Loeffel, J., A. Martin, B. Simon and A. Wightman, 1969, Phys. Lett. **30B**, 656.

Magnen, J., and V. Rivasseau, 1986, The Lipator Argument for ϕ_3^4 Perturbation Theory, to appear.

Magnen, J., and R. Sénéor, 1977, Commun. Math. Phys. **56**, 237.

Nagamachi, S., and N. Mugibayashi, 1976a Commun. Math. Phys. **46**, 119.

Nagamachi, S., and N. Mugibayashi, 1976b, Commun. Math. Phys. **49**, 257.

Nagamachi, S., and N. Mugibayashi, 1986, J. Math. Phys. **27**, 832.

Newman, C., 1981, Commun. Math. Phys. **79**, 133.

Parisi, G., 1975, Nucl. Phys. **B10**, 368.

Parisi, G., 1977, Cargese Lectures, Hadron Structure and Lepton–Hadron Interactions, eds M. Levy, J.-L. Basdevant, D. Speiser, J. Weyers, R. Gastmans and J. Zinn-Justin (Plenum, New York) pp. 665–686.

Parisi, G., 1979a, Phys. Rev. **49**, 215.

Parisi, G., 1979b, Nucl. Phys. **B150**, 163.

Pauli, W., 1933, Die allgemeinen Prinzipien der Wellenmechanik, Handbuch der Physik, Zweite Aufl. 24-1 (Springer, Berlin).

Petermann, A., 1953, Helv. Phys. Acta **26**, 291.

Poincaré, H., 1882, Oeuvres, **I**, XXXVI.

Rivasseau, V., 1984, Commun. Math. Phys. **95**, 445.

Rosenfeld, L., and J. Solomon, 1931, J. de Physique (7) **2**, 129.

Ruijsenaars, S., 1983, J. Math. Phys. **24**, 922.

Sakato, S., H. Umezawa and S. Kamefuchi 1952, Prog. Theor. Phys. **7**, 377, 551.

Salam, A., 1951a, Phys. Rev. **82**, 217.

Salam, A., 1951b, Phys. Rev. **84**, 426.

Schroer, B., 1964, J. Math. Phys. **5**, 1361.

Schroer, B., and T. Truong, 1977, Phys. Rev. **D15**, 1684.

Schwartz, L., 1952, Théorie des Distributions I and II (Hermann, Paris).

Schwinger, J., 1951, Phys. Rev. **82**, 664.

Schwinger, J. (ed.), 1958, Selected Papers on Quantum Electrodynamics (Dover, New York).

Siegel, C.L., 1952, Nach. Akad. Wiss. Göttingen Math. Phys. Klasse, p. 21.

Simon, B., 1970, Ann. Phys. (USA) **58**, 76.

Skyrme, T.H.R., 1961, Proc. Roy. Soc. London **A262**, 237.

Sokal, A., 1980, J. Math. Phys. **21**, 1583.

Sokal, A., 1982, Ann. Inst. H. Poincaré **A37**, 317.

Speer, E., 1968, J. Math. Phys. **9**, 1404.

Steinberger, J., 1949, Phys. Rev. **76**, 1180.

Streater, R., and I.F. Wilde, 1970, Nucl. Phys. **B24**, 561.

Symanzik, K., 1973a, Lett. Nuovo Cimento **6**, 77.

Symanzik, K., 1973b, Lett. Nuovo Cimento **8**, 771.

Symanzik, K., 1975, Commun. Math. Phys. **45**, 79.

't Hooft, G., 1979, Can we make sense out of quantum chromodynamics, in: The Whys of Subnuclear Physics, Lectures at "Ettore Majorana", Int. School of Subnuclear Physics Erice 1977, ed. A. Zichichi (Plenum, New York).

't Hooft, G., 1982, Phys. Lett. **119B**, 369.

Thirring, W., 1953, Helv. Phys. Acta **26**, 33.

Tomonaga, S., 1948, Phys. Rev. **74**, 224.

Wick, G.C., 1950, Phys. Rev. **80**, 268.

Wightman, A., 1977, in: Cargese Lectures 1976, eds M. Levy and P. Mitter (Plenum, New York) pp. 67–77.

Wightman, A., 1979, in: The Whys of Subnuclear Physics, Lectures at "Ettore Majorana", Int. School of Subnuclear Physics, Erice 1977, ed. A. Zichichi (Plenum, New York). pp. 983–1026.

Wightman, A., 1981a, Physica Scripta **24**, 813.

Wightman, A., 1981b, Adv. Math. Suppl. Studies **7B**, 769.

Zimmermann, W., 1969. Commun. Math. Phys. **15**, 208.

Zinn-Justin, J., 1982, The Principles of Instanton Calculus: A Few Applications, Les Houches 1982, Session XXXIX, eds J.B. Zuber and R. Stora (North-Holland, Amsterdam, 1984) p. 39.

Discussion, session chairman A. Salam

Salam: This extension of field theory to non-renormalizable cases is surprising. I remind you that the success of the standard model was due to the fact that it *was* renormalizable.

Johnson: What has worried people about non-renormalizable theories is not that they don't exist, but rather that there are infinitely many of them for a given Lagrangean. So my question is: what about the uniqueness of these theories?

Wightman: Even with conventional methods it is difficult to specify a theory uniquely. Suppose you solve a theory defined by a functional integral with a cut-off and find that the appropriate axioms are satisfied. How can you answer the question, "What problem did you solve?" The answer would be "I solved the problem of showing that certain limits existed and that they had certain properties". But you never write down any condition which fixed the theory you were talking about.

We argue, of course, that conventional renormalized theories are fixed by choosing coupling constants and masses. In a non-renormalizable model these parameters are usually infinite in number. But in the procedure I have discussed you only have to fix the same number of constants that you would fix in a lower number of dimensions. In this way you decide which terms in the Lagrangean are fixed by the renormalization procedure and which are fixed by the normalization.

Lehmann: You mentioned the work of Gawedzki and Kupiainen. I think that they have made real progress, by showing that the $2 + \epsilon$ dimensional Gross–Neveu model is well defined with Green functions bounded polynomially by non-integer powers. This is the first time that any non-renormalizable theory has been treated rigorously. In addition this theory is non-asymptotically free, so we have a direct proof that such theories can exist.

Wightman: Yes, this is very interesting, but this theory is not unitary since the dimensionality is non-integer. The $d = 3$ case would be much more interesting.

Lee: There seems to be two types of theories, those which confine and those which do not. From the axiomatic point of view, how does one make the distinction?

Wightman: There are some general theorems proved by the group in Zürich in the early 1960s which say you cannot get an increasing potential out of a relativistic positive metric theory. Gauge theories have positive metric only for non-covariant gauges, so you can easily see the possibility of an increasing potential.

It has now been shown that the Higgs model has a continuum limit in three space–time dimensions. This theory [with a compact Lie group broken to U(1)] is supposed to confine. Perhaps confinement can be proved in this concrete continuum model.

Nielsen: Is it possible in an abstract manner, that is, without constructing examples, to tell whether there exist many non-renormalizable theories?

Wightman: I don't know, but there is the grand program of Symanzik to classify all non-trivial theories (which is by no means completed).

The Lesson of Quantum Theory, edited by J. de Boer, E. Dal and O. Ulfbeck
© Elsevier Science Publishers B.V., 1986

Particles, Fields, and now Strings

Steven Weinberg * **

Physics Department, The University of Texas
Austin, Texas, USA

We are celebrating the 100th birthday of a great man, so it is natural here to think back over the history of his and our field. In thinking about any sort of history it is often a useful approximation to see it in terms of the struggle between great opposites, whether Revolution and Reaction, or Cross and Crescent, or even Guelf and Ghibelline. (Somehow historians always manage to find opposites that start with the same letter.) In our field a fair amount of the history, if not of the last hundred at least of the last fifty years, can be seen as the struggle between two grand views of the world: as a world made of particles, or as a world made of fields. This opposition crystallized in the 1950s into a more specific distinction, between theoretical physics as quantum field theory, and as S-matrix theory. Just in the last year or so this old oscillation between quantum field theory and S-matrix theory has entered a new cycle, in the recent revival of what are called string theories. I will come back to string theories later in this talk, but it will be useful first to try to set the historical stage.

As a theory of everything, I suppose quantum field theory starts with the pair of papers written by Heisenberg and Pauli in 1929. Before that there was only a quantum field theory of the electromagnetic field, limited to photons. The Heisenberg–Pauli formalism was further developed in 1934 in papers by Pauli and Weisskopf and Furry and Oppenheimer. Quantum field theory presented a grand view: The fundamental ingredients of our universe are fields, and the particles are mere quanta of energy and momentum of the fields. But despite the grandness of this vision, quantum field theory did not really catch on the 1930s. I wasn't doing physics then, but my impression is that there was great doubt about the future of quantum field theory for two particular reasons. One is that a more dualistic approach, especially associated with Dirac, survived and for many purposes was as good as quantum field theory. In Dirac's view (which I think he kept to the end) there are fields, but only the electromagnetic fields, and there are particles, electrons of both positive and negative energy, which are eternal but which can be raised from negative energy to positive energy states, giving the appearance of pair creation.

---

* Author's note: This talk is largely based on one I gave at the close of the American Physical Society Division of Particles and Fields meeting in Eugene, Oregon, on August 15, 1985. I was asked at Eugene to assess future prospects for elementary particle physics in the light of its past history, and I thought that nothing else would be so appropriate for me to talk about at the Bohr Centenary.
** Discussion on p. 238.

This Dirac hole theory was for most purposes as good as quantum field theory. A notable exception of great importance in the history of physics was Fermi's theory of beta decay, in which for the first time a physical process was described in a way that could not have been understood within the old hole theory. Pais emphasized the importance of Fermi's theory of beta decay in the development of quantum field theory in a talk I heard him recently give in Princeton. It is surprising that the success of Fermi's theory did not convert more of the theorists of the 1930s into enthusiasts for quantum field theory.

Another reason was that although quantum field theory worked well in electrodynamics and was beginning to give a reasonable description of the weak interactions, it had some obvious failures. Most important of all were the ultraviolet divergences, discovered when quantum field theory was applied to the electromagnetic self-energy of the electron by Oppenheimer and independently by Waller in 1930. These infinities have been on our minds ever since. Also there were problems in the application of quantum electrodynamics to electromagnetic cosmic ray showers, which were actually somewhat of a red herring, because these problems are now believed to have been due to the production of mesons. There were also some wrong calculations done, which never helps. Many of these problems were of course equally problems for hole theory and quantum field theory, and it didn't seem very important to distinguish the two, as both were failing.

The first revival of quantum field theory took place in the arena of quantum electrodynamics in the late 1940s through the work of Feynman, Schwinger, Tomonaga, Dyson, Salam and others. The success was brilliant. In a few short years it was learned how to develop a Lorentz invariant perturbation scheme for quantum electrodynamics, how to carry out calculations in such a way that the infinities were not only swept under the rug, but dealt with in a physically satisfying way by renormalization of physical constants, and how these calculations not only could be done, and gave physically sensible answers, but gave answers that agreed with experiment to large numbers of decimal places.

It was, I suppose, as exciting a time as theoretical physicists have had since the 1920s. I heard about all this when I was an undergraduate a few years later, and it seemed to me obvious that what I wanted to work on was this brilliant, successful new field, quantum field theory, which had been so beautifully vindicated a few years earlier. I came to Copenhagen as a first-year graduate student in 1954, knowing nothing whatever about quantum field theory, but all keen to learn the technology of Feynman diagrams and renormalization. I was very disappointed and surprised to learn that already a disillusionment had set in, and that the theorists at Copenhagen, some of the leading spirits of the age, were already beginning to have renewed reservations about the future significance of quantum field theory in physics. These reservations we all remember. The first was that the renormalization theory which had worked so wonderfully when applied to quantum electrodynamics didn't work at all when applied to weak interactions in the only theory that was then extant, the Fermi theory of beta decay. Also, although there weren't necessarily any problems with renormalizability, the strong interactions were too strong to allow any use of perturbation theory.

There was another problem which is perhaps less direct, but which I think was also important. Quantum field theory has built into it the sense of an elementary particle as being different from something like an atom or a piece of chalk. An elementary particle in quantum field theory is a particle whose field appears in the fundamental Lagrangean or field equations. This seemed reasonable as long as the elementary particles were just a few in number, the electron, the photon, the proton, and then a little bit later the neutron and the neutrino. But beginning in the 1950s and then with a rush in the 1960s, the old distinction between elementary and composite particles began to look less and less physical. Particles were discovered like the ρ meson; in what sense could you say that it was not elementary and the proton was? It really didn't seem to make any sense to choose a few particles which just happen to be the ones that were historically discovered first, the proton, the neutron, and so on, and make a fundamental field theory out of those particular particles.

In the mid-1950s in Copenhagen and then in Princeton I began to hear about an alternative to quantum field theory. Already in 1937, John Wheeler had introduced the S-matrix, and then independently Heisenberg in 1943, and Christian Møller here in Copenhagen in 1945 had developed the beginnings of what might be called S-matrix theory, a theory in which the fundamental object of interest is not the evolution of a state vector in time, but rather the probability amplitude that given an initial state at $t = -\infty$, a particular final state will be observed at $t = +\infty$.

In this early work, up to the end of the 1940s, there was no real idea what sort of dynamical scheme might be used to calculate the S-matrix without referring back to a quantum field theory. But then, picking up the earlier work on electromagnetism of Kramers, Kronig, Toll and Wheeler, in the mid-1950s Chew, Goldberger, Gell-Mann, Low, Nambu, Thirring and others began to apply these ideas to the strong interactions, and there developed the beginnings of a self-contained S-matrix theory of elementary particle dynamics. The starting idea was that the unitarity principle would relate the imaginary part of any scattering amplitude to the total cross-section, which involves squares of scattering amplitudes, and the causality principle would relate the real and the imaginary part of the scattering amplitude to each other, and so in this way one would have a closed self-consistent set of non-linear equations, which one might either solve perturbatively, or perhaps in some nonperturbative way. This approach became quite popular, in part I think because it satisfied a deep philosophical preconception that many physicists have shared since earliest times.

Physics of course deals with observables, and in the end we always have to relate our theories to doable experiments. But there is the more stringent demand, which from time to time emerges in the history of physics, that not only must the end product of our calculations relate to observables, but that every ingredient in our theories must be directly observable. This requirement was very important in Bohr's thinking. We remember the work that Bohr and Rosenfeld did on the measurability of the electric field, a homework problem that had been given them by Landau and Peierls. Bohr regarded it as a matter of great importance to establish that the electromagnetic field was in fact as measurable as the commutation relations allowed it to be. In S-matrix theory one could feel satisfied that one was again dealing only with observables, whereas that was not the case in quantum field theory.

This last remark may seem a little strange, after what I've just said about Bohr and Rosenfeld showing that the electromagnetic field was observable. Remember, however, they were not working in what we now call quantum field theory, but rather in the old hole theory, in which the only fields were the electromagnetic fields. In a quantum field theory, the electron field is hardly observable, because it is massive and charged and a Fermi field. Quantum field theory deals with fields that we are not going to measure; they only appear in the theory as parts of the machinery for calculating the S-matrix. And so, by turning away from quantum field theory toward S-matrix theory, the physicists of the 1950s could enjoy an act of purification, an expelling of the unobservables.

It soon became clear that in order to develop S-matrix theory in this way, one had to go far beyond the forward-scattering dispersion relations that were first studied. After all, unitarity relates every physical process to every other physical process. A full fledged S-matrix theory dealing with processes of arbitrary complexity began to be developed at the end of the 1950s, especially at Berkeley in a group including Mandelstam and Stapp, centered around Geoffrey Chew.

The group in Berkeley, and many others, felt that S-matrix theory was the most interesting thing going on in physics, but I found it forbidding for various reasons, and did not get into it myself. One thing I did do that was motivated by S-matrix theory I will mention here because it comes up later on in this lecture. I tried to see how the well-known properties of electromagnetism and gravitation could be understood in a purely S-matrix context without any ideas about quantum field theory. I showed that any mass zero, spin one particle would, according to the axioms of S-matrix theory, have to behave pretty much the way we know photons do, and any mass zero, spin two particles would have to behave the way we believe gravitons do, as was also remarked by Feynman. I will come back to this point later.

The S-matrix theory which began in the mid-1950s had by the mid-1960s already collapsed, at least as a grand program for solving the problems of physics. This was for a number of reasons. (There are always a number of reasons for everything.) One reason is that in order to understand how to deal with physical processes involving anything more complicated than elastic scattering, it was necessary to understand the analytic behavior of functions of many complex variables, to understand functions that have simultaneous cuts in more than one variable at a time. In fact, this is necessary not just to do calculations, but even to state the axioms of S-matrix theory. This kind of mathematics of many complex variables is notoriously difficult; it may or may not be beyond the capacity of the human brain to understand the analytic structure of the full multi-particle S-matrix, but I was able to show quite rigorously that it was beyond me.

A second problem with S-matrix theory was that there never was any rational approximation scheme; there was no small parameter in the theory. And a third problem was that despite the wonderful, logical-positivistic flavour of S-matrix theory, it really wasn't very successful in predicting anything much about the real world. The forward-scattering dispersion relations and Regge pole systematics, of course, were and continue to be very important, but beyond this very little about the strong interactions was successfully predicted by the S-matrix theories. In particular, although they concentrated very much on pion scattering, they missed the important point (revealed through the application of current algebra in the mid-1960s) that pion interactions at low energy are quite weak.

However, before the end of S-matrix theory, there was a last curious spurt of activity. Since it was so difficult even to formulate exactly what the axioms of S-matrix theory were, let alone apply them, a number of theorists began just to guess at candidate scattering-amplitudes which would satisfy a number of our general ideas about the properties of physical scattering amplitudes, ideas having to do with crossing symmetry, Regge asymptotic behavior, and so on. Veneziano, in particular, discovered a remarkable formula for a candidate meson–meson scattering amplitude. This amplitude can be expressed *either* as a Breit–Wigner sum of an infinite number of resonances, *or* as the sum of an infinite number of exchanged resonances (which give it the desired Regge asymptotic behavior at high-energy and fixed-momentum transfer), a property known as "duality".

Great effort went into the exploration of these dual models, starting with the paper of Veneziano. They were very rapidly generalized to multiparticle processes, by Bardacki and Ruegg, Goebel and Sakita, Chan and Tsun, and Koba and Nielsen in 1969, and it was then pointed out by Nambu, Nielsen, and Susskind that the problem to which the dual scattering amplitudes were the solution was the problem of the motion of a relativistic string. That is, the infinite number of particles whose exchange produces the Regge trajectory are in one to one correspondence with the modes of vibration of a relativistic string. For the Veneziano model in its original form the string is open, with free ends which travel at the speed of light. The strings have a certain tension, and since this was all in the context of strong interaction physics, this string tension was imagined to be about a pion mass per fermi, or roughtly $(100 \text{ MeV})^2$. There were other models developed; another model called the Virasoro–Shapiro model was found to be based on a closed string, with a string tension that again was taken to be about $(100 \text{ MeV})^2$. Spinors were incorporated into string theories by Neveu and Schwarz, Ramond, and others, and it was this formalism that led Wess and Zumino in 1974 to their development of supersymmetry.

It is really remarkable, now looking back at the period of the late 1960s and early 1970s, how much work went into these string theories without the slightest encouragement from experiment. In fact, the string theories incorporated features that were not only not confirmed by experiment, but were in gross contradiction with experiment. One of these features was that these theories contained massless particles, and given the background of these theories, these were taken to be massless strongly interacting particles, which clearly would have already been observed. In the open string theory, there was a massless spin one particle, and in the closed string theory massless particles of spin two and also spin zero. Another problem was that when one tried to take into account the unitarity corrections to the S-matrix elements, it was found that these theories really only made sense in 26 dimensions, or if you included fermions in the theory, then in 10 dimensions. That was quite embarrasing. Schwarz and Scherk in 1974 suggested that these theories should not be thought of as describing the strong interactions; rather, the mass-zero spin-two particle that appeared in the closed-string theory should be identified as the graviton. In other words, the string tension would not be of order $(100 \text{ MeV})^2$, but would be something like the square of the Planck mass, $(10^{38} \text{ GeV})^2$. One does not assume general covariance or the equivalence principle here, but these theories are Lorentz covariant, and thus as I mentioned earlier, any mass-zero spin-two particle in them would have to interact like the graviton of general relativity. However, this proposal received little attention, I think above all because it was embarrasing to be working on theories that made sense only in 10 or 26 dimensions.

However, perhaps even more influential than any failure of the dual models or string theory was a second revival of quantum field theory, which began at just about the time of the development of string theory. I need not dwell on this, as I suppose it is familiar to everyone here. First, it was shown by 't Hooft and others that the earlier suggestion of a spontaneously broken renormalizable gauge theory of electroweak interactions would indeed work mathematically. Very soon thereafter, the discovery of neutral currents and the measurement of their properties showed that not only could an electroweak theory be constructed along these lines, but the already constructed $SU(2) \times U(1)$ theory was in fact the correct theory of electroweak interactions. At the same time, there was also the development at last of a theory of strong interactions, based on much the same Yang–Mills mathematical structures, in which instead of the gauge symmetry being spontaneously broken, it provides a trapping mechanism which hides the underlying constituents, the gluons and quarks, from our view. We had in the early 1970s a fully developed standard model of weak, electromagnetic, and strong interactions, which apparently was capable of describing everything to do with accessible particle physics. Exciting experiments then later in the 1970s gave a final confirmation to these theories, showing that in fact the neutral currents did behave in the way they were supposed to, the quarks and leptons really are what we think they are (except that there are more of them), and the W and the Z are really there.

Once again it became possible to think of a quantum field theory as being a fundamental theory of nature. In particular it made sense once again to talk about particles as being elementary, because we no longer had to think of the proton or the neutron or the ρ meson as elementary particles. The elementary particles had been reduced to a fairly manageable set; leptons, quarks, gluons, W, Z, the photon, and maybe a few Higgs bosons. One thing that especially excited me, was the fact that at last in these theories we could understand in a natural way why symmetries like parity and strangeness and isotopic spin were symmetries of some interactions and not of others. All the mysterious facts that my generation of physicists had had to learn as empirical rules when we were graduate students suddenly began to make sense rationally.

I tried for this talk to find a couple of quotes in the literature to exemplify the mood of enthusiasm in the mid-1970s, and then the sense of disillusionment that set in a few years after. After a good deal of searching I found two perfect quotes, both by me. In 1977 in an article in Daedalus I had the effrontery to say,

> "If something like a set of ultimate laws of nature were to be discovered in the next few years [and thank goodness I said in parentheses 'an eventuality by no means expected'], these laws would probably have to be expressed in the language of quantum field theory."

Just four years later in 1981 I was at another anniversary, the 50th Anniversary of the Lawrence Berkeley Laboratory, and I gave a talk that also dealt with the oscillation between *S*-matrix and field theory. At the end of the talk I concluded with the words,

> "These have been exciting times. Quantum field theory is riding very high, and one might be forgiven for a certain amount of complacency with it. But perhaps we will now see another swing away from quantum field theory. Perhaps that swing will be back in the direction of something like *S*-matrix theory."

Well, I was at Berkeley, and I wanted to say nice things about *S*-matrix theory, but I had two serious reasons for this sense of caution about the future of quantum field theory.

One reason is that theorists had failed to make further progress in explaining or predicting the properties of elementary particles, beyond the progress that was already well in hand by the early 1970s. The standard model has many loose ends; mass ratios, coupling constants, a whole menu of quarks and leptons, and we have simply not succeeded in explaining it. Several attractive ideas were tried: grand unification, technicolor, preons, supersymmetry, Kaluza–Klein theories, and all that, but despite so much clever mathematical work, almost nothing has come out in the way of concrete numbers that could be compared with experiments. Perhaps the only success in that hard quantitative sense was the grand-unification prediction of $\sin^2 \theta$, a prediction that does seem to be quite robust, and also to agree with experiment. It soon became clear that in trying to make the next step beyond the standard model we would probably have to understand physics at or near the Planck scale, partly because that is where whatever grand unified group there might be would break down, and also because after all gravity exists, and gravity becomes a strong interaction at the Planck scale. Unfortunately, throughout the 1970s most of us saw no hope for a quantum field theory of gravity.

The second reason that I gave in 1981 for being skeptical about the future of quantum field theory is that we could understand its successes in the low-energy range, up to a TeV or so, without having to believe in quantum field theory as a fundamental theory. There is a folk theorem (a term of Wightman, meaning something which is generally known to be true although it has not been proved) that says that any theory which satisfies the axioms of *S*-matrix theory, and contains only a finite number of particles with mass below some M, will at energies below M look like a quantum field theory involving just these particles. That is, quantum field theory by itself has no content; it is just a way of calculating the most general scattering amplitudes that obey the axioms of *S*-matrix theory. Of course, one might argue that the quantum field theories that we have developed, like quantum chromodynamics, are not just any old quantum field theories, but simple, even beautiful, field theories. However, another folk theorem tells us that in the effective field theory, to which essentially any theory reduces at sufficiently low energy, the non-renormalizable interactions are all suppressed by powers of the underlying fundamental mass scale, which one might imagine is the Planck mass scale. Thus, the physics we see at accessible energies should be described by a renormalizable effective field theory, and we know that the interactions in such theories are always limited in number and complexity. (This is a view of the rationale for renormaliz-able field theory that is somewhat different from the one described at this meeting by Arthur Wightman.) To see anything else we would have to do experiments at the Planck scale, except that a few interactions though very weak may be detectable for special reasons, like for example the special circumstance that gravity adds up coherently, so that although the gravitational interaction is fantastically weak, we can still measure its effect for macroscopic bodies.

To summarize: quantum field theory has not done much for us lately, and what it had done for us earlier we can understand without having to believe that quantum field theory is in any sense fundamental.

When I made these remarks in 1981, I had no idea what direction S-matrix theory would take as a possible replacement for quantum field theory. In fact, Schwarz had been forcefully advocating string theory as the only hope for a quantum theory of gravity ever since his 1974 work with Scherk, but he was ignored by almost everyone (myself included). In 1980 Green and Schwarz proved the space–time supersymmetry of superstring theory. In the following two years they developed a new supersymmetric formalism for superstrings and used it to invent the Type II superstring theories and to prove their finiteness at one loop. At the 1982 Solvay Conference in Austin, Zumino mentioned this work of Green and Schwarz as the natural candidate for a finite theory of gravitation. Then in 1983, at the Fourth Workshop on Grand Unification, Witten gave an influential talk about $d = 10$ superstring theory, and other theorists began to take such theories seriously as a promising approach to quantum gravity. The time was ripe for this suggestion, as it had not been in 1974, partly because we were all so frustrated within everything else, and also because several years of work on Kaluza–Klein theories had made us comfortable with the idea that space–time might really have more than 4 dimen-sions, with all but 4 wrapped up in a compact manifold of very small circumference.

Witten in his talk also pointed out the theoretical obstacle, having to do with a hexagon anomaly, that impeded this development. Later, with Alvarez–Gaumé, he discovered the cancellation of this anomaly in one example of a superstring theory, that unfortunately seemed phenomenologically unpromising. Then the great breakthrough came last year when Green and Schwarz demonstrated that there were a few potentially realistic superstring theories, with very specific gauge groups, in which the anomalies that Witten had worried about cancelled. This started an explosion of interest in string theory, which has not yet even peaked.

I suppose that this should be scored as a victory for the S-matrix approach. String theory grew out of S-matrix theory, but in a sense it has some of the features of both S-matrix theory and quantum field theory — the experts have not yet settled down in their view of what string theory really is. Indeed, this is one of the things that makes the theory hard to learn; not everyone will tell you the same thing about what it is you are supposed to be learning.

On one hand there is a view of string theory which takes seriously that these are theories of strings. Instead of quantum fields that are time-dependent functions of the position in space of a particular particle, you have quantum fields that are time-dependent functionals of the configuration in space of a moving string. This second quantized quantum field theory of strings unfortunately does not yet exist, but many of the leading experts in this area are working very hard to develop it. The particles are the normal modes of these strings, so when you calculate an S-matrix element (which in the end is what you always have to do) you imagine a string in a particular normal mode colliding with another string in another normal mode, and perhaps two strings joining together to make a single string, and then that single string breaking apart to be two other strings, which finally wind up in two other normal modes. Calculations are not actually done that way. The description I just gave is often presented in public talks about string theories by the experts, but as far as I can tell they do not actually do calculations that way. It is too hard, and the formalism has not been fully developed yet.

There is another approach to string theory, which is the one that almost everyone actually uses. In this other approach, one starts with the observation that a string moving through space sweeps out a two-dimensional surface in space–time. You can describe the string by giving the space–time coordinates x^μ as functions of two parameters. One parameter, σ, tells you where along the string you are, and the other parameter, τ, tells you how long the string had to move. So the string theory can be regarded as a quantum field theory in *two* dimensions, the "fields" being taken as the d quantities $x^\mu(\sigma, \tau)$, (where $d = 4$ or 10 or 26 or whatever) with perhaps some spinors $\psi(\sigma, \tau)$ as well. The interpretation of this two-dimensional field theory in terms of physical processes in d space–time dimensions has to be accomplished by asking what sort of quantum averages in the two-dimensional world have the unitarity and Lorentz transformation properties that we require for the S-matrix in d-dimensional space–time, so we have a curious blend here of quantum field theory and S-matrix theory.

It is natural to write the Lagrangean for the two-dimensional theory so that it is independent of the choice of the parameters σ and τ, which of course requires the introduction of a two-by-two metric tensor. As emphasized by Polyakov, it turns out then that the old string theory is not only a generally covariant two-dimensional field theory, but is invariant as well under conformal transformations, in which the metric is multiplied with an arbitrary function of the two-dimensional coordinates. This may sound like I am getting into technicalities here, but the addition of conformal invariance to two-dimensional general covariance and d-dimensional Lorentz invariance has an overwhelmingly important consequence: the string Lagrangean must be that of a *free* field theory in two dimensions, with the non-triviality of the S-matrix in flat d-dimensional space–time arising not from interaction terms in the Lagrangean but from the non-trivial topology of the Riemann surface described by the 2×2 metrics. So here we have the realization of an old dream of what a fundamental theory ought to be: interactions are not something we insert in a more-or-less arbitrary way into a Lagrangean (and might if we wished be left out altogether), but are inevitable consequences of the nature of the theory's degrees of freedom.

The differences between these two views of string theory can be illustrated by considering how each would deal with a "one-loop" calculation of the S-matrix for two-particle scattering in a closed string theory. In the two-dimensional field theory approach, one imagines the two-dimensional space to form a torus, and carries out a free-field quantum average of a product of four "vertex functions" of position on the torus, one for each incoming or outgoing particle. On the other hand, in the second-quantized string theory approach, one imagines two closed strings in different normal modes approaching, joining to form one string, then breaking up again into two closed strings, then joining again to form one closed string, and finally breaking apart again to form two closed strings in definite normal modes. This gives one more of a sense of the physical reality of strings, but it is not a very elegant way to describe a torus.

I have not been entirely impartial here in drawing the contrast between the two leading approaches to string theory. My preference for the two-dimensional field theory approach may be due in part to the fact that this is the only approach that I have so far been able to learn. Certainly one should try to understand all possible approaches. Also, it may be, as often argued, that the second-quantized field theory of strings offers the best hope for an understanding of non-perturbative effects. Nevertheless, since string theory is supposed to be better than quantum field theory, it does not seem clear to me that the best strategy is to make string theory look as much as possible like field theory, only with strings instead of particles, which I take is the spirit of the second-quantized approach.

In the last few minutes, I want to take up the question that has doubtless been on the minds of all those of you who have not yet become string mavens (meyvn, mevinim, Jiddish for expert; eds.). The question is: Is it safe to ignore string theory, and hope that it will go away? For a theorist, the question takes the form whether it is necessary to learn all about automorphic functions, Riemann surfaces, Virasoro algebras, and all that, or just bypass all this effort and wait for the next fashion in theoretical physics. For the experimentalist, the question is whether it is worthwhile beginning to think of possibly testing these theories?

In trying to answer these questions, I must say right away that there is not the slightest shred of experimental evidence for string theory. The same was also true of the other theories that we developed in our desperate attempt to go beyond the standard model, in particular for supersymmetry and Kaluza–Klein theories, which have now been incorporated into superstring theory. Never has so much brilliant mathematics been done by physicists with so little encouragement from experiment. Furthermore, just as for supersymmetry and Kaluza–Klein theories, the string theories have still not settled down so that they could make very definite predictions which could be tested experimentally. However, in this respect I think string theories are really different from the Kaluza–Klein and supersymmetry theories. Supersymmetry is a symmetry like Lorentz invariance; it allows a tremendous variety of possible dynamical theories. Likewise, Kaluza–Klein theory is just a general idea, that there might be some higher number of dimensions which are compactified; again, this idea allows a great many specific theories. String theories, on the other hand, are very rigid. There are almost no string theories at all, and of the few possibilities only one at present (the "heterotic" superstring of Gross, Harvey, Martinec and Rohm) seems at all promising phenomenologically. As I discussed earlier, what you are really doing in string theory is studying two-dimensional conformal gravity (actually supergravity), which is a free-field theory, so that the only interactions are those that arise from the topology of the two-dimensional manifolds. Also, the topology of a two-dimensional manifold is completely specified by the number of handles that you put on it (assuming it an orientable closed surface). And so there is nothing you can tinker with in these theories; they are either right or wrong as they stand. We do not have any experimental evidence for string theories, and I cannot really tell the experimentalists what they should look for, but the string theorists in the next few years should be able to come up with definite predictions, which can then be tested. Already, there is an indication that any realistic string theory when compactified down to four dimensions will probably contain at least an extra U(1), so that it is worth looking for one more gauge symmetry in addition to the SU(3) × SU(2) × U(1) of the standard model. But the precise features of this extra U(1) are certainly not yet predicted. The biggest gap that will have to be crossed before such predictions can be made is in understanding the dynamics of the compactification from 10 to 4 space–time dimensions. Candelas, Horowitz, Strominger, Witten and others have made some progress in understanding the general features of this compactification, but much remains to be done.

I have remarked that there is no evidence for string theories, but there are other criticisms of a more fundamental nature. Georgi has argued to me that it really would be very unlikely that string theory should provide a fundamental theory of gravity and everything else, in part because after all string theories developed out of the original guess by Veneziano of a scattering amplitude for strong interaction meson–meson scattering. Why in the world should we believe that mathematical structures that grew out of the attempt to understand the strong interactions should be applicable not to the strong interactions but to everything, including gravity? I do not agree with this argument, because I do not agree with its historical basis. The string theories and the dual models which preceded them did not grow out of an attempt to understand the detailed empirical facts of the strong interactions; in fact, they never were much good at that. They grew out of an attempt to find some solution to the problem of constructing scattering amplitudes that satisfy the basic axioms of analyticity, unitarity, and so on. And, these are of course the same problems that we are all solving, whether at 100 MeV or at the Planck scale. There is a good chance that the way of satisfying these fundamental S-matrix principles that is provided by string theory is unique, at least if we want to include gravitation. The string theorists of the late 1960s and early 1970s guessed that the kind of dual model which they were proposing would turn out (when suitably unitarized) to be the more or less unique solution of the axioms of S-matrix theory. And maybe it is, not as applied only to the strong attractions, but as applied to everything, including gravitation. Maybe we really do now know the laws of nature, and the only thing that is left is to work hard for the next few years trying to figure out how the ten dimensions get compactified to four, and then finding out what low-mass quarks and leptons and gauge bosons are left over, calculate their mass ratios, check to see whether they agree with experiment, and then go home.

Relying on a general sense that nothing ever in this life works out the way we want it to ("we" here means theorists), my guess is that there are many surprises that will be provided to us both by imaginative theorists and by enterprising experimentalists before we finally get to the solution to our problem. Nevertheless, I would argue that it is not safe to ignore string theories and wait for the next change in fashion. These theories are much too promising and too beautiful for us not to take them very seriously and explore their consequences for as long as it takes.

Discussion, session chairman A. Salam

Nielsen: There is at least one aspect of the program in which uniqueness seems doubtful. That is the compactification of the six extra dimensions. Of course it may be determined dynamically.

Weinberg: It is true that you have to be cautious concerning uniqueness. I should have mentioned that when you compactify the free two-dimensional theory (whose fields $x^\mu(\sigma, \tau)$ lie on a ten-dimensional manifold) to something which is four-dimensional space–time and something else, you find that the two-dimensional string theory is not free and perhaps strongly interacting.

One can say that the fundamental actions are either stated at the ten-dimensional level or at the four-dimensional level. I lean towards the first point of view, and hope that compactification is a dynamical afterthought. Perhaps this happens in a unique way, but perhaps there are still a few parameters which are undetermined (which would be disappointing since then quantities like α could not be calculated without knowing the initial conditions of the universe).

Johnson: Early in your talk you pointed out that renormalizable theories are theories in which the details of the cut-off are irrelevant. We have a similar situation in condensed matter physics, where the type of cut-off does not affect critical exponents. Why then do you need a fundamental theory if it is irrelevant? How can you tell what the theory is on the basis of physics at our own scale in which we only know what the order parameters are?

Weinberg: The fundamental theory is not really irrelevant in that it tells us where the parameters in the effective theory come from. By way of analogy with critical phenomena, there is certainly universality for critical exponents, but the Curie point of a metal must be determined from the microscopic theory. Here we want to see not only the 17 or so free parameters of the standard model but also the 3, 2 and 1 in $SU(3) \times SU(2) \times U(1)$ and the dimensionality of space–time coming out of the final theory.

I think that in the next few years the theory will prove to be either stunningly successful in explaining things we already know or else turn out to be quite wrong.

Bleuler: Will a general unification explain the enormous gap between gravity and other interactions and also why there are four types of interaction?

Weinberg: That is the right question to ask. We have learned, however, to ask it another way: why are all the mass scales, such as particle masses, so small compared to the Planck mass? There is a natural way of getting such hierarchies with supersymmetry which can protect the masses of certain particles from becoming of the order of the Planck mass.

Of course, the superstring formalism is so restrictive it may fail — but that's what is good about it.

Bleuler: Would there be a change in the definition of four-dimensional space?

Weinberg: Yes, a radical change, as the x^μ's are now dynamical variables in a two-dimensional theory.

Fowler: When John Schwarz talks about superstrings, he describes photinos, etc. Why is there not some hope that these can be discovered experimentally?

Weinberg: Actually, there is hope. The difficulty in finding superpartners is that they are not protected by gauge symmetry from getting large masses (whereas the particles we see are protected). Some theories say that photinos, quarks, etc. are right around the corner, whereas others say that they're at the Planck scale. In the former case there is a good chance that photinos make up the dark matter in the universe.

Lee: You made a remark about this theory being either proved right or dying, in the next few years. I don't think it will do that. Very rarely in our experience have either people or nature not been ingenious enough to keep theories workable for long periods of time. Perhaps we will neither prove it right nor wrong and it will reappear in a few decades.

Weinberg: Yes, and we may have a proliferation of string theories, but then I say to hell with it.

Salam: I do not think we should take this attitude. I am recalling a remark by Niels Bohr in 1948, when I first heard him. He was speaking of the fundamental length. He wanted to have a fundamental length at 1 GeV. Now we do have a fundamental length with a local field theory, which is an amazing statement to be made. We have a theory of which my student Chris Isham said that he joined the quantum-gravity bandwagon in order to get quantum physics understood in terms of general covariance. That is to say, Einstein would be supreme and Planck would come out of it. What has happened is exactly the opposite: you get the theory of Einstein as a very special case of the string theories and Planck's constant has to be put in by hand. These are very important remarks which have come out of these string theories. I do not agree with T.D. Lee. But even if the theories do not satisfy the uniqueness principle advocated by its adherents, it will have a very strong place.

I would like to end by a remark by Witten when he said that 57 years of point quantum field theory have come to an end. We shall now in the future have string field theories of which point quantum field theories will be special cases.

The Lesson of Quantum Theory, edited by J. de Boer, E. Dal and O. Ulfbeck
© Elsevier Science Publishers B.V., 1986

Collective and Planetary Motion in Atoms

R. Stephen Berry

Department of Chemistry and The James Franck Institute
The University of Chicago, Chicago, Illinois 60637, USA

Contents

Abstract

The evolution of the conception of electron correlation is sketched, particularly the ideas that emerged in the late 1970s. Those ideas have led to a deep reexamination of the behavior of electrons sharing a valence shell. Doubly-excited helium was the first case in which it could be clearly established that the electronic states exhibit collective rotations and vibrations, rather than predominantly independent-particle-like behavior. More recently, it has appeared that the ground states and most but not all of the low-lying excited states of the alkaline-earth atoms are also much more like collective rotor-vibrators than like quantum analogues of solar systems. The appearance of such molecule-like characteristics for the electrons in atoms leads to a search for independent-particle-like behavior for atoms in highly excited vibrational states of small molecules such as H_2O, NH_3 and CH_4. Together, the two kinds of systems potentially exhibiting characteristics traditionally associated with the other suggest trying to find a more unified formulation of few-body problems that makes collective and independent-particle behavior into related but complementary manifestations of some more general characterization of the states of few-body systems.

1. Introduction

The dazzling success of the planetary model of Niels Bohr's 1913 trilogy, in reproducing the known spectral lines of hydrogen and predicting others and in explaining the lines found by Pickering (1897) in the spectrum of ζ-Puppis and by Fowler (1912) in laboratory discharges, stimulated attempts to interpret the spectrum of neutral helium atoms with an extension of the same model. The solution to this problem was far subtler than suspected [except perhaps by Einstein (1917)] and was not solved in terms of Bohr–Sommerfeld quantization until the work by Leopold and Percival (1980), and then only for the ground state. [See also Percival (1977) for a review, and Coveney and Child (1984) for the next development.] However by 1920, in his address to the German Physical Society, Bohr expressed an expectation whose fulfillment is beginning to emerge with the ideas we shall explore here. [See Nielsen (1976).] Bohr said, regarding the attempt to extend the results for the hydrogen atom to other atoms and to molecules,

> "It appears no longer possible to justify the assumption that in the normal states the electrons move in orbits of special geometric simplicity like 'electron rings'. Considerations relating to the stability of atoms and molecules against external influences and concerning the possibility of the formation of an atom by successive addition of the individual electrons compel us to claim, first that the configurations of electrons are not only in mechanical equilibrium, but also possess a certain stability in the sense required by ordinary mechanics, and secondly that the configurations employed must be of such a nature that transitions to these from other stationary states of the atom are possible. *These requirements are not in general fulfilled by such simple configurations as electron rings and they force us to look for possibilities of more complicated motions.*"

These sentences presage not only the shell structure of atoms; we can even read into them the seeds of our contemporary ideas regarding collective behavior of electrons in atoms, and it is to this topic that this essay is primarily devoted. We shall, however, look a bit beyond, toward some generalizations to which those ideas inevitably lead us.

Certainly the accepted picture today of atomic structure supposes that the best first approximation is the independent-particle model in which each electron has an energy and an angular momentum of its own. The orientations of those angular momenta cannot be even modestly good constants of the motion, but we suppose that one-electron energies and angular momenta can be treated as approximate constants of the motion with corresponding "pretty good" quantum numbers. This behavior is in sharp contrast to the collective rigid-body rotations and normal-mode vibrations we use as the first approximation for describing the behavior of atoms in simple molecules. Bohr himself recognized this difference in 1913, in the third paper of the trilogy (Bohr 1913).

We shall see here how some of those distinctions between the two classes of systems break down when we scrutinize them with the tools now available to us. As the distinctions begin to blur, we find atomic states that are characterized by molecule-like quantization and perhaps molecular states for which independent-particle atom-like quantum numbers are appropriate. The systems begin to seem like the armadillos of Rudyard Kipling (1946), neither tortoises nor hedgehogs but a blending of the two. Nuclear physics has long had to confront systems that exhibit both sorts of characteristics and concepts from this field have been important in influencing what has happened recently in atomic physics. It is tantalizing to conjecture that the cross fertilization might lead eventually to a formulation of the few-body problem that puts nuclei, atoms and molecules on a common footing, so that we can recognize their commonalities as easily as their differences.

The slowness with which the interpretation of electron correlation in atoms has emerged might be associated with the history of computation. In the earliest days, back-of-the-envelope hand calculations were all one could do, and sufficed very nicely for the original model of Bohr. With the advent of hand-operated desk calculators, it became possible to do self-consistent field calculations at the level of Hartree and Hartree–Fock methods, and to interpret the results of those calculations. When electronic computers became available, elaborate, many-electron variational wave functions could be and were generated, for example of the multiconfiguration type. However, during the period when these functions, often quite accurate, were first being generated, their interpretation was apparently a computation problem just a step more complicated than their generation, so that the interpretive phase for wave functions of present accuracy waited until the powerful microprocessor-based computers of today. Ironically, the task of extracting the relevant information to interpret even very complicated wave functions turns out to be not much more difficult than carrying out the calculations that produce those functions.

The recent history relevant here begins with the approaches that treated the electrons of doubly-excited helium atoms like almost-hydrogenic electrons interacting weakly via their mutual repulsion. The special symmetry of the hydrogen atom that makes its nth level n^2-fold degenerate is the symmetry of rotation of a sphere in four dimensions, O(4), so that the starting point used for calculating energy levels and wave functions of doubly-excited helium takes the zero-order Hamiltonian as O(4) × O(4). This is just the approach Herrick and Sinanoglu (1975) followed and was also followed independently by Wulfman (1973). The electron–electron interaction breaks this symmetry; the problem was to find what smaller group gives a proper picture of the approximate symmetry of the full Hamiltonian with the electron repulsion included. The answer they found is a single O(4) whose extra invariant is the square of the length of the difference of the two electrons' Runge–Lenz vectors, A_1 and A_2. Since each Runge–Lenz vector is essentially the semi-major axis of the classical Kepler ellipse, the invariance of $|A_1 - A_2|^2$ indicates that the two classical ellipses precess together. Wulfman (1973) showed that diagonalizing the operator corresponding to this invariant is approximately but not exactly equivalent to diagonalizing the operator $1/R_{12}$, so things seemed to be starting to fall into place. However, the quantum-mechanical counterpart of the precession of the ellipses was very much a mystery; it did suggest that helium in a state well described by the O(4) model might show some tendency toward maintaining a geometric structure to its probability distribution.

Pursuing this idea, Rehmus, Kellman, Roothaan and the present author in 1978 (Rehmus et al. 1978a, b) presented suitable probability distributions from the wave functions generated by Herrick and Sinanoglu to look for spatial correlations, particularly angular correlations that would give the atom a persistent "shape". There was little question of what was wanted: mean deviations of the symmetry of the probability distribution with respect to the mean interelectronic angle θ_{12} would be the very least. Fortunately it was possible to extract vastly more information by not reducing the data nearly so far. Beginning with the full probability distribution for the two electrons $|\Psi(R_1, R_2)|^2$, a function of six variables, one can remove the dependence of this distribution on its orientation in the space of the laboratory

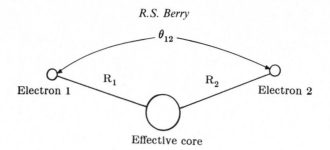

Fig. 1. The intrinsic coordinates for two-electron and quasi-two-electron atoms; for He and iso-electronic ions, the effective core is simply the nucleus, while for complex atoms such as the alkaline earths, various effective potentials may be used.

coordinates by integrating it over the Euler angles that specify that orientation. This leaves a reduced distribution in three variables; what three variables ought one to use? Clearly one must be θ_{12}; for the interpretive purposes we had in mind, the others ought to be the distances R_1 and R_2 of the two electrons from the nucleus. Figure 1 shows these variables. But this is still not quite sufficient; one cannot readily graph a function of three independent variables. However, one can graph a function of two independent variables and even represent such a graph as a projection on a sheet of paper. The reduction is natural: one reduces the already reduced probability distribution or density $\rho(R_1, R_2, \theta_{12})$ to the conditional probability distribution $\rho(R_2, \theta_{12}; R_1 = a)$ which is the probability distribution for the distance R_2 and the angle θ_{12}, provided that the distance R_1 has the value a. With the reduced density or probability distribution based on these variables, it is straightforward to construct the distribution from a variety of conventional forms for wave functions, and it is equally straightforward to interpret the distributions. (See Rehmus et al. 1978a, b and Rehmus and Berry 1979.) We discovered a bit later that exactly this choice had been made by Shim and Dahl (1978) to help them interpret the physical basis of Hund's rule, and in 1959 by Munschy and Pluvinage (1963) to explore the correlation in the ground state of helium. Computational tools were not yet powerful enough in 1963 to make the appropriate computations feasible for more than one simple graph. Now one can construct the graphs readily, and much of the subsequent discussion focuses on examining and interpreting such graphs. Figure 2 illustrates the probability distributions for the ground and first two excited states of the helium atom, specifically the distributions for R_2 and θ_{12} when R_1 takes on its most probable value. (Strictly, all the distributions shown here are multiplied by the factor R_2^2 from the Jacobian; without this factor, it is difficult to see clearly the asymmetry in the lower states, particularly the S states.) These and some of the subsequent illustrations are done in a Cartesian representation; others are done in cylindrical polar representation. The former spreads out the region near the origin, which is helpful for seeing that region. However, the polar representation seems closer to reality because it shows the nucleus as a singular point at the origin, not as a line at the bottom of a trough. Both fig. 2a and 2c exhibit some polarization of the distribution toward $\theta_{12} = \pi$; that is, electron 2 does indeed show some tendency to be on the side of the nucleus away from electron 1. However this tendency is not strong. We can expect more correlation of the electron distribution

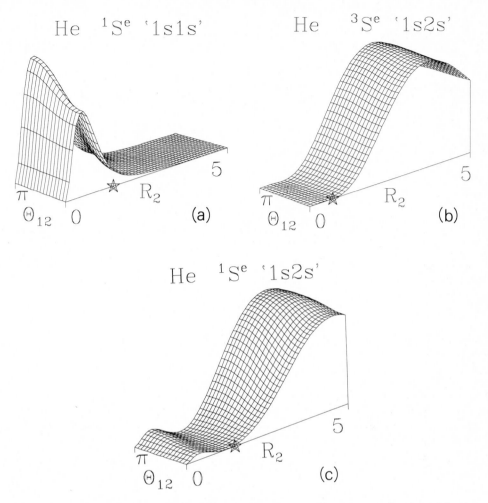

He $^1S^e$ '1s1s'

He $^3S^e$ '1s2s'

He $^1S^e$ '1s2s'

Fig. 2. Cartesian representations of the conditional probability distributions $\rho(R_2, \theta_{12}; R_1 = a)$ for the three lowest states of the helium atom with a at the most probable value of R_1. In all three cases (a, b, c) some angular correlation is evident, but the large amplitude of the distributions near $\theta_{12} = 0$ is indicative that the behavior is dominated by independent-particle character. The star indicates the value of R_1.

in coordinate space in the two-electron H^- ion. Figure 3 shows the conditional probability distribution for this species, again in Cartesian representation, for several values of R_1, from a very small and improbable value through the most probable region out to a very large and very improbable value. The H^- ion does show more asymmetry in θ_{12} than do the states of He in fig. 2, but still electron 2 can appear at any value of θ_{12} and R_2, as one expects of an electron that can be moderately well described as occupying its own atomic orbital.

Section 2 reviews the current concept of correlation in doubly-excited helium, beginning with the recognition of rotor series and then supermultiplets. Section 3 describes some of the recent findings regarding the atoms of the alkaline earth

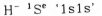

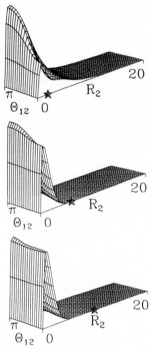

Fig. 3. The conditional probability distribution for the ground (and only bound) state of the H^- ion for three values of the "fixed" electron–nuclear distance. At the closest distance shown, the angular correlation is more marked than in the ground state of He, but if one electron is far from the nucleus, as in the lower two graphs, the other electron has a nearly spherical distribution.

elements. Section 4 peeks a bit through some of the doors opened by the new conceptions described in sections 2 and 3.

2. Doubly-excited helium atoms

In 1978, Kellman and Herrick (1978) pointed out that the observed and calculated levels of helium in which both electrons have the same quantum number (greater than 1) contain sets of levels that correspond remarkably well to rotor series, That is, in each of the known groups of levels for which $n_1 = n_2$, there is a terminating set of levels with angular momentum quantum numbers $J = 0, 1, 2, \ldots$, alternating even and odd, with energies approximately $\hbar^2 J(J+1)$ above the level of the lowest for which $J = 0$ in that group. This was followed by three much longer papers in 1980 (Herrick and Kellman 1980, Herrick, Kellman and Poliak 1980 and Kellman and Herrick 1980) in which the authors showed that the rotor series could be fit into supermultiplets corresponding to symmetry broken from that of the O(4) of the two interacting electrons. Kellman and Herrick, in the third of the 1980 papers, showed

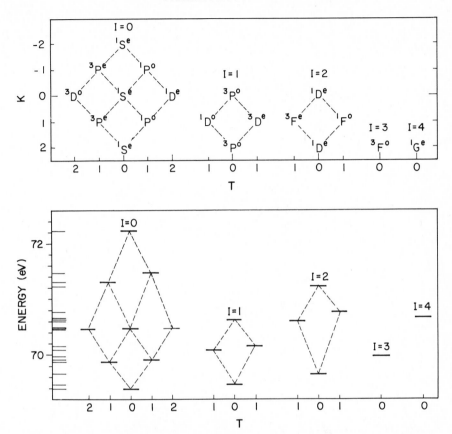

Fig. 4. The supermultiplet pattern developed by Herrick and Kellman (1980) as manifested by the states of He with $n_1 = n_2 = 3$; the upper box is the pattern imposed by the supermultiplet structure, with the spin, orbital angular momentum and parity of each state indicated. The lower box shows all the states of this manifold, on the ordinate at left in order of their energies without classification, and in the body of the figure arranged by quantum numbers into the supermultiplet pattern. In all cases here, the quantum numbers correspond to those demanded by the pattern as shown in the upper box.

that the particular supermultiplet pattern of He**, the doubly-excited helium atom, is the $O(3) \times SU(2)$ of a rotor in three dimensions and a two-dimensional harmonic oscillator. The ideal supermultiplet pattern and the corresponding states of He** for $n_1 = n_2 = 3$ are shown in fig. 4. The papers culminate with the identification of this symmetry as equivalent to that of the rotations and bending vibrations of a linear e-He-e rotor-vibrator like a linear ABA triatomic molecule. Kellman and Herrick speculated that the level pattern might well reveal symmetric and antisymmetric stretching vibrations as well. This analysis, it must be remembered, was done from phenomenological and symmetry arguments; the interpretation in terms of molecule-like collective rotation and vibration came afterward, and was not put intentionally into the model. The molecular interpretation came only when the analysis was complete.

We had of course been in fairly close touch with Herrick and Kellman as this work took shape, and it did indeed stimulate our thinking. If their assignments were

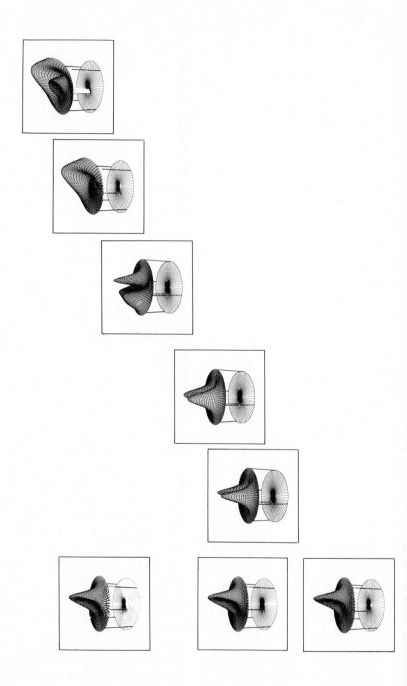

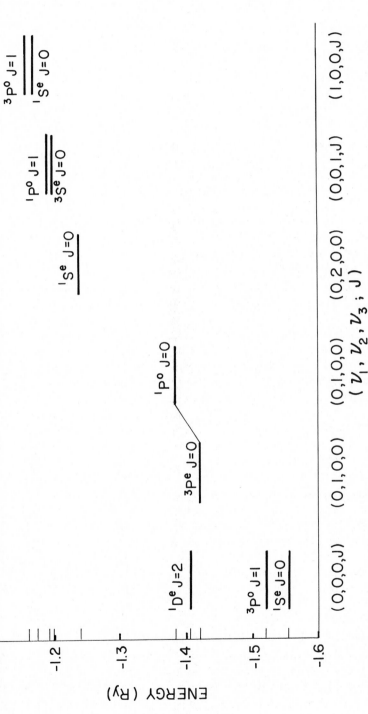

Fig. 5. (a, bottom) The levels of the helium atom for which $n_1 = n_2 = 2$ and, at the right upper corner, the four lowest levels of the manifold with $n_1 = 2$, $n_2 = 3$, arranged according to the quantum numbers ν_1, ν_2, ν_3, J of the linear three-body system, corresponding respectively to the symmetric stretch, the doubly degenerate bend, the antisymmetric stretch and the total rotational angular momentum (spin is neglected in the designations at the bottom). The spin, orbital angular momentum and parity are indicated for each level. The quantum number I of fig. 4 corresponds to J here; K of fig. 4 is $\nu_2 + 2$ of this figure, and T of fig. 4 denotes the magnitude of angular momentum along the figure axis associated with the doubly degenerate bending mode. (b, top) Cylindrical polar representations of the conditional probability distributions for the levels whose energies are shown in fig. 5a), the two highest P levels excepted. The distributions shown here exhibit very strong angular correlation, of the degree expected for strongly collective behavior; for example, the three states of the rotor series at the left differ only very slightly when viewed in the intrinsic coordinate frame this way, although they obviously differ in their overall rotational distributions in the laboratory frame. Similarly, the partners of the degenerate bending mode have very similar distributions not expected for independent-particle behavior. In this polar representation, the position of the "fixed" electron is shown by the vertical that falls from the surface of the distribution to a point within the indicator circle below. The other four "legs" indicate the positions on that circle of $\theta_{12} = 0$, $\pi/2$, π and $3\pi/2$.

correct, then we reasoned that the conditional probability distributions for the states of He** in the supermultiplets should look like the distributions for rotors and vibrators, not like those of independent-particle systems. In the intrinsic coordinates R_1, R_2, θ_{12}, the members of a rotor series should have distributions that look very much alike; the differences among them should be primarily in their behavior with respect to the Euler angles whose time variation corresponds to rotation in the laboratory frame. By the same token, the degeneracies of the ideal harmonic two-dimensional oscillator should be reflected in the distributions of the corresponding nearly-degenerate states of the supermultiplet model. And there should be states corresponding to combinations of rotational and vibrational excitation, rotational ladders built on excited vibrational states.

Yuh et al. (1981) constructed probability distributions of the electrons of the $2s^2$ $^1S^e$, $2s2p$ $^3P^o$, $2s2p$ $^1P^o$ and $2s3s$ $^3S^e$ states of He**, using a functional basis with the electron–electron distance R_{12} as one of the variables. The first two of these ought to be the first two members of a rotor series—the third should be the $2p^2$ $^1D^e$—the $^1P^o$ should be one of the two partners of the first excited level of the bending vibration, and the last should be the first excited state of the antisymmetric stretching mode. (The designations such as 2s2p refer of course to an independent-particle picture and are strictly inappropriate labels for states characterized by collective behavior. However, there is no ambiguity here in using the independent-particle labels and the observed levels are generally designated in the literature that way. Hence, we shall refer to the states in terms of the orbital labels for convenience with the understanding that the single-particle quantum numbers must not be taken literally.) Indeed, the distributions do have the expected forms. This was strong persuasive support for the collective molecular model but its validity depended on all the states of the manifold with $n_1 = n_2 = 2$ fitting the molecular pattern.

The full picture was established for the states of this lowest set of states of He** by the calculations of Ezra and Berry (1983). The energy level pattern for these states, organized according to the quantum numbers of the normal modes of vibration ν_1, ν_2 and ν_3 (symmetric stretch, bend and antisymmetric stretch, respectively) and J, the quantum number for rigid-body rotation, is shown in fig. 5a. (An additional approximate quantum number λ, designating angular momentum along the figure axis associated with the doubly degenerate bending vibration, is also useful, but is not indicated in fig. 5a.) The conditional probability distributions for the corresponding eigenstates are shown in fig. 5b, for one electron at its most probable distance from the nucleus. The three members of the rotor series—the "$2s^2$" $^1S^e$, the "$2s2p$" $^3P^o$ and the "$2p^2$" $^1D^e$—do appear very much the same in this intrinsic-coordinate picture, although in the independent particle model there is no particular reason why they should be similar. The partner states corresponding to one quantum of bending—the "$2s2p$" $^1P^o$ and the "$2p^2$" $^3P^e$—are also very much alike despite their differences in (presumed) configuration, spin and parity. The "$2p^2$" ^1S state corresponds to two quanta of bending vibration but no angular momentum. The first states above this manifold are also shown; the "$2s3s$" $^3S^e$ does correspond to the one-quantum state of the antisymmetric stretch mode and the corresponding singlet, to the one-quantum state of the symmetric stretch mode. The molecular model seems well justified, at least for some of the doubly excited states of helium.

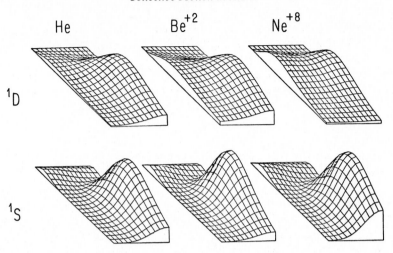

Fig. 6. Cartesian representations of the conditional probability distributions of the "$2s^2$" ^{1}S (lower row) and "$2p^2$" ^{1}D levels (upper row) of He, Be^{+2} and Ne^{+8}, respectively from left to right. The ^{1}S levels retain their strong angular correlation as the nuclear charge Z increases but the distributions for the ^{1}D levels become progressively more symmetrical about $\theta_{12} = \pi/2$, so that in the helium-like neon ion, the distribution indicates that each electron carries its own nearly constant angular momentum.

The model does have its limits. Nikitin and Ostrovsky (1976, 1978) showed that if the principal quantum numbers n_1 and n_2 are very different, then as intuition suggests, the independent-particle model becomes the more appropriate one. Furthermore, because it is the large Coulomb repulsion of the electrons *relative to their kinetic energy* that causes the extreme molecule-like collective behavior, one suspects that if the kinetic energy were made large enough, the strong collective behavior would give way to independent-particle behavior. This can be done by increasing the nuclear charge; figure 6 shows conditional probability distributions for the $2s^2$ ^{1}S and $2p^2$ ^{1}D levels of He, Be^{+2} and Ne^{+8}. The ^{1}S levels retain their highly correlated form, independent of the nuclear charge Z, but the ^{1}D levels transform from rotor-like in He to independent-particle-like, almost symmetrical about $\theta_{12} = \pi/2$, in Ne^{+8}. The explanation for the persistence of correlation in the $2s^2$ (and $2s2p$ ^{3}P as well) has been interpreted by Navaro and Freyre (1971) and much more systematically by Ho and Wulfman (1983) and by Wulfman and Levine (1984). In effect, the increase of kinetic energy in these states arising from increases in Z is compensated by the accompanying decrease in the spacing and ease of mixing of the configurational levels that must take place to generate the highly correlated rotor character. The ^{1}D levels, on the other hand, do reflect the uncompensated effect of increasing Z by becoming truly $2p^2$-like as Z increases. Hence the correlated, molecule-like picture clearly has only a limited range of validity; it is by no means a universal model for states of many-electron atoms. It might even seem presumptuous at this point to suppose that the collective molecular model is applicable to anything except doubly-excited two-electron species.

A technical point needs to be mentioned here. All the doubly-excited states of helium are indeed at energies above the first ionization limit and are therefore strictly resonances, not bound states. The method used to calculate the states described above is a variational procedure with only square-integrable functions in the basis. Hence, the calculation was equivalent to one based on a Feshbach projection in which the continuum was projected away. The coupling to the continuum can then be made, e.g. by Fano's bound state-continuum mixing procedure (Fano 1961); Rehmus and Berry (1981) used just such an approach to evaluate the lifetime of two of the ^{1}S states of He**.

3. The atoms of the alkaline-earth elements

The next natural step was to ask whether any states of other atoms exhibit any tendencies toward the kind of collective behavior found in He**. Are there any more common species, easier to study in the laboratory than the exotic, short-lived doubly excited helium, in which collective rotations and vibrations might be found? The obvious targets of this question were the quasi-two-electron atoms and ions, those with two electrons in their valence shell outside a closed core. This meant studying the alkaline-earth atoms Be, Mg, Ca, Sr and Ba, and the negative ions of the alkali atoms, Li$^-$ through Cs$^-$. To carry out systematic calculations of the full electronic structure for even a modest number of states for all of these would be a major project. Moreover, such an effort would develop far more information than is needed to answer the questions at hand. It would suffice to have the results of a model calculation for those states of interest: provided the results are well-converged and robust to small changes in the core potential, it would be enough to study the electron distributions of the two valence electrons in the effective potential due to the field of the nucleus and all the core electrons. It is only the correlation between the valence electrons that we investigate at this point, not the valence-core correlations, which are certainly far the smaller in the alkaline-earth atoms and alkali negative ions. Following this course, Krause and Berry (1985a, b) constructed two-electron wave functions for the ground states and a variety of bound excited states of all the alkaline-earth atoms, and the ground states and stable excited states or resonances of the alkali negative ions, using effective core potentials of several types, most extensively those of Bachelet et al. (1982), but also those of Weeks and Rice (1968) and of Barthelat et al. (1977). Figure 7 shows conditional probability distributions of the ground states of the alkaline earths, for four values of the "fixed" electron–nucleus distance. The distributions are a little broader in θ_{12} than those of the rotor states of He** but are clearly far more like those rotor states than like the ground state of the helium atom.

Collective behavior seems to be the rule for most of the low-lying excited states of these atoms as well. Figure 8 contains distributions for all the alkaline-earth atoms with one electron–nuclear distance at approximately its most probable value, for most of the states analogous to the He** states with $n_1 = n_2$, plus the two states of each corresponding to the "2s3s" ^{3}S and ^{1}S levels of He**. As with He**, the ^{1}S^e, ^{3}P^o and ^{1}D^e levels in the three left columns have the similar shapes expected of the members of a rotor series; the next ^{1}P and ^{3}P levels, odd and even respectively, have their maxima at values of θ_{12} much less than π and have zeros at $\theta_{12} = \pi$, as the first

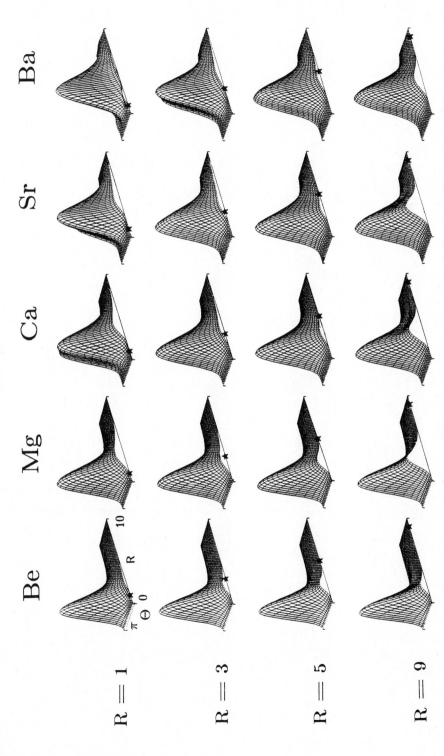

Fig. 7. Cartesian representations of the conditional probabilities of the valence electrons of the alkaline-earth elements in their ground states, for four values of the "fixed" electron–nuclear distance. While broader in the direction of θ_{12} than the corresponding distributions for the "$2s^2$" state of He, these all exhibit much more angular correlation and therefore more collective character than the distributions for the bound excited states of He or the ground state of H^-.

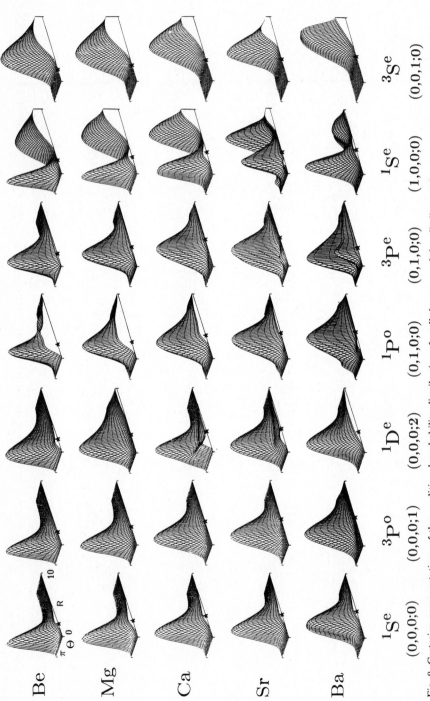

Fig. 8. Cartesian representations of the conditional probability distributions for all the states of the alkaline-earth atoms (analogous to those of He** shown in fig. 5b), the analogues of the "$2p^2$" ^1S excepted. The collective quantum numbers v_1, v_2, v_3, J are indicated below, so that the first three columns correspond to the rotor states and the next two to the first excited states of the bending mode. The ^1P levels of Be and Mg show moderately large probability densities at $\theta_{12} = 0$, and the $^3P^e$ level of Ba exhibits a distribution intermediate between the single maximum characterizing a first-excited bending-state and the symmetrical double maxima of a d^2 $^3P^e$ state which would correspond to independent-particle behavior. All these distributions are constructed with R_1 at its most probable value, indicated by a star.

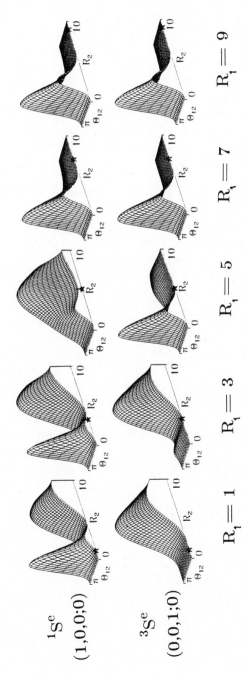

$R_1 = 1$ $R_1 = 3$ $R_1 = 5$ $R_1 = 7$ $R_1 = 9$

$1S^e$
$(1,0,0;0)$

$3S^e$
$(0,0,1;0)$

Fig. 9. Cartesian representations of the conditional probability distributions for the S states of Mg that correspond to the symmetric stretch ($^1S^e$) and the antisymmetric stretch ($^3S^e$) state, respectively, for five values of the electron–nuclear distance R_1, indicated by a star, showing how these do indeed correspond to the appropriate classical models. The symmetric stretch picture fails if R_1 is very small or very large; in the former case, its effective force is almost as if it is part of the nucleus, and in the latter case, it behaves like a Rydberg electron. Both these extreme cases correspond to situations of very low probability.

excited partner states of a doubly-degenerate bending vibration should. The $^1P^o$ levels of Be and Mg are somewhat more independent-particle-like than the "2s2p" $^1P^o$ level of He, it seems, from their moderately high probability densities at $\theta_{12} = 0$. The probability density of the lowest $^3P^e$ level of Ba does not look at all like that of a first excited state of a bending vibration; we shall return to this state shortly. The ^3S levels all have the forms expected of antisymmetric stretching vibrations. This is even more apparent in fig. 9 where the probability densities for the presumed excited stretching-mode states of Mg are shown for several values of the fixed electron–nucleus distance. The ^1S state of this pair seems a bit less of a pure stretching mode than the triplet; however, one must keep in mind that if one electron is very close to the nucleus, the other electron is subject to a nearly centrosymmetric potential, and if one electron is very far from the nucleus, it must behave as a Rydberg-like, independent particle. Hence, we should expect symmetric-stretch character for the ^1S state only for intermediate values, the most probable values of course, of the fixed electron–nucleus distance. And this is what we find from the calculations.

The energy levels of the low-lying states of the alkaline-earth atoms seem to show patterns characteristic of three-particle rotor-vibrators, with rotor series and bending-mode states, according to Kellman's (1985) analysis. The splittings of the vibrational partner-states are considerably larger than in He**, and in a few instances, states identified by Krause and Berry (1985a, b, c) as belonging to the vibrator-rotor manifold are not the same as those Kellman assigned to the same niche in the manifold. Nevertheless, the energy-level pattern is largely interpretable in terms of the collective model.

The distributions for the ground states of the alkali negative ions are very similar to those of their iso-electronic counterparts in the alkaline earths. The "np^2" $^3P^e$ states of the alkali negative ions are, with the exception of Li⁻, stable with respect to spontaneous detachment, according to the calculations of Norcross (1974). These states have energies below the energies of the corresponding excited ^2P levels of the neutral alkalis, and have no allowed decay mechanism to the ground states of the neutrals, plus a free electron. These triplets, like the "np^2" $^3P^e$ levels of the alkaline earths, have nodes at $\theta_{12} = \pi$ and maxima at angles corresponding to bent structures. The alkali negative ions have been discussed by Krause and Berry (1985c).

A few additional points need to be made to give a balanced assessment of the case for collective, molecule-like behavior in two-electron atoms. First, one of the most persuasive points is the make-up of the ^1D functions whose distributions appear in fig. 8. For Be and Mg, these are, respectively, dominated by the $2p^2$ and $3p^2$ configurations. For the heavier elements of this group, sd configurations are more important; for Sr and Ba, respectively, the dominant configurations in their lowest ^1D levels are 5s4d and 6s5d. Despite this major difference, the spatial distributions of the full variational wave functions are extremely similar. This inevitably suggests that a common characteristic responsible for that similarity is more fundamental to the behavior of the ^1D states than is the dominant configuration. The rotor character seems to serve as that common characteristic.

In some cases, notably the "np^2" $^3P^e$ and $ns(n + 1)s$ $^3S^e$ levels, approximately the same distribution emerges whether the independent-particle or collective picture is used. In both these kinds of states, the constraints imposed by particle symmetry, the exclusion principle and the spatial symmetry force the distribution to be like a first excited bending state in the case of the ^{3}P and like an antisymmetric stretch-state in the case of the ^{3}S. One can only say from these examples that they do not contradict the expectations based on the collective model. By contrast, some few states conform to neither model. The lowest ^{3}P^e level of Ba is not based on the $6p^2$ configuration, but on the $5d^2$, in an independent-particle picture or in a configuration-based variational calculation. A pure bending-vibration form for this state would give a single maximum in the probability distribution at some angle θ_{12} less than π; a pure d^2 ^{3}P^e level would give a distribution with two equal maxima at $\pi/4$ and $3\pi/4$. The distribution shown for this level in fig. 8 has two maxima of very unequal heights. This state seems to fall between the two extreme models.

All in all, while the collective molecular picture is clearly not universally applicable for the description of the states of two-electron and quasi-two-electron atoms, it seems to be a more appropriate description for many of the states of these species than is the alternative independent-particle model. The physical basis for this behavior in the quasi-two-electron systems is not very different from that in He**. In both kinds of systems, the valence electrons are kept from the region near the nucleus, so that they never have very high kinetic energy, and their spatial distributions are consequently very much affected by the electron–electron repulsion. In the alkaline-earth atoms, the ion core keeps the valence electrons away from the nucleus; in He** and the iso-electronic ions, orthogonality of the 1s orbitals to all the higher orbitals serves the same function. In both kinds of systems, we see a kind of behavior that can be simply characterized and is in no way particularly startling, yet is quite different from what we have traditionally associated with the character of two-electron atoms.

Some unanswered questions are obvious at this stage. Is it possible to quantify the extent of validity of the molecular model by projecting well-converged variational functions onto simple vibrator-rotor functions to determine the amplitude and total contribution of the basic collective states to the stationary states? Would variational expansions in series of rotor-vibrator functions converge significantly faster than series expansions in the traditional configurational, independent-particle basis? The effective force constant and moment of inertia of a rotor-vibrator model can be treated as variational parameters to maximize the overlap of a rotor-vibrator function with an accurate function; are the optimizing values of those parameters sensible, and consistent with the properties of atoms? Do the valence electrons of the atoms of the third and fourth groups of the periodic table exhibit collective, molecule-like behavior? Do all the electrons with the same principal quantum number participate on an equal footing in the correlation in the valence shell, or does the separation of s- and p-subshells emerge in the form of a separation or distinction between the probability distributions of ns- and np-subshells? If the collective, molecular picture is correct at least for the ground states of some atoms, then should we not think of these atoms as having internal geometric structure which only needs to be oriented in order to form directional chemical bonds? Traditionally, we envision directional bonds being formed by the distortions and polarizations of the atomic charge distribution due to the fields of neighboring atoms or ions. Whether there would be observable consequences associated specifically with this phenomenon remains to be investigated.

4. Independent-particle behavior in small molecules

Suppose we turn around the question "Can electrons in atoms exhibit collective rotations and vibrations like those of atoms in small molecules?" and ask instead, "Can the atoms in a small molecule such as the hydrogens in H_2O or CH_4 exhibit independent-particle-like behavior comparable to that of electrons in the ground state of the helium atom?" Could we find states in which the hydrogens of H_2O each have their own pretty good orbital angular momentum quantum numbers? Why do the hydrogens in a water molecule not have independent orbital angular momentum? The reason, crudely but accurately put, is that they bump into each other. If they could be kept out of each other's way, then instead of bending vibrations, H_2O would show hindered but rather independent rotation of the hydrogen atoms.

It has been apparent since the work of Henry [see Henry (1977) for a summary and review of the early development] that when enough quanta are put into a bond, it can happen that the quanta remain there in a *local* stretching mode. Lawton and Child (1979, 1980, 1981) and then Child and Lawton (1981) showed how a water molecule excited by four or five or more quanta in its stretching modes would be capable of exhibiting local O–H stretching; the benzene molecule exhibits local C–H stretching when it absorbs a quantum of visible radiation from a helium–neon laser, a phenomenon found by Swofford et al. (1976). Of course the effective potential for a local stretching mode is much like a Morse potential, steeply repulsive at short distances, with a deep well, and tailing out to a weakly attractive long-distance portion as the bond stretches toward dissociation. In a highly excited state in such a well, the probability distribution is much more concentrated in the long-distance parts where the kinetic energy is low than in the region over the deep well. In other words it is much more probable to find a bond stretched nearly to its classical turning point in such a state than at a distance near its equilibrium distance. It is imaginable that a water molecule, highly excited in a local stretching mode so that the two hydrogen atoms are most probably at very different internuclear distances, could also be excited in its bending mode. If the distances of the two hydrogens from the oxygen are different enough, they would not "bump into each other." Rather, they would be capable of passing so that the bending mode would have turned into an internal rotation of the inner hydrogen about the oxygen nucleus.

It is not yet known whether or not such a transformation from bending, stretching and overall rotation into independent atomic motions can take place in highly excited vibration–rotation states of H_2O. Some very simple calculations by Berry, Ezra and Natanson (1983) were based on fixing one O–H distance at its equilibrium value and the other at classical turning points for several local stretching modes. Then the wave function for the angular motion of this system was determined with two different potential surfaces then available for the H_2O mole-

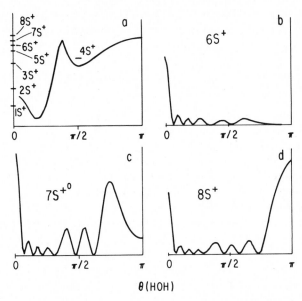

θ(HOH)

Fig. 10. The effective potential (upper left) for motion in the θ_{12} direction with one O–H distance fixed at its equilibrium value and the other at its classical outer turning point for the eighth excited local stretching mode, based on the potential surface of Sorbie and Murrell (1978), with energy levels of some of the eigenstates of the bending mode indicated on the left ordinate; and probability distributions for the highest three of those states in the upper right, lower left and lower right, in succession. The $6S^+$ has a sharp maximum probability for a linear H–H––O geometry; the $7S^+$ has a maximum at this geometry too but has, in addition, a high probability of an obtuse H–O–H with some tunnelling through the barrier around $\theta_{12} = \pi$, and the $8S^+$ state has its highest probability at the linear H–O–H configuration, with free passage over the barrier.

cule, one by Sorbie and Murrell (1978) and the other by Murrell, Carter, Mills and Guest (1981). The former potential supported independent-particle rotations below the dissociation limit; the latter did not. Figure 10 shows the effective potential for the angular motion and three of the probability densities $\rho(\theta_{12})$ for S states, one with energy well below the barrier, one also with energy below the barrier but close enough that tunneling is important, and one with energy above the barrier; these are of course based on the Sorbie–Murrell potential. More sophisticated methods are nearly ready to be applied to this problem, such as that of Natanson et al. (1984, 1986), and a better answer should be available soon. However it is more likely that the greatest bottleneck to getting an unambiguous theoretical prediction of this phenomenon is the generation of a reliable potential surface for the water molecule.

5. Concluding remarks

What can we expect to observe of these phenomena, of molecule-like atoms and independent-particle-like molecules? There are several possibilities. Very nonrigid molecules appear to have the characteristics of liquids, with well-defined diffusion coefficients based on the rate of increase with time of mean displacements; with liquid-like pair distribution functions; with spectral distributions containing modes at frequencies close to zero. Hence measurements capable of showing these properties, particularly measurements showing angular distributions, may tell us of non-rigid, independent-particle-like character in small molecules. Diffraction experiments may well give us strong persuasive indications, but diffraction results do not distinguish disordered rigid forms from fluids. We shall need to study excited molecules with other techniques such as Raman–Brillouin scattering to probe for soft modes, and perhaps examination of angular and energy correlations of fragments from Coulomb explosions of molecules previously excited to known states of interest. Mean angles will be less interesting in this context than their next moment, the mean deviations of the bond angles from their means.

To look for strong correlations of electrons, several techniques come to mind. Electron diffraction is now capable, in favorable cases, of yielding mean values of the interelectronic distance (strictly, of its inverse); with a little more resolving power, it may be possible to evaluate the mean deviation of $1/r_{12}$, which is a measure of the correlation of the two electrons. A method more tractable now but requiring some new theory for its interpretation is the measurement of angular and energy distributions of photoelectrons from atoms photoexcited to the state of interest. Still another method, in a sense the counterpart of the Coulomb-explosion experiment for molecules insofar as it is a sudden process, is an (e, 3e) double ionization with very fast incoming electrons. This method will give a snapshot of the two-electron distribution in the atom, and since the alkaline earths seem to exhibit molecule-like correlation of their valence electrons, coincidence experiments to detect the angular correlation of the two slow electrons following a fast (e, 3e) process can be carried out with vapor of Ca, Sr or Ba. Here, however, as with the Coulomb explosions, the mean deviation in the angular distributions will be much more interesting than the mean angles.

These speculations are meant to give assurance that the phenomena of correlation and independent-particle character are more than games for a computer exerciser. They are observables, and we are only beginning to conceive laboratory probes to test their extent and importance. The experiments just suggested may well be more difficult or more complex than others that readers can invent.

Finally, we may ask whether it might be possible to find a unifying way to describe few-body systems that will encompass electrons in atoms, atoms in molecules, nucleons in nuclei and perhaps other few-body systems as well. Can we find a formulation from which the collective or independent-particle character of each state will emerge naturally before, rather than after each problem is solved on an ad hoc basis? Niels Bohr was right that we must look for possibilities of more complicated motions than co-planar orbits; the search to find those motions gives us the stimulus to look for a new level of unity for our ideas concerning the dynamics of simple systems.

Acknowledgements

The research described here that was conducted by my own group at The University of Chicago was carried out by the efforts of several collaborators: Michael Kellman,

Francois Amar, Paul Rehmus, Gregory Ezra, Grigory Natanson, Julius Jellinek, Jeffrey Krause and Thomas Beck. The research was supported by the National Science Foundation.

References

Bachelet, G.B., D.R. Hamann and M. Schlüter, 1982, Phys. Rev. **B26**, 4199.
Barthelat, J.C., P. Durand and A. Serafini, 1977, Mol. Phys. **33**, 159.
Berry, R.S., G.S. Ezra and G. Natanson, 1983, in: New Horizons of Quantum Chemistry, eds P.-O. Löwdin and B. Pullman (Reidel, Dordrecht) p. 77.
Bohr, N., 1913, Phil. Mag. **26**, 1, 476, 857.
Child, M.S., and R.T. Lawton, 1981, Faraday. Disc. Chem. Soc. **71**, 1.
Coveney, P.V., and M.S. Child, 1984, J. Phys. **B17**, 319.
Einstein, A., 1917, Verh. Dtsch. Phys. Ges. **19**, 82.
Ezra, G.S., and R.S. Berry, 1983, Phys. Rev. **A28**, 1974.
Fano, U., 1961, Phys. Rev. **124**, 1866.
Fowler, A., 1912, Mon. Not. R. Astron. Soc. **73**, 62.
Henry, B.R., 1977, Accts. Chem. Res. **10**, 207.
Herrick, D.R., and M.E. Kellman, 1980, Phys. Rev. **A21**, 418.
Herrick, D.R., and O. Sinanoglu, 1975, Phys. Rev. **A11**, 97.
Herrick, D.R., M.E. Kellman and R.D. Poliak, 1980, Phys. Rev. **A22**, 1517.
Ho, Y. and C. Wulfman, 1983, Chem. Phys. Lett. **103**, 35.
Kellman, M.E., 1985, Phys. Rev. Lett. **55**, 1738.
Kellman, M.E., and D.R. Herrick, 1978, J. Phys. **B11**, L755.
Kellman, M.E., and D.R. Herrick, 1980, Phys. Rev. **A22**, 1536.
Kipling, R., 1946, The beginning of Armadillos, in: Just So Stories, reprint (Doubleday, New York).
Krause, J.L., and R.S. Berry, 1985a, Phys. Rev. **A31**, 3502.
Krause, J.L., and R.S. Berry, 1985b, J. Chem. Phys. **83**, 5153.
Krause, J.L., and R.S. Berry, 1985c, Comments At. Mol. Phys., submitted.
Lawton, R.T., and M.S. Child, 1979, Mol. Phys. **37**, 1799.
Lawton, R.T., and M.S. Child, 1980, Mol. Phys. **40**, 773.
Lawton, R.T., and M.S. Child, 1981, Mol. Phys. **44**, 709.
Leopold, J.G., and I.C. Percival, 1980, J. Phys. **B13**, 1037.
Leopold, J.G., I.C. Percival and A.S. Tworkowski, 1980, J. Phys. **B13**, 1025.
Leopold, J.G., I.C. Percival and D. Richards, 1982, J. Phys. **A15**, 805.
Munschy, G., and P. Pluvinage, 1963, Revs. Mod. Phys. **35**, 494.
Murrell, J.N., S. Carter, I.M. Mills and M.F. Guest, 1981, Mol. Phys. **42**, 605.
Natanson, G., G.S. Erza, G. Delgado-Barrio and R.S. Berry, 1984, J. Chem. Phys. **81**, 3400.
Natanson, G., G.S. Erza, G. Delgado-Barrio and R.S. Berry, 1986, J. Chem. Phys. **84**, 2035.
Navaro, O., and A. Freyre, 1971, Mol. Phys. **20**, 861.
Nielsen, J.R. (ed.), 1976, N. Bohr, address to the German Physical Society, 27 April, 1920, in: Niels Bohr, Collected Works, Vol. 3 (North-Holland, Amsterdam) p. 241.
Nikitin, S.I., and V.N. Ostrovsky, 1976, J. Phys. **B9**, 3141.
Nikitin, S.I., and V.N. Ostrovsky, 1978, J. Phys. **B11**, 1681.
Norcross, D.W., 1974, Phys. Rev. Lett. **32**, 192.
Percival, I.C., 1977, Adv. Chem. Phys. **36**, 1.
Pickering, E.C., 1896, Astrophys. J. **4**, 369.
Pickering, E.C., 1897, Astrophys J. **5**, 92.
Pickering, E.C., 1898, Astrophys J. **8**, 119.
Pickering, E.C., 1901, Astrophys J. **13**, 230.
Rehmus, P., and R.S. Berry, 1979, Chem. Phys. **38**, 257.
Rehmus, P., and R.S. Berry, 1981, Phys. Rev. **A23**, 416.
Rehmus, P., M.E. Kellman and R.S. Berry, 1978a, Chem. Phys. **31**, 239.

Rehmus, P., C.C.J. Roothaan and R.S. Berry, 1978b, Chem. Phys. Lett. **58**, 321.
Shim, I., and J.P. Dahl, 1978, Theor. Chim. Acta **48**, 165.
Sinanoglu, O., and D.R. Herrick, 1975, J. Chem. Phys. **62**, 886.
Sorbie, K.S. and J.N. Murrel, 1978, Mol. Phys. **29**, 1387.
Swofford, R.L., M.E. Long and A.C. Albrecht, 1976, J. Chem. Phys. **65**, 179.
Weeks, J.D., and S.A. Rice, 1968, J. Chem. Phys. **49**, 22741.
Wulfman, C., 1973, Phys. Lett. **23**, 370.
Wulfman, C., and R.D. Levine, 1984, Phys. Rev. Lett. **53**, 238.
Yuh, H.-J., G.S. Ezra, P. Rehmus and R.S. Berry, 1981, Phys. Rev. Lett. **47**, 497.

Discussion, session chairman T.A. Bak

Broglia: Can one view the results you have shown as being the transition densities associated with some of the elementary modes of excitation of the few-electron systems under study?

Berry: Not the pictures I have shown here; they are all truly reduced densities of either stationary states for the alkaline earths and the ground and singly-excited states of He, or projected quasi-stationary parts of the scattering resonances of doubly excited helium. However Paul Rehmus and I have examined a few transition densities for bound-free auto-ionizing transitions in doubly-excited He in order to find what parts of configuration-space and what operators are most important for generating the transitions amplitude. This was an extremely productive line, and we are planning to study transition densities systematically for the atoms and states I have been discussing here.

Temmer: Has the time not come to do experiments which, in nuclear physics, are called "pickup" and "stripping" reactions to highlight the single-particle aspects of these molecules? I mean reactions of the type $H^- + X \rightarrow H^0 + X^-$, where X is the atom or molecule whose excited "single-particle" states we wish to characterize? The negative hydrogen ion is merely an example of the type of ion one can use. The spectroscopy is done on the outgoing H^0 with high resolution which is now technically possible.

Berry: Charge-transfer processes have, of course, been studied for a long time in atomic collision experiments, but most of these used positive ions on neutrals. The analogue of your suggestion would be, for example, Be (instead of H^-) + $He^+ \rightarrow$ $Be^+ + He$. I cannot recall whether Be has been used for such experiments but I believe Ca has been. The (negative ion + neutral) electron-transfer reactions are more difficult because some collisions $A^- + B \rightarrow A + B^-$ are dominated by detachment processes, $A^- + B \rightarrow A + B + e$. However, one could study $F^- + Cl$, or $H^- + Na$, or $Li^- + Na$, to pursue the process you suggest; one might even be able to make the predicted $^3P^e$ state of Na^- by colliding excited Na(3p) with Cs^- or with Cs in a high Rydberg state.

The Lesson of Quantum Theory, edited by J. de Boer, E. Dal and O. Ulfbeck

The Origin and Evolution of Large-Scale Structure in the Universe

Wallace L.W. Sargent

Palomar Observatory, California Institute of Technology
Pasadena, California, USA

Contents

1. Introduction

Observational cosmology today is concerned with several questions. Perhaps the most exciting developments are the theoretical attempts to link physics of the early

Universe to grand unification theories of the fundamental forces of nature. A particularly influential recent development of this kind has been Guth's (1981) proposal that the very early Universe experienced a phase of exponential expansion, picturesquely called "inflation" by its author. Guth's idea provides an explanation of why the Universe is so homogeneous on large scales despite the fact that different regions did not appear to have been in causal contact (the "horizon" problem). It also explains at the same time why the Universe is geometrically "flat". Finally, the "inflationary Universe" leads to a prediction that the spectrum of initial density perturbations in the Universe, generated ultimately by vacuum fluctuations, should have equal power on all scales—the "Harrison–Zeldovich spectrum" (Guth and Pi 1982). These developments have justifiably exited considerable optimism that the way is open to understanding the physical history of the Universe from the Planck time (10^{-44} s) to the indefinite future.

However, we should recall Landau's remark that "cosmologists are often wrong but never in doubt" and remember that this dictum applies mainly to theoreticians. Thus we recognize that ideas like the "inflationary Universe" which points the way to the solution of several fundamental problems are ultimately of little value unless they are subject to observational or experimental test. Therefore, I shall focus attention on a particularly interesting general problem which is susceptible to observational attack using a variety of techniques and approaches. This is the problem of the origin and evolution of large-scale structure in the Universe. In my view this general topic illustrates the connection between the microscopic and macroscopic worlds and the connection between theory and observation in ways that are both beautiful and unexpected.

The study of the early Universe suddenly gained respectability as a serious science in 1965 with the surprising discovery of the microwave background radiation by Penzias and Wilson (1965). It is true that considerable theoretical work on a hot, compressed phase of the Universe had already been done by a few intrepid pioneers —in particular by Gamow and his collaborators who had predicted the existence of the background radiation (Alpher and Herman 1948, 1949). However, the discovery of the 2.7 K background showed directly that the Universe had evolved from an early, hot phase and, as Gamow had inferred on theoretical grounds, it had been initially radiation dominated in the sense that most of the mass-energy lay in the radiation and not in the matter. (No other plausible source for such a high photon density and no plausible mechanism for thermalizing the radiation from such hypothetical sources have been suggested.) Also, the background gave a value for the Universal photon/baryon ratio which is required for primordial nucleosynthesis calculations. This discovery was made three years after Bohr died; it was certainly the most important discovery in cosmology in the second half of the century and one of the most important in the whole of science because at one stroke it transformed the study of the nature and evolution of the early Universe into an observational science.

Historians used to speak of the "Dark Ages"—the Early Medieval period which lasted from the fall of the Roman Empire until the ninth century. The period was characterized by a retreat from urban life and was a time of relative chaos in European affairs after the order imposed by the Romans. In a similar vein there is at present a long dark age in our understanding of how the Universe evolved. It lasts from roughly 1 million years after the "Big Bang" when the microwave background-radiation was emitted until 2 billion or so years later when the first quasars formed. Like the terrestrial "Dark Ages", this period was also one of chaos —although its origin is still a mystery. More importantly it was the period in which the chaos developed into the large-scale structure which is so evident today. This epoch in the development of the Universe is literally dark as far as we are concerned because the gas in the Universe was too cool to shine and, according to current wisdom, no stars or galaxies with their nuclear sources of energy had yet formed.

Partly as a result of the existence of the Dark Ages, there are serious problems in understanding how structure formed in the Universe at all. The problem is that the theory of infinitesimal perturbations, worked out by Jeans early in the century, and which is successful in explaining how stars form from the static interstellar medium, runs into difficulties in the expanding Universe; the perturbations at the time of recombination (redshift $z = 1000$) are predicted on the most naive application of the theory to be larger than is consistent with the limits on angular fluctuations in the microwave background-radiation.

Fortunately, the problem is susceptible to observational attack. I shall describe several lines of work which are aimed at tracing the evolution of clustering in the Universe as far back as possible and at searching for the first objects to appear. At the other extreme, studies of the microwave background and of the relative abundances of the primordial elements are providing information about the physical state of the Universe as far back as a few seconds after the Big Bang.

2. The distribution of galaxies and quasars

2.1. Galaxies

The existence of galaxies outside our own Milky Way system was only established in 1925 when Hubble discovered Cepheid variable stars in the Andromeda nebula. Four years later Hubble discovered his redshift–distance relation which, following the earlier theoretical work by first Friedmann and then de Sitter, was interpreted in terms of an expanding Universe. The existence of clusters of nebulae had already been noticed as early as 1903 by Max Wolf; it was therefore immediately clear as soon as extragalactic astronomy was born that the galaxies were clustered, at least on scales of 1 Mpc or so. Later, it became evident that clustering extended the way from loose groups of galaxies such as our own Local Group up to compact clusters of hundreds or thousands of bright galaxies such as the nearby Virgo, Coma Berenices and Perseus clusters. The existence of clustering on still larger scales was a controversial subject for many years. Abell produced his catalogue of 2500 clusters of galaxies in 1957 using the plates of the first Palomar Sky Survey. The Abell clusters have a typical redshift of 0.15 (distance 1000 Mpc) and are exceptionally rich in galaxies—the Virgo cluster would not qualify for inclusion. Abell (1958) analyzed the distribution of his clusters on the sky and concluded that there was a probability of only 1 part in 10^{60} that they were not clustered. Abell also identified several examples of putative "superclusters." Nevertheless, a more sophisticated analysis of the same data by Yu and Peebles (1969) reached the conclusion that there was no significant clustering of the clusters. De Vaucouleurs (1956) studied the distribution of nearby, bright galaxies on the sky and concluded that the Local Group lies on the outskirts of an elongated, flattened structure called the Local supercluster or the Virgo supercluster, roughly centered on the Virgo cluster. Astronomers were slow to accept the importance or even the existence of the Local supercluster, but in recent years evidence has accumulated that superclusters are basic structures in the Universe. A recent review of their properties has been given by Oort (1983), who has played a particularly influential role in persuading astronomers that superclusters are important. The exploration of the Universe in three dimensions has also been facilitated by the very large redshift surveys carried out in the last few years. (As a measure of the rapid progress in this area, we recall that in 1956 Humason, Mayall and Sandage assembled a catalogue of redshifts measured up to that time; it contains 580 entries. A catalogue which is maintained by J. Huchra at the Center for Astrophysics currently contains redshifts for 7500 galaxies.) It is now possible to construct three-dimensional diagrams of the spatial distribution of nearby galaxies as inferred from their redshifts and their positions on the sky. The Local supercluster appears clearly in such diagrams with a density of galaxies which is about twice the mean smoothed-out density and, in general, the picture is one of elongated filaments containing the large clusters and enclosing apparently empty volumes. A similar picture emerges from deeper redshift surveys which have been conducted over restricted areas of the sky. There are almost no galaxies over a substantial range in redshift: the large-scale distribution of galaxies is once more revealed as taking the form of elongated structures separated by immense "voids". Note that the redshift gives a reliable indication of the distance of a galaxy if the peculiar motion is small. Galaxies in large clusters have velocity dispersions of about 1000 km s^{-1}. This produces a distortion of the inferred distribution along the line of sight, the so-called "fingers of God" effect, which distorts the true galaxy distribution. As a corrolary, the fact that the voids remain in such diagrams shows directly that the peculiar motions in the general field of a supercluster are small. There is now a growing realization that large voids in which at least intrinsically bright galaxies are sparse, are a basic feature of the distribution of galaxies. A large void discovered recently in the constellation Bootes by Kirshner et al. (1981) is about 100 Mpc in diameter.

Several years ago, Peebles and Hauser (1974) generated a remarkable map of the distribution on the plane of the sky of fainter and more distant galaxies from the Shane–Wirtanen counts. The Shane–Wirtanen catalogue goes out to a redshift of about 0.2 or to a distance of 500 Mpc. About 1 million galaxies were counted manually to produce this map; it will be some time before redshifts have been measured for all of them in order to obtain a deep two-dimensional map of the galaxy distribution over the whole northern sky! Nevertheless, even on the two-dimensional map, one gets the impression that the large-scale galaxy distribution has a cellular character.

Clusters of galaxies have recently been detected out to a redshift of about $z = 1$ and a few individual radio galaxies have been identified with redshift up to $z = 2$. A single galaxy with $z = 3.15$ has been found close to a quasar with the same redshift. Nothing is known about the clustering of galaxies at distances beyond a few hundred Mpc. Moreover, there is no direct evidence as to how galaxy clustering evolves in cosmic time.

2.2. Quasars

A typical quasar is about a factor of 10 brighter intrinsically than the brightest galaxies observed at the current epoch; it is likely that galaxies were brighter in the past, but by what factor is unknown. Thus at present most of our knowledge of the most distant parts of the Universe back to the end of the Dark Ages comes from studies of quasars. The relatively nearby quasars all appear to be in the centers of galaxies, probably mostly in galaxies of the spiral type. Thus, as has long been suspected, quasars appear to be scaled-up versions of the "Seyfert nuclei" which locally are found in a few percent of spiral galaxies. The highest quasar redshift found so far is $z = 3.78$ for the radio object PKS 2000-330 (Peterson et al. 1982). There is some doubt about whether quasars exist with much higher redshifts. A barrier (the "edge of the Universe") was suspected to exist at a redshift of $z = 2$ several years ago and then suddenly after a number of years of stasis, objects were quickly found with redshifts of $z = 3.44$ and 3.53, respectively. Hazard and the present author have made an intensive study of the redshift distribution of quasars in a particular field and have found that the number of quasars per unit redshift range begins to fall off at $z = 2$. Extrapolating our numbers beyond $z = 3.5$, we find that at $z = 4$ one would only expect one object per 30 square degrees on the sky. Although this means that there should be 1000 quasars with redshifts beyond $z = 4$ in the whole easily accessible sky, they would be hard to find and could well have escaped detection in the surveys carried out so far. It is of course very important to know if quasars suddenly switched on at a redshift of around $z = 4$ because this epoch, whenever it was, was likely to have been the epoch of galaxy formation as well.

Schmidt and others (see Schmidt and Green 1982) have shown that the co-moving space density of quasars has declined rapidly since the epoch corresponding to a redshift of $z = 2$—roughly by a factor of 1000. Moreover, it appears that the decline in number to the present day has been chiefly manifested by the intrinsically bright objects. Similar cosmological evolution is shown by the extragalactic radio sources (galaxies and quasars). It is interesting that the evolution in the number of these objects has occurred on the same time scale as the expansion of the Universe and not on a much longer or shorter time scale.

Quasars are not observed to be clustered; however, they are such rare objects even at high redshifts that this observation gives no interesting limit on the clustering of the associated galaxies at early times.

2.3. Clustering at large redshifts and the Lyman-alpha clouds

Our main quantitative information concerning the clustering of galaxies comes from the two-point correlation function of galaxies. The correlation function is defined as follows: Given a galaxy G1 we investigate the probability dP that there is a second galaxy G2 in a volume dV, distance r from G1. It is found empirically from studies of the distribution of nearby galaxies that

$$dP = dV\, N_0 [1 + \xi(r)], \tag{1}$$

where N_0 is the mean number density of galaxies per unit volume and the function $\xi(r)$ represents the enhanced effects of clustering. (For randomly scattered galaxies we would have $dP = N_0\, dV$.) Peebles found that the function $\xi(r)$ is a power law of the form (see Peebles 1980)

$$\xi(r) = \left(\frac{r}{r_{\mathrm{c}}}\right)^{-1.77}, \tag{2}$$

where

$$r_{\mathrm{c}} = 10\left(\frac{50}{H_0}\right) \text{Mpc}. \tag{3}$$

Since the two-point correlation function depends only on r by hypothesis, it gives no information on the shapes of galaxy clumpings. Moreover, many different kinds of distributions are consistent with such a simple one-parameter description. Nevertheless, the two-point correlation function serves as a crude measure of clustering in the Universe which we can attempt to study at earlier epochs.

A particular interesting approach to the study of clustering at earlier times and large distances is afforded by the sharp absorption lines observed in the spectra of quasars. These lines can be divided into two distinct types (Sargent et al. 1980). Those redshifts containing lines of heavy elements as well as the Lyman series lines of H are believed to originate in the interstellar gas in galaxies distributed along the line of sight to the quasar. These "heavy element" redshifts are not common, having a density $dN(z)/dz$ of about 1 per unit redshift range ($\Delta z = 1$) at a redshift of $z \sim 2$. Such a rate of interceptions is compatible with the known space density of galaxies and the inferred sizes of their gaseous halos. A second category of quasar absorption lines contains only the Lyman lines of hydrogen with no detectable lines of heavier elements. These redshifts are about 50 times more common than the "heavy element" redshifts and produce a "forest" of Lyman-alpha absorption lines which extends from the Lyman-alpha emission line (i.e. from the redshift of the quasar itself) down to shorter wavelengths. Unlike the "heavy element" redshifts, the lines of the Lyman-alpha forest exhibit a strong redshift dependence—greater than that expected for objects of fixed cross section and with a fixed co-moving density which move apart as the Universe expands. Moreover, the lines of the "forest" are found to be completely randomly distributed on smaller scales (from 30–$30\,000$ km s^{-1} or from 0 to 0.05 Hubble radii) while the "heavy element" lines show considerable clustering on small scales (< 200 km s^{-1}). Primarily, because of their lack of heavy elements it is currently supposed that the Lyman-alpha forest lines are produced by primordial intergalactic clouds. They appear to be confined by the pressure of an external medium rather than by the self-gravity and can be used to obtain important limits on the physical state of the intergalactic medium at large redshifts (see section 4).

For cosmologically distributed intervening objects, a clumping in redshift can be directly related to clumping in space even at large redshifts where the objects responsible for the absorption lines cannot be observed. The Lyman-alpha forest lines are so plentiful that they can be used to place a very fine limit on the value of r_c in the correlation function $\xi(r)$. The best limit so far obtained is $\xi \leq 0.2$ Mpc at $\langle z \rangle = 2.44$ (Sargent et al. 1982). It is not known how galaxy clustering evolves in time; however, on the hierarchial clustering model developed by Peebles and his associates (Peebles 1980), the correlation function retains the form $\xi(r) \sim r^{-1.77}$ provided the clustering is non-linear ($\delta\rho/\rho > 1$). If this is the case it is easy to show that r_c changes with redshift according to the relation

$$r_c(z) = r_c(z = 0)(1 + z)^{-5/3}. \tag{4}$$

Thus, on this simple hypothesis, it is expected that r_c should be 1.07 Mpc at $z = 2.44$. Therefore, the observed clustering of the Lyman alpha clouds at $z = 2.44$ is much weaker than expected. There are several possible explanations of this observation. One is that the Lyman-alpha clouds are confined to the voids in the galaxy distribution and so cluster much less strongly than galaxies at all epochs. Another is that the galaxy clustering developed late and that the Lyman-alpha clouds share the distribution of galaxies. Future observations from the Space Telescope of Lyman-alpha clouds at low redshifts will solve the problem of their distribution. In the meantime it is noteworthy that the most distant objects that we can study show no signs of clustering.

3. The cosmological density parameter Ω

3.1. Definitions

In addition to observations of the galaxy distribution, information on the overall distribution of mass in the Universe can be obtained from studies of dynamics of aggregates of galaxies on all scales on which they are clustered. The Friedmann equations lead to the notion of a critical mean smoothed out density in the Universe

$$\rho_c = \frac{3H_0^2}{8\pi G}, \tag{5}$$

where H_0 is the present value of the Hubble expansion constant. The value of H_0 is about 50 km s^{-1}/Mpc (Sandage and Tammann 1984); its value is in dispute in the range $40 < H_0 < 90$ km s^{-1}/Mpc (Rowan-Robinson 1985). It is usual to define $\Omega_0 = \rho_0/\rho_c$ where ρ_0 is the actual density of the Universe at the present epoch. As is well known, if $\Omega_0 < 1$ the Universe is open and will expand forever, if $\Omega_0 > 1$ it will eventually reverse its expansion and experience the "Big Crunch." A Universe with

$\Omega_0 = 1$ is geometrically flat and is the value which appears as the natural outcome of the "inflationary" scenario (Guth 1981) and its subsequent modifications (Albrecht and Steinhardt 1982) which attractively explain the Universe's present day isotropy and its close approach to flatness.

In general, Ω changes with time as the Universe expands. In a model with zero cosmological constant

$$\Omega(z) = \frac{\Omega_0(1+z)}{1 + \Omega_0(z)}.$$
(6)

Thus for $z \gg 1$

$$\Omega - 1 = \frac{1}{z}\left(1 - \frac{1}{\Omega_0}\right).$$
(7)

As many authors have remarked, it is strange that Ω_0 is so close to unity at the present epoch; it must have been very close to unity in the past.

Current estimates of Ω_0 lead to the result that the Universe contains a preponderance of "dark matter"; considerations which we shall outline later indicate that this must be primarily in the form of non-baryonic matter.

3.2. Mass-to-light ratios

A summary of our present knowledge of Ω_0 is as follows. Astronomers find it convenient to discuss masses and mass densities in terms of the mass-to-light ratio, M/L, in solar units. Dynamical studies of the central parts (say the inner 10 kpc) of galaxies, both spiral and elliptical, lead to values $M/L = 6$. This appears as reasonable, because stars similar in type to the sun (with $M/L \sim 1$) appear to dominate the spectrum of most galaxies.

In the solar neighborhood studies of the motions of stars lead to $M/L = 3$, although only about 50 percent of the mass can be accounted for in visible stars (Oort 1960).

It has been shown in the last few years that almost invariably spiral galaxies exhibit flat rotation curves in which the rotational velocity $V_r \sim r$ out to as far as it can be observed (approaching 100 kpc radius in some cases). Such a flat rotation curve leads to an inferred mass distribution in which the mass inside radius r, $M(<r) \sim r$, i.e., the total mass diverges logarithmically. On the other hand, the projected luminosities of spiral galaxies are observed to fall off exponentially,

$$L(r) \sim e^{-\alpha r},$$
(8)

with a scale length $1/\alpha$ which is typically a few kpc. Thus, the inferred M/L ratios of spiral galaxies rise with increasing radius and approach values $M/L \sim 100$ in their extreme outer parts (100 kpc scale). Accordingly, spiral galaxies are now considered to be luminous condensations in extended "dark halos" of non-luminous matter.

The dynamical studies of the motions of galaxies in small groups of a few galaxies and in large clusters of hundreds of objects also lead to mass-to-light ratios in the range $M/L = 100$ to 300. This is on scales of a few hundred kpc to a few Mpc.

Recently studies have been made of the local anisotropy of the local "Hubble-flow" introduced by our relative proximity (15 Mpc) to the Virgo cluster of galaxies: this also leads to a value of $M/L \sim 300$ on scales of tens of Mpc.

3.3. Dark matter

The mean luminosity density in the local vicinity is $L = 8 \times 10^7 \, L_\odot \, \mathrm{Mpc}^{-3}$. This leads to a relationship between Ω_0 and M/L of the form

$$\frac{M}{L} = 840 \; \Omega_0 \frac{H_0}{50} \,. \tag{9}$$

Accordingly, if $M/L = 300$ locally, then $\Omega_0 \sim 0.3$. However, we have seen that the M/L ratios of the luminous parts of galaxies are small, $M/L \sim 6$. Thus the visible matter in the Universe has $\Omega_0 \sim 0.01$ and therefore contributes only a small fraction of the total mass density. Whatever its form, most of the mass in the Universe is dark. It will be observed from the preceding discussion that M/L and Ω_0 appear to increase with the scale on which they are measured from a few kpc to tens of Mpc.

In general, dynamical analyses cannot distinguish mass distributions which are smooth on a scale larger than the scale of the test particles being used for the mass determination. Accordingly, attempts are now under way to determine M/L and Ω_0 on scales as large as 100 Mpc. A powerful observation in this regard is that the direction indicated by the dipole anisotropy in the microwave background radiation is now well established to correspond to a velocity of $V = 600 \pm 50 \, \mathrm{km \; s^{-1}}$ in a direction 45° away from the nearest large mass concentration, the Virgo cluster. Part of this velocity is due to our "infall" into the Virgo cluster at 200–300 km s^{-1}: this leads to a value of $\Omega_0 = 0.2$. It now appears that the Local Group of galaxies and the entire Virgo cluster are in turn moving towards a more distant large concentration of galaxies in Hydra-Centaurus at 400 km s^{-1}. If confirmed this would lead to a very large value of $\Omega_0 \sim 0.5$ on a scale of 100 Mpc. The details of these large scale motions are currently being investigated via mapping of the local velocity field through large redshift surveys.

3.4. Primordial nucleosynthesis

The discovery of the microwave background radiation made it possible to begin accurate calculations of conditions in the Universe; a particularly fruitful outcome has been detailed calculations of primordial nucleosynthesis yields as a function of the assumed baryon density. The isotopes D, He^3, He^4, and Li^7 are all synthesized in the first 1000 seconds of the expansion of the Universe. The resulting abundances of these isotopes are sensitive in varying degrees to the mass density in the form of baryonic matter Ω_b^0. For example, the higher values of Ω_b^0 result in more He^4 and less D. A recent critical review of the observed abundances led to values of Ω_b^0 in the range 0.1 to 0.14 (Steigman and Boesgaard 1986). There are considerable difficulties in measuring the abundances of all of these elements and in extrapolating back to the "primordial" values before nucleosynthesis by stars became important. Nevertheless, the present abundance measurements (all of which have difficulties) are consistent with the same value of $\Omega_b^0 \sim 0.1$.

3.5. Summary

Dynamical studies of the motions of galaxies on supercluster scales (~ 30 Mpc) indicate that the cosmological density parameter has a value $\Omega_0 \sim 0.3$ at the present epoch. Visible matter only contributes $\Omega_0 \sim 0.01$. The primordial nucleosynthesis studies show that the baryonic contribution to Ω_0 cannot exceed 0.15. Thus, if $\Omega_0 = 1$ as is demanded by the "inflationary" Universe scenario (Guth 1981), then it cannot be in the form of baryons. Moreover, if the dark matter is present in sufficient amount to close the Universe it must be distributed on scales larger than that of superclusters.

4. The intergalactic medium

The Lyman-alpha forest of absorption lines in the spectra of quasars has been interpreted as being due to intergalactic clouds of very low heavy-element content. We introduced these clouds in section 2.3 where we also discussed their clustering properties. The Lyman-alpha clouds appear to be of galactic dimensions; a beautiful observation of the similarity in Lyman-alpha absorption lines in the spectra of two images of the gravitationally lensed quasar Q2345 + 007 shows that they have a diameter of about 10 kpc (Foltz et al. 1984). Theoretical studies show that the clouds are not supported by their own self-gravitation, but instead must be confined by a general intergalactic medium whose properties can be estimated from the inferred physical state of the Lyman-alpha coulds themselves (Sargent et al. 1980, Ostriker and Ikeuchi 1983). In summary, the condition that the clouds are ionized by the metagalactic quasar flux and the requirement that they are not ablated on a short time scale by the surrounding hot medium leads to the conclusion that the clouds themselves have masses of about 10^8 $M_\odot$ (of the same order as dwarf galaxies) and are highly ionized with an ionization fraction for hydrogen N(H-II)$/n$(H-I)$ = 10^5$. The clouds have a temperature of $T_c = 3 \times 10^4$ K and a density of $n_c = 10^{-4}$ electrons and ions per cm^3. They only contribute a negligible fraction to the total mass density of the Universe—$\Omega_c \sim 10^{-3}$. The general intergalactic medium required to confine the clouds has a density $n_c^M = 10^{-5}$ cm^{-3} and a temperature of $T_M = 3 \times 10^5$ K at a redshift $z = 2.4$. Since the Universe is expanding, this general intergalactic gas must be cooling and becoming more tenuous with cosmic time. It can easily be estimated that at the current epoch $T_M^0 \sim 10^3$ K and $n_M^0 \sim 2 \times 10^{-7}$ cm^{-3}. The intergalactic gas contributes a larger fraction of the mass density of the Universe than the Lyman-alpha clouds, namely $\Omega^M \sim 0.1$ for $H_0 = 50$ km s^{-1}/Mpc

—comparable to the visible galaxies. (Note that the densities involved are so low that the intergalactic gas stays highly ionized as it cools because the recombination time is larger than the Hubble time.) At the low temperatures and densities quoted, it will be very difficult to make direct observations of the intergalactic gas. It is noteworthy, however, that if the inferred properties of the Lyman-alpha clouds are anywhere near the true situation, these objects could not co-exist with a hot, dense intergalactic medium ($T^M \sim 10^8$ K; $\Omega^M \sim 1$) which is required to explain the diffuse X-ray background.

The Lyman-alpha clouds thus enable us to establish that there cannot be a large baryonic contribution to the cosmological density parameter in the form of hot intergalactic gas. Also, the process of galaxy formation must have been relatively efficient, with ~ 50 percent of the initial gas being converted into galaxies.

5. The microwave background radiation

The discovery of the 2.7 K microwave background radiation has had an enormous impact on cosmology since it enables us to make direct observations of the state of the Universe at a redshift of $z = 1000$ and at a cosmic time only 10^5 years after the Big Bang.

The radiation peaks at a wavelength of about 1 mm; however, it is most easily studied from the ground at wavelengths of 1 cm and above, because of variable atmospheric transmission problems at shorter wavelengths. A wavelength of about 1 cm is ideal because the Galactic background radiation is falling as $\nu^{-0.7}$ while the microwave background radiation spectrum is rising as ν^2 in this wavelength region. Thus, particularly for studies of the angular fluctuations in the background, a wavelength of about 1 cm is optimal.

The scales of interest for studies of early galaxy formation are obtained from the following considerations. A scale of 1 min corresponds to $\sim 10^{11}$ $M_\odot$, the size of a galaxy. A scale of 10 min corresponds to $\sim 10^{14}$ $M_\odot$, about the size of a cluster of galaxies. Finally, 1° corresponds to $\sim 5 \times 10^{16}$ $M_\odot$, the scale of a supercluster. A radio telescope with a diameter of 40 m has a beam with a FWHM size of 1 min at 1 cm. Thus, such a telescope is ideal for studies of angular fluctuations on scales of a few minutes of arc which correspond to the scales of clusters of galaxies.

The only anisotropies in the microwave background radiation which have been discovered so far are the "dipole" anisotropy due to the Solar System's peculiar motion in the Universe and the smaller scale anisotropy seen in the direction of clusters of galaxies due to the Sunyaev–Zel'dovich effect. (In this last effect the hot electrons in the intergalactic gas in a cluster of galaxies scatter the microwave background radiation to high frequencies, resulting in a diminution or "cooling" of the radiation in the Rayleigh–Jeans part of the spectrum.) The dipole anisotropy, $\Delta T/T \sim 2 \times 10^{-3}$, implies that the Local Group of galaxies is moving at a velocity of ± 600 km s^{-1} in a direction some 45° away from the Virgo cluster: the implications of this result for Ω were discussed in section 3.4.

On smaller scales, in directions away from clusters of galaxies, increasingly sensitive measurements have so far not succeeded in detecting any anisotropy. Such measurements present a difficult technical challenge. Even at 1 cm, variable absorption by atmospheric water vapor is a problem. Also the side lobes of the antenna see the surrounding ground which is a temperature of 300 K. Hence, the observations must be made at a cold, dry site and the antenna moved as little as possible to eliminate variable ground spill-over. A particular useful technique is to point an altazimuth mounted telescope at the N pole and to alternate it between two directions in the sky close to the pole by moving it only in azimuth. At the same time two feeds are Dicke-switched. Such measurements have been reported by Uson and Wilkinson (1984) and are currently being carried out by Readhead, Sargent, and Moffet (1986). In typical measurements, the Dicke switching is done between two points 5–7 min apart in azimuth, while an annulus 1.5 min wide is traced out by the rotation of the earth in a 24-hour period. The best limits are $\Delta T/T < 3.8 \times 10^{-5}$ on a scale of 7.1 min (corresponding to about 10^{14} $M_\odot$ at $z = 1000$.) It is hoped that eventually the limits obtained by this technique can be lowered by a factor of 10 on the ground. In the meantime, the present limits already pose serious problems as shall be seen in section 6.3.

6. Theory of fluctuations

6.1. Adiabatic and isothermal fluctuations

It is supposed that the present clumpiness in the distribution of galaxies is the result of the evolution of primordial fluctuations, since the Universe is known to have been in a gaseous form at the time when matter and radiation decoupled at $z = 1000$. For several years it has been hoped that a theory of galaxy formation would be devised in which infinitesimal primordial perturbations in the density of matter would give rise eventually to galaxies. In fact, this approach leads to conflicts with the observations.

Two types of primordial fluctuations have been considered. In "adiabatic fluctuations" both matter and radiation are clumped. In "isothermal" fluctuations only matter is clumped. Adiabatic fluctuations are thought to be more natural because it is expected that a fundamental theory of elementary particles would lead to the prediction of a specific ratio $n_{\text{photon}}/n_{\text{baryon}}$ which would be the same everywhere at some early stage of the Universe.

The first-order question in the theory of galaxy formation is: which came first—superclusters or galaxies? On the hierarchial clustering scenario which has been explored particularly by Peebles and his associates, galaxies, or even globular clusters, form first and are subsequently aggregated by gravity into clusters and superclusters. This kind of picture would result from isothermal initial perturbations.

The other extreme view is that supercluster-sized objects (called "pancakes" by Zel'dovich) fragmented to form galaxies. This scenario would result from adiabatic perturbations.

The spectrum of the initial fluctuations is important in determining the range of masses which are eventually produced. The Fourier spectrum of density fluctuations

$\delta = \delta\rho/\rho$ as a function of wave number κ is assumed to be of the power-law form

$$|\delta_\kappa|^2 = \kappa^n. \tag{10}$$

This translates into a mass spectrum

$$S = \kappa M^{-\alpha}, \tag{11}$$

where κ is a constant whose value is $\kappa < 10^{-4}$ (from the absences of a quadrupole to the anisotropy of the microwave background) and where $n = 6\alpha + 1$.

6.2. Growth of fluctuations

The value $\alpha = 0$, or $n = 1$ is known as the Harrison–Zel'dovich or "constant curvature" spectrum. This particular fluctuation spectrum is predicted to emerge from quantum vacuum fluctuations in the "inflationary" scenario. In a matter dominated Universe infinitesimal density fluctuations are found to grow with time relative to the smoothly expanding background as

$$\delta = \delta\rho/\rho \sim t^{2/3}, \tag{12}$$

exactly the same time dependence as the scale factor $R = R_0/(1 + z)$. On the other hand, in a radiation dominated Universe it is found that

$$\delta \sim t. \tag{13}$$

In an adiabatic fluctuation

$$\delta_R = \tfrac{4}{3}\delta_B. \tag{14}$$

The first result of the discovery of the microwave background was to give a value for the ratio of photons to baryons,

$$n_P/n_B \sim 10^9, \tag{15}$$

a quantity which is conserved in the expansion of the Universe. (The number density of photons produced by stars, etc., is lower than the primordial number by about 5 orders of magnitude.)

It can be shown that a growing fluctuation continues to expand, although at slower rate than the background Universe until it has reached a maximum size which, for $\Omega = 1$, is 5.6 times its initial size. It then begins to collapse to a galaxy, cluster, or supercluster depending on the scale of the initial fluctuation. However, the analysis of which fluctuations can grow is more complicated in the case of the expanding Universe than in the case of the interstellar medium. The first point is that fluctuations can only grow when they are causal—that is when they are smaller than ct at time t after the Big Bang. This happens for a galaxy-sized fluctuation at about one year and a supercluster at about $t = 10^5$ years. Secondly, fluctuations can only grow if they are larger than the Jeans length

$$\lambda_J = \frac{2\pi}{\kappa} = \left(\frac{\pi\kappa T}{\mu G}\right)^{1/2}.$$ (16)

The corresponding Jeans mass is

$$M_J = \frac{4}{3}\rho\left(\frac{\lambda_J}{2}\right)^3,$$ (17)

During the radiation-dominated era, the cosmic fluid is relativistic so that

$$M_J \sim M_H.$$ (18)

Between the onset of the matter-dominated era and recombination M_J levels out at $\sim 10^{17}$ solar masses—which is larger than the mass of a supercluster. Before this period fluctuations cannot grow, but oscillate like acoustic waves. However, photon diffusion damps small adiabatic fluctuations (this is known as "Silk damping"). The critical damping length is at time t,

$$d_S \sim \left(\frac{ct}{n_e \sigma_T}\right)^{1/2},$$ (19)

where n_e is the electron density and σ_T is the Thomson scattering cross-section. The corresponding "Silk mass", below which fluctuations are damped out is

$$M_S = \frac{4}{3}\pi\rho_m d_S^3,$$ (20)

where ρ_m is the baryon density or

$$M_S \sim 1.3 \times 10^{12} \left(\Omega h^2\right)^{-3/2} M_\odot.$$ (21)

Thus, for $\Omega_b = 0.1$, $h = 1/2$, $M_S = 3 \times 10^{14} M_\odot$.

As we have seen, between t_{eq} and the recombination time t_r, M_J levels out at about $10^{17} M_\odot$. Thus under the adiabatic picture only large initial structures can form, i.e. those with masses in the range 10^{14}–$10^{17} M_\odot$. At recombination the Jeans mass suddenly falls to $10^6 M_\odot$; however, small fluctuations have already been damped out.

Thus, adiabatic fluctuations lead to large structures which form first. These are then supposed to fragment into galaxies. This attractive picture is the basis of Zel'dovich's "pancake" theory for the formation of galaxies via superclusters.

6.3. Problems with the simple adiabatic picture

As we have seen, fluctuations $\delta = \delta\rho/\rho$ grow linearly with the scale factor $R = R_0/1 + z$ while the Universe is matter-dominated. However, only fluctuations with $M > M_J \sim 10^{17} M_\odot$ can grow until the time of recombination when the matter temperature suddenly falls. The subsequent growth of an initial density fluctuation δ slows when the Universe begins to expand freely: this happens at $z \sim \Omega^{-1}$. Accordingly, in a baryon dominated Universe with $\Omega_b = 0.1$, δ only grows from $z = 1000$ (the value at recombination) to $z = 10$, i.e. by a factor of 100.

Now the value of $\Delta T/T$, to be expected in the microwave background at recombination when the radiation experiences its last scattering, is related to the size of the corresponding density fluctuations by

$$\frac{\Delta T}{T} = \frac{1}{3}\frac{\Delta\rho}{\rho}.$$

(22)

Thus we expect that, since $\Delta\rho/\rho \sim 1$ at the present epoch it must have been $\Delta\rho/\rho = 10^{-2}$ at recombination. The corresponding value of $\Delta T/T \sim 3 \times 10^{-3}$ should be observed on scales larger than the Silk mass $M_S \sim 10^{14}\ M_\odot$; this corresponds to an angle on the sky of a few arcminutes.

Observations of the microwave background show that fluctuations on this scale are less than 3×10^{-5}—a factor of 100 smaller than is expected on the simplest picture based on adiabatic perturbations.

There are several possible loopholes in the above argument. It is possible that isothermal fluctuations which suffer no Silk damping and which lead to the formation of globular cluster-sized ($M \sim 10^6\ M_\odot$) primordial objects were generated on scales much larger than the causal horizon at early times. A second possibility is that matter was re-ionized some time after recombination. If this occurred at $z > 10$ it would have washed out the fluctuations $\Delta T/T$ produced by proto-pancakes. However, no source of ionizing radiation is known earlier than $z = 3.8$ where the earliest quasars are observed.

The most likely hypothesis is that non-baryonic dark matter preponderates in the Universe. Such matter, if it interacts weakly with baryonic matter, could begin to clump early, before the baryons were released from the radiation field. The baryonic matter would then fall into the potential wells created by the dark matter. Such a scenario would avoid the conflict with observations of the microwave background which is seen with the simple adiabatic picture in a baryon-dominated Universe.

7. Simulations of galaxy clustering

7.1. Types of dark matter

We have seen that studies of the motions of galaxies on large scales have indicated that "dark" matter predominates the mass density of the Universe. Active research is under way at the present time to try and determine the nature of the dark matter. A particularly promising avenue of approach is the study of the evolution of galaxy clustering in the Universe under various hypotheses concerning the nature of the dark matter.

The dark matter candidates can be divided into three broad categories: "cold", "warm" and "hot".

Hot dark matter is the term given to abundant, light particles, weakly interacting with baryonic matter, which remain relativistic as the Universe expands until slightly before the era of recombination. A candidate for such a particle would be a massive neutrino with mass in the range $10 \text{ eV} < m_\nu < 100 \text{ eV}$. It can be shown that freely streaming relativistic particles erase density fluctuations on a scale smaller than the cosmological horizon: for massive neutrinos, fluctuations would have damped out if their present scale is less than the critical damping scale λ_c which, at the present epoch translates to

$$\lambda_c = 41\left(\frac{m_\nu}{30 \text{ eV}}\right) \text{ Mpc}. \tag{23}$$

Warm dark matter consists of particles which interact more weakly than neutrinos, which are less abundant than neutrinos and which have a mass of around 1 keV. Candidates for such particles are the as yet hypothetical gravitinos and photinos. Warm dark matter is able to erase fluctuations on (present day) scales of < 1 Mpc, i.e. galaxy sized and below.

Cold dark matter comprises weakly interacting particles which become non-relativistic early and which therefore can diffuse only a negligible distance. The axion is a possible candidate for such a particle.

There are theoretical problems with scenarios for galaxy formation involving hot dark matter. Since small-scale fluctuations are erased, the formation process must occur through the fragmentation of supercluster-sized clouds into galaxies. However, studies of the dynamical collapse of superclusters indicate that they formed relatively recently at redshift z_{sc} in the range $0.5 < z_{sc} < 2$, while galaxies and quasars exist beyond $z = 3$. There are other problems with hot dark matter, including the difficulty of understanding how galaxies acquired massive halos. In particular, there is good observational evidence that dwarf galaxies with masses as low as 10^6–$10^8\ M_\odot$ have massive halos. It is hard to see how they can be retained, since initial fluctuations on this small scale would have been erased.

The cold dark matter scenario of galaxy formation suffers few problems of principle. Since the dark matter is decoupled from baryonic matter and from the radiation field which dominates the mass density, fluctuations in the dark matter can start to grow before the Universe becomes matter-dominated and before recombination. On the other hand, the baryonic matter is held by the smoothly distributed radiation until the era of decoupling. After recombination, the amplitude of baryon fluctuations grows rapidly to match those of the cold dark matter fluctuations which have had longer to grow. Thus, smaller mass fluctuations grow to non-linearity ($\Delta\rho/\rho > 1$), virialize and cluster within successively larger and larger bound systems. It can be shown that ordinary baryonic matter in gravitationally bound systems of mass 10^8–$10^{12}\ M_\odot$ cools within their dark matter halos to form galaxies while larger mass fluctuations form clusters of galaxies. The essential feature is that the dark matter forms potential wells into which the baryonic matter falls when it is released from the radiation field.

7.2. Numerical simulations

Simulations of the evolution of galaxy clustering in various model Universes dominated by cold dark matter have been calculated recently by Davis et al. (1985). They assumed a "constant curvature", $n = 1$, spectrum of initial fluctuations in the distribution of the dark matter and calculated models for $\Omega_0 = 1$ and $\Omega_0 < 1$, as well as one model with a finite, positive value of the cosmological constant Λ.

The distribution of baryonic matter was approximated by distributing 32 768 particles in a box of size 32.5 $(\Omega_0 h^2)^{-1}$ Mpc, where $h = H_0/100$ is the present value of the Hubble constant. Two-dimensional projections of the particle distribution, i.e. the distribution of visible matter, of the resulting simulations at various phases in the expansion of the Universe were examined and compared with similar plots of the real galaxy distribution at the present epoch. In addition to making visual comparisons of the simulated galaxy distributions with the observed distributions, Davis et al. also calculated the two-point correlation function and random velocity distribution of the galaxies at various epochs in order to make quantitative comparisons with the observations.

The remaining results of the cold dark matter simulations were as follows:

(1) While there is a superficial resemblance with the observed galaxy distribution, models with $\Omega_0 = 1$ are inconsistent with observations if galaxies are assumed to be unbiased tracers of the underlying mass distribution. The random velocities of galaxies are predicted to be higher than is observed. In addition, it is not possible to simultaneously obtain the correct shape and amplitude of the galaxian two-point correlation function at the present epoch.

(2) For $\Omega_0 = 0.2$ the agreement with observations is better than with $\Omega_0 = 1$, but is still not adequate. A model with a positive, finite value of Λ resembles an open model with the same Ω.

(3) Accordingly, if galaxies sample the mass distribution no simulated cold dark matter simulation matches the observations.

(4) If galaxies only form at the peaks of the underlying mass distribution then it is possible to obtain consistency with the observations, both with $\Omega_0 = 0.2$ and $\Omega_0 = 1$. (The way in which this effects the simulations follows from the fact that the peaks in a random distribution are more highly correlated than random points. This may be seen by imagining a spectrum of white noise: the larger high-frequency fluctuations will tend to arise from places where the base provided by the low-frequency fluctuation is high.)

Thus the results of numerical simulations of the galaxy formation on the cold dark matter scenario forces us to the idea of biased galaxy formation in which galaxies form preferentially at places where the underlying density is high.

8. Conclusions

8.1. Summary

Largely from the results of large galaxy redshift surveys, we are steadily obtaining a clearer picture of the large-scale distribution of galaxies in space. These objects (and the quasars) seem to have formed at a redshift $z \sim 4$ (about 2 billion years after the Big Bang) when the scale of the Universe, measured by $1 + z$, was not orders of magnitude different from its present scale. The process of galaxy formation was very effective; not much intergalactic gas remains. The cosmological density parameter measured at the present epoch has a value $\Omega_0 \sim 0.3$; the evidence on the baryonic component Ω_b derived from primordial nucleosynthesis shows that $\Omega_b < \Omega_0$ and that the mass density of the Universe is dominated by some form of dark matter. There is evidence that Ω_0 increases with the scale over which it is measured: it may be as high as $\Omega_0 = 0.5$ on scales of 50–100 Mpc. However, there is no observational evidence that $\Omega_0 = 1$, as the attractive "inflationary" Universe scenario demands.

The galaxies in the Universe around us at $z = 0$ are strongly clustered, with density fluctuations $\Delta\rho/\rho > 1$. We have no empirical information on how galaxy clustering has evolved, but the Lyman-alpha clouds (which may or may not be distributed like the galaxies) show no observable clustering in the redshift range $1.8 < z < 3.8$ over which they can be observed. Also, there is no detectable small-scale anisotropy in the microwave background radiation down to a level $\Delta T/T \sim 10^{-4}$ to 10^{-5}.

Although the existence of superclusters of galaxies points to an origin of galaxies in the form of "pancakes" and adiabatic fluctuations, the studies of absorption lines in quasars do not reveal gaseous pancakes before they have fragmented into galaxies. Moreover, the observed limits on microwave background fluctuations on scales of arcminutes are incompatible with adiabatic fluctuations unless the initial fluctuation spectra were very much steeper than that countenanced by theory.

Adiabatic fluctuations as the origin of large-scale structure in the Universe can in principle be saved by invoking "cold" dark matter as the dominant constituent of the Universe. However, detailed calculations show that this scenario only works if most of the mass of the Universe is non-baryonic and galaxies form preferentially in regions of high-matter density and not randomly.

Accordingly, the present view must be that not only is most of the matter in the Universe invisible, but that the visible galaxies are poor tracers of the mass distribution in the Universe.

8.2. The future

With larger optical telescopes, it would be possible to undertake deeper redshift surveys and directly observe the evolution of the galaxian correlation function at earlier times—say, out to a redshift of $z = 0.5$. The limits on the angular fluctuations in the microwave background must be pursued further on scales of arcminutes to degrees. It appears from current work that the limits on arcminutes scales could be pushed down to $\Delta T/T \sim 5 \times 10^{-6}$ by paying careful attention to sources of systematic errors. The clustering of the "heavy-element" absorption redshifts in quasar spectra could be used to investigate the distribution of galaxies at high redshifts and to settle the critical question of whether the Lyman-alpha clouds (which show no measurable clustering) are distributed in the same manner as the galaxies. Finally, on a more local scale, massive redshift surveys for galaxies out to a distance of 100 Mpc would enable us to evaluate Ω_0 on this scale.

8.3. Epilogue

When Niels Bohr was born, the science of cosmology as we know it today did not exist. The nature of the Universe of galaxies was demonstrated when Bohr was 40 and the expansion of the Universe was discovered when he was 44. At about this time, Lemaitre and Gamow began their bold speculations about the physics of the early Universe and identified many of the problems with concern us today—the problem of the formation of galaxies, the formation of the heavier elements and the creation of the relict radiation. It was the accidental discovery of this radiation by Penzias and Wilson (1965) three years after Bohr's death which unleashed the present activity in the study of the early Universe.

I believe that our present serious attempts to explore the evolution of the Universe back to at least the first few seconds represents one of the most remarkable developments in scientific history. Moreover, the possibility that the large-scale structure can ultimately be directly traced back to primordial quantum fluctuations would no doubt have pleased Bohr. And yet the most interesting question for Bohr's 200th birthday would surely be: "was the dark matter that cosmologists believed in during the late 20th century as ephemeral as the phlogiston of 200 years earlier?"

Acknowledgements

This paper was begun during a visit to the European Southern Observatory, Garching bei München; I wish to thank L. Woltjer for his hospitality. The work was also supported by the National Science Foundation under Grant AST84-16744.

References

Abell, G.O., 1958, Astrophys. J. Suppl. **3**, 211.
Albrecht, A., and P.J. Steinhardt, 1982, Phys. Rev. Lett. **48**, 1220.
Alpher, R.A., and R.C. Herman, 1948, Nature **162**, 774.
Alpher, R.A., and R.C. Herman, 1949, Phys. Rev. **75**, 1089.
Davis, M., G. Efstathiou, G. Frenk and S.D.M. White, 1985, Astrophys. J. **292**, 371.
de Vaucouleurs, G.H. 1956, Vistas Astron. **2**, 1584.
Foltz, C.B., R.J. Weymann, H.-J. Röser and F.H. Chaffee, 1984, Astrophys. J. Lett. **281**, L1.
Guth, A.H. 1981, Phys. Rev. **23**, 347.
Guth, A.H., and S.-Y. Pi, 1982, Phys. Rev. Lett. **49**, 1110.
Humason, M.L., N.U. Mayall and A.R. Sandage, 1956, Astrophys. J. **61**, 97.
Kirshner, R.F., A. Oemler, P.L. Schechter and S.A. Shectman, 1981, Astrophys. J. Lett. **248**, L57.
Oort, J.H. 1960, Bull. Astr. Inst. Netherlands **15**, 45.
Oort, J.H. 1983, Ann. Rev. Astron. Astrophys. **21**, 373.
Ostriker, J.H., and S. Ikeuchi, 1983, Astrophys. J. Lett. Ed. **268**, L63.
Peebles, P.J.E., 1980, The Large-Scale Structure of the Universe (Princeton University Press, Princeton, NY).
Peebles, P.J.E., and M.G. Hauser, 1974, Astrophys. J. Suppl. **28**, 19.
Penzias, A.A., and R.W. Wilson, 1965, Astrophys. J. **142**, 419.
Peterson, B.A., A. Savage, D.L. Jauncey, and A.E. Wright, 1982, Astrophys. J. Lett. **260**, L27.
Readhead, A.C.S., W.L.W. Sargent and A. Moffet, 1986, Astrophys. J., in press.
Rowan-Robinson, 1985, The Cosmic Distance Ladder (Freeman).

Sandage, A.R., and G.A. Tammann, 1984, Large-scale structure of the universe, cosmology and fundamental physics, in: 1st ESO–CERN Symposium.

Sargent, W.L.W., P.J. Young, A. Boksenberg and D. Tytler, 1980, Astrophys. J. Suppl. **41**, 42.

Sargent, W.L.W., P.J. Young, and D.P. Schneider, 1982, Astrophys J. **256**, 374.

Schmidt, M., and R.F. Green, 1982, in: Astrophysical Cosmology, eds H.A. Brück, G.V. Coyne and M.S. Longair (Vatican Observatory, Vatican City, Italy).

Steigman, G., and A.M. Boesgaard, 1986, Ann. Rev. Astron. Astrophys. **24**, in press.

Uson, J.M., and D.T. Wilkinson, 1984, Astrophys. J. **283**, 471.

Yu, J-.T., and P.J.E. Peebles, 1969, Astrophys. J. **158**, 103.

Discussion, session chairman T.A. Bak

Bleuler: You assumed a perfectly new kind of matter (so-called dark matter) which (apart from gravitation) has no interaction whatever with visible matter. Is that not too much freedom undermining the basic assumption of understanding the universe through known physical laws?

Sargent: The existence of the dark matter has been inferred for more than 40 years and the evidence for it is becoming stronger. Apart from gravitation, the particles have to interact weakly with matter after the first second or so in the expansion of the Universe: there are several theoretical candidates for such particles.

Ginzburg: In the main part of your presentation you have assumed that the cosmological "constant" Λ is equal to zero. I am, however, convinced that there is at present no reason to assume that $\Lambda = 0$. What, in your argumentation and conclusions, depends on the hypothesis that $\Lambda = 0$?

Sargent: A finite Λ was introduced into some of the numerical simulations of the evolution of clustering in a Universe dominated by "cold" dark matter that I described. It did not help to resolve the problems that I outlined.

Jones: The only problem with introducing a cosmological constant is that we need to explain the remarkable coincidence that we are living at the time when $\Lambda \sim 3 H_0^2$ (to within 90%). Appeal to anthropic principles is hardly a scientific explanation.

The Lesson of Quantum Theory, edited by J. de Boer, E. Dal and O. Ulfbeck

Molecular Replication and the Origins of Life

Leslie E. Orgel

The Salk Institute for Biological Studies
San Diego, California, USA

Contents

1. Introduction

All living organisms have a great deal in common. The interior environment of their cells is a concentrated aqueous solution containing a characteristic set of organic macromolecules. They are surrounded by membranes composed of a different set of organic molecules called lipids. This similarity in composition results from an even more striking similarity in organization. The universal genetic system of all forms of life is dependent on a very complicated interplay between two sets of macromolecules—proteins and nucleic acids. The proteins are made up by joining 20 standard amino acids together in specific linear sequences, while the nucleic acids are formed in a similar way from 4 standard nucleotides. Membranes are somewhat more variable, but all are constructed according to the same general plan. They consist, in the main, of an impermeable bilayer of lipid molecules in which specific protein-carrier molecules and channels are embedded.

It became clear very early in the history of chemistry that nothing at all like the components of living systems occurs on the earth in a non-biological context—hence, the traditional division of chemistry into organic and inorganic subdisciplines. The chemists' interest in the problem of the origins of life is mainly concerned with the origin of organic material on the primitive earth before the appearance of life and with the sequence of events that led to its self-organization into a living system.

In retrospect one can see that Wöhler's synthesis of urea from ammonium cyanate in 1828 [1] was an important step toward our understanding of the origins of life. Prior to this discovery it was thought to be impossible to produce organic products from inorganic starting materials without the assistance of a "vital force". Since urea is organic and ammonium cyanate can be produced from inorganic sources, the publication of Wöhler's results made it clear that organic material could originate in an inorganic environment through normal abiotic processes.

The investigation of reactions that produce important biochemicals from elementary constituents known to be present in planetary atmospheres was initiated by the speculations of Oparin [2] and the experimental studies done by Miller in Professor Harold Urey's laboratory [3]. This subject has been reviewed repeatedly [4]. The basic finding is that mixtures containing some or all of the elementary atmospheric constituents, hydrogen, water, nitrogen, ammonia, methane, carbon monoxide and carbon dioxide, when subjected to "high energy" sources such as ultraviolet irradiation or electric discharges, yield complex mixtures of products including amino acids, nucleotide bases and a variety of other biochemicals. These laboratory studies are complemented by the observation of a great variety of small organic molecules in the dust clouds where new stars form [5] and by the discovery of amino acids and other organic molecules as indigenous constituents of certain stony meteorites [6].

At the present time, the sources of the organic material that must have accumulated on the primitive earth before the emergence of life is obscure. There are too many possibilities. Synthesis in the atmosphere or in volcanoes or deep-sea vents is one possibility. Accumulation as constituents of the material from which the earth accreted, or from meteorites falling on the primitive earth, is another. It is, however, abundantly clear that organic material, including a surprising number of important contemporary biochemicals, is formed in the cosmos without the intervention of living organisms, intelligent or otherwise. The origin of the organic chemicals in the prebiotic environment is now a subject for detailed study, but does not present major conceptual difficulties.

Direct evidence concerning the steps leading from the mixture of organic compounds that accumulated on the primitive earth to an organized living system has not survived in the geological record. Thus we can turn only to laboratory experiments and to the properties of contemporary living organisms for relevant information. The obstacles to the formulation of a detailed and plausible model, given these limitations, are formidable. It is not surprising that there is at present no coherent, generally accepted model for the origins of life.

Attempts to recapitulate in the laboratory the actual events that occurred on the primitive earth are doomed to failure. We do not know how long it took for life to evolve once the conditions necessary for its appearance were present. Although the upper limit placed by the geological record is about one thousand million years, the evolution of life could have taken a very much shorter time. However, it is unlikely to have taken a time as short as a human lifetime. We are forced, therefore, to search for model systems which provide useful information in days about processes that may have taken millenia. Similarly, the scale of human experimentation is measured, at most, in litres, while the volumes of lakes or tide pools where life is thought to have originated must have been measured at least in cubic kilometers.

The first simplification which is made in almost all studies relevant to the origin of biological organization is to work with a simple mixture of pure organic compounds. In our own work we use D-nucleotides, other investigators use, for example, single L-α-amino acids or mixtures of L-α-amino acids. Prebiotic syntheses rarely, if ever, produce single substances or simple mixtures. Even the most plausible reactions, such as Miller's synthesis of amino acids in an electric discharge, produce numerous "non-natural" analogues and, of course, produce D- and L-enantiomers in equal amounts. Spontaneous reactions leading to sugars, nucleotides, etc. are far less specific. We know in many cases that the unwanted materials would interfere seriously with the reactions which we are able to demonstrate in the laboratory using pure reagents. Do these simple considerations invalidate all laboratory work on the origins of biological organization?

This is a hard question to answer. The environment on the primitive earth must have permitted extensive enrichment of particular classes of organic compounds on the basis of solubility, thermal stability, tendency to adsorb on inorganic surfaces, etc. Furthermore, reactions that could be relevant to the origins of biological organization vary widely in their sensitivity to inhibition by substrate analogues. However, even when these favourable considerations are taken into account, belief in the relevance of most laboratory experiments and most theoretical studies to the origins of life on the earth, involves, consciously or unconsciously, a good deal of faith.

At least one group of researchers, appalled by these difficulties, has proposed a radical solution—an inorganic origin of life based on self-replicating clays [7]. Any experimental demonstration of the main claim of the theory—that there are clay structures that act as primitive catalysts with the specificity of enzymes and that are also able to replicate accurately—would command the greatest respect. The absence of any scrap of experimental evidence twenty or more years after the first publication of the theory is one of the reasons why I find the new theory even less plausible than the conventional one that it is designed to replace.

2. Replication (theoretical)

Even the simplest forms of life are so complicated that their spontaneous appearance in a mixture of prebiotic organic compounds would constitute a miracle. We must conclude that contemporary cellular life was preceded by a series of systems of gradually increasing complexity. Natural Selection is the only mechanism that could have generated such a series of intermediates.

The theory of Natural Selection was, of course, conceived by Darwin and Wallace as an explanation of the origin of new species on the earth. However, the central dogma of natural selection—that those who reproduce most successfully eliminate all competitors—must apply to any family of objects that are capable of sufficiently but not perfectly accurate replication.

The objects that we will be interested in are macromolecules made up by arranging a set of related but distinguishable small molecules (monomers) in a determined order, for example, proteins and nucleic acids. A macromolecule replicates when it brings about the synthesis of a new macromolecule sharing the parental sequence. Variant progeny macromolecules that differ from the parental molecule at one or a few positions in the sequence are called mutants. Their appearance is inevitable, even if at very low frequency, and provides the variability on which natural selection depends.

The eventual evolution of complexity by natural selection in complicated chemical systems is not inevitable. If, for example, formaldehyde, a highly reactive and versatile molecule, is fed into a continuous flow reactor and the products are sampled from time to time, it is found that the product mixture soon reaches a steady state. The steady-state mixture is certainly complicated, but it shows none of the characteristic organization that interests us [8]. No mechanism for achieving very complex organization by natural selection seems to exist in this system.

In fact, the only method that we know to be able to generate increasing molecular complexity through natural selection is residue-by-residue replication. There may be other ways that do not involve residue-by-residue replication, for example through complex cycles of reactions in which each cycle generates molecules that play a part in other cycles. However, no detailed description of such a system of cycles has ever been offered, and I am sceptical that such systems are possible.

The arguments presented so far suggest that experimental study of molecular replication might provide models of an important step in the origins of life. But where should one start? One can draw on paper a very large number of potentially self-replicating macromolecules, and many of them have special features of interest to the chemist. What considerations guide the choice of a particular experimental system?

The student of the origins of life (on earth) should, I believe, require minimally that:

(1) The monomeric components of the system (or sufficiently close analogues) can be synthesized under prebiotic conditions;

(2) The proposed replication mechanism is compatible with and, if possible, supported by the established chemistry of the components.

It is also desirable, but perhaps not essential, that there be a reasonably close relation between the proposed primitive mechanism and contemporary nucleic-acid replication.

Even with these restrictions there remain a number of possibilities. The most conservative is the hypothesis that the self-replicating genetic molecule has always been a nucleic acid. Another possibility is that the first replicating polymer was related to a nucleic acid but had a different and simpler structure. It is possible, but in my opinion less likely, that life originated with a self-replicating protein or carbohydrate, or with something even more bizarre.

We have chosen to work with the standard nucleotides. In making this choice we have in part been guided by our belief that this is a very plausible first choice. However, there is another and far more important reason. The chemistry of nucleotides and polynucleotides is well-developed. A variety of starting materials are available commercially and, most importantly, enzymes can be used to synthesize new starting materials and analyze reaction products. These factors allow one to avoid the very extensive preliminary chemistry that is needed when one modifies the nucleic-acid structure in any way. Like the man who lost his watch on a dark night, we are forced to look where the light is; fortunately it seems a very good place to start.

3. Template-directed synthesis

DNA and RNA replication as they occur in living systems are very complicated enzymatic processes. However, they always depend on the same basic chemistry. A single-stranded region of a preformed nucleic acid directs the synthesis of a complementary antiparallel strand. The four bases, A, U, G and C in the case of RNA direct the incorporation of the complementary bases U, A, C and G, respectively (fig. 1) via Watson–Crick base-pairing (fig. 2). For an excellent description of this fundamental aspect of molecular-biology the reader is referred to J.D. Watson's textbook [9].

We have attempted to imitate this template-directed reaction without using enzymes. I will try to summarize very briefly the conceptual background of these experiments and the progress that we have made. However, to make the treatment accessible to the non-chemist reader, it will be necessary to neglect many important details.

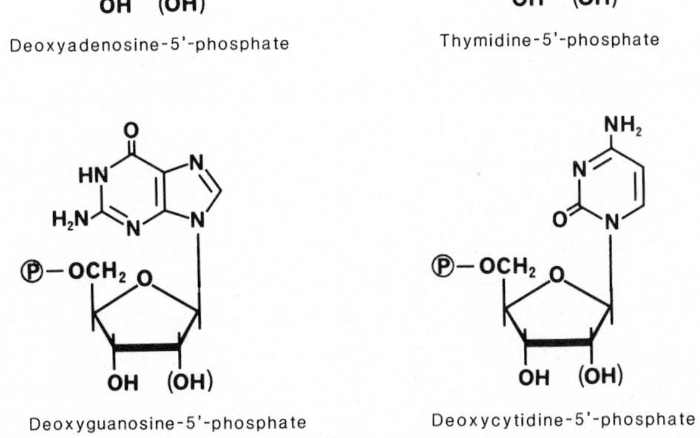

Fig. 1. The structure of the monomeric components of the nucleic acids. The figure illustrates the deoxynucleotides, components of DNA. The corresponding ribo-nucleotides (components of RNA) have slightly different structures. The differences are indicated in parentheses.

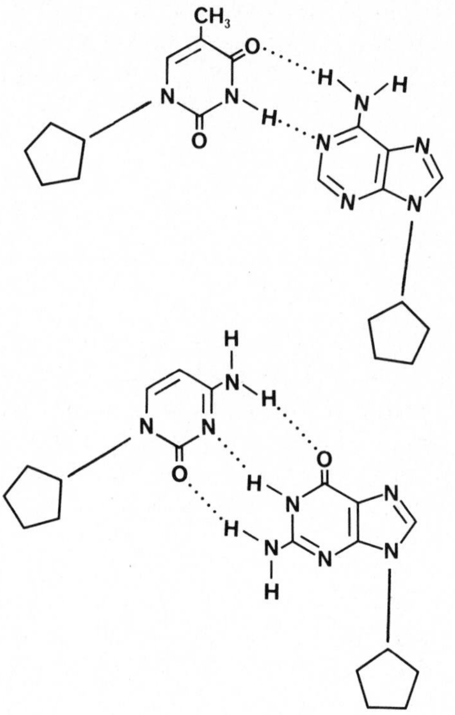

Fig. 2. The Watson–Crick base-pairs for DNA. Virtually identical pairing occurs between the bases in RNA double-helices and in RNA–DNA hybrids.

Todd Miles and Paul Ts'o and their co-workers discovered that a preformed strand of poly(U) or poly(C) was able to organize the complementary monomeric base or one of its derivatives into a helix with a structure closely related to that of DNA or RNA (fig. 3). This helix is stable only at low temperatures. The important point is that poly(U) organizes A derivatives into a helix while ignoring G derivatives, while poly(C) organizes G derivatives and ignores A derivatives. Poly(A) and poly(G), for well-understood reasons, do not organize U or C into helices.

The principle of template-directed synthesis is very simple. If the G derivative organized on poly(C), for example, is activated, it might zip-up rapidly if the arrangement in the helix brought the correct parts of the two adjacent monomers close together. Of course, activated derivatives also condense together in the absence of a template, but in dilute solution and without the orienting effect of a template this reaction is known to be inefficient and non-specific with respect to the nature of the bases involved.

It proved easy to demonstrate that poly(U) does bring about the polymerization of a variety of A derivatives, that poly(C) facilitates the reactions of G derivatives, and that neither template has any influence on non-complementary bases. The Watson–Crick pairing rules are obeyed [10]. However, it has proved much more

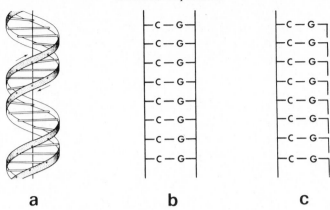

Fig. 3. The complementary structures formed between two polynucleotide strands or between one polynucleotide strand and complementary monomers. (a) Diagramatic illustration of the double-helices; (b) simplified view of a poly(G):poly(C) double helix; (c) corresponding diagram of a helix formed by poly(C) with a monomeric derivative of G.

difficult to achieve efficient copying of templates, particularly of those containing more than one base.

The activated nucleotide substrates in all enzymatic replications are tri-phosphates such as ATP. These substances are not suitable for laboratory investigation of non-enzymatic reactions because they react too slowly. Instead we use the imidazole derivative I, and the corresponding derivatives of the other bases (fig. 4). When I is used as a substrate, in the presence of poly(C), a very efficient polymerization occurs. The products are oligomers of G up to at least 30–40 units long that are identical to the naturally occurring substances. The reaction is highly selective—if one incubates a mixture of I and the corresponding derivative of another base with poly(C), the G derivative is incorporated 100–500 times more efficiently than the second base [11].

This is a good example of a template-directed reaction. In the absence of poly(C), the oligomerization of the G derivative is inefficient and yields a very complicated mixture of low molecular-weight products. When poly(C) is present long complementary oligomers are obtained in excellent yield. All bases except G are rejected by a poly(C) template.If one could extend this reaction to any arbitrary template, the

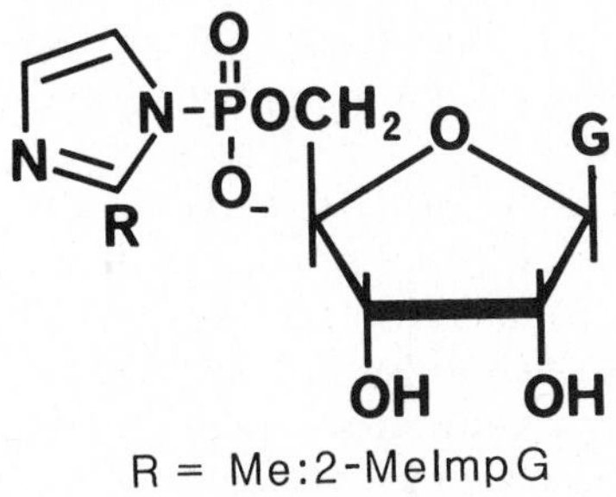

R = Me:2-MeImpG

Fig. 4. The imidazole derivate I; R = Me:2-MeImpG.

problem of non-enzymatic replication would be well on the way to solution. Unfortunately, this has not proved possible as yet. If one uses a template containing C and one or more other bases, one does indeed obtain oligomeric products that contain only the bases complementary to those in the template. However, the reaction is efficient only as long as C is the major component of the reaction mixture. For a variety of reasons, polymers containing less than 60% of C do not act as good templates in our system [12].

It has also proved possible to demonstrate "information-transfer" from template to product unambiguously in a few cases. The template CCGCC, for example, directs the synthesis of GGCGG without producing detectable amounts of isomers such as GGGCG. Unfortunately, the efficiency achieved so far never exceeds 20% [13]. Similarly the template $d(C_3GC_3GC_3GC_3)$ directs the synthesis of complementary oligomers up to $G_3CG_3CG_3CG_3$. The longer oligomers, however, are formed in yields well below 1% [14].

While these studies show unambiguously that the Watson–Crick pairing scheme is applicable to non-enzymatic synthesis of oligonucleotides, they also uncover a number of difficulties. First they show that the initiation of product synthesis is not restricted to the end of the template. A significant proportion of the products on a C_7GC_7 template, for example, are of the form G_6CG_n and G_5CG_n. A second difficulty is the stable self-structure formed by many templates which prevents further synthesis. The oligomer C_3GCGC_3, for example, has no template activity because it oligomerizes to form a stable mini-helix with four GC base pairs. A related difficulty is the stability of the template-product helix which must be dissociated before further reaction can occur. Finally, in our system, templates work efficiently only if they contain more than 60% of C. The complements, therefore, contain at most 40% of C and cannot act as templates. This is probably a technical difficulty that can be overcome by modifying the chemistry slightly.

4. What do experiments on template-directed synthesis suggest about the origin of nucleic-acid replication?

The overwhelming impression gained by studying a wide range of template-directed reactions is one of detailed, idiosyncratic, chemical complexity. No two templates behave in the same way; the products of template-directed synthesis even on a simple template are always complex mixtures containing oligomers of different lengths, isomers with different linkages etc. The principle of complementary base-pairing is obeyed, as in enzymatic synthesis, but beyond that all of the precision of the biological system is lost.

These difficulties might, of course, be due to an unfortunate choice of substrates, but I do not think this likely. It seems probable that the complexity of the product mixtures reflects the variety of chemical reactions that can occur between molecules as complicated as activated nucleotides. The enzymatic process of replication is neat and tidy only because it greatly accelerates one single sequence of chemical reactions while suppressing all other reactions. In the absence of enzymes it is unreasonable to expect a precise end-to-end copying of a template strand.

A more realistic picture is obtained by considering what would happen in a steady-state chemical reactor into which activated nucleotides are fed at a slow, steady rate, while a random mixture of substrates and products is extracted at an equivalent rate. At first a complex, "zero-order" mixture of short oligonucleotides with varying base-sequence and isomeric structure would begin to form, necessarily by non-template reactions. Among these oligomers would be some that are able to direct complementary synthesis. The composition of the mixture would then begin to drift away from that of the "zero-order", non-templated product mixture.

There seem to be two possible end points for this drift. Pessimistically, one might expect the system to settle down to produce a steady-state mixture not very different from the "zero-order" mixture; non-template-directed synthesis would remain dominant, and small amounts of inconsequential additional material would be produced by template-directed processes. The outcome expected by the optimist is very different—the reaction mixture might gradually change until template-directed synthesis became dominant. The final, steady-state product would consist of a complex mixture of oligomers very different from the "zero-order" mixture. No individual molecule would necessarily have the capacity to replicate accurately, but the whole family of molecules, through template-directed synthesis, could sustain itself with a composition very different from that of the "zero-order" mixture.

What reactions would be essential for the maintenance of a steady-state self-replicating system? Clearly one must incorporate new material into longer oligomers to compensate for the longer oligomers that are removed at random from the system. The primary source of new material, under prebiotic conditions is likely to be activated monomers, since most primary activation processes do not work efficiently on oligomers. Thus, template-directed incorporation of monomers is likely to be one fundamental reaction.

Since template-directed reactions do not always generate full length copies, it also seems necessary to have a second elongation process to ensure that the average length of the product molecules is maintained. The most plausible mechanism is the template-directed joining-together of short oligomers, a process called "ligation" in biochemistry. The reaction has been demonstrated as a non-enzymatic chemical reaction, for example by Naylor and Gilham [15] who showed that two T_6 molecules can be joined to make a T_{12} molecule on a poly(A) template. The intermediates in biochemical ligation, oligonucleotides capped with a pyrophosphate bonded A residue, are also likely prebiotic ligation intermediates, formed in secondary reactions from oligomers and activated monomers.

In principle, these two reactions are sufficient to maintain a "steady-state" system. However, an additional process is likely to have been important, namely the sequence-specific hydrolysis that generates free ends. The amount of new synthesis that can occur must depend in part on the presence of free ends, so hydrolysis that specifically produces free ends suitable for further elongation could have been very important for the evolution of an efficient replicating system.

Finally, one should mention the possibility that reaction cycles were important for prebiotic replication. Templates are likely to form stable complexes with their products which block further reaction. Repeated cycles in which the temperature is temporarily raised, and then lowered again, for example, could melt these complexes and allow synthesis to re-initiate. Cycles involving periodic changes in the pH, salt concentration, etc. could have a similar effect.

The final picture of replication that emerges does not have the elegant simplicity of modern replication. Instead, we have a picture of a complex mixture of macro-molecules, sub-sequences of which are templates for the synthesis of intermediate-sized oligomers. These latter oligomers are mobile and can move to new sites where they can be extended by incorporation or joined by ligation. From time to time an oligomer breaks in such a way as to generate two rapidly growing fragments. A system of this type seems to me as close an approximation as one can expect to a modern replication mechanism in the absence of more specific catalysts.

5. RNA catalysts—Ribozymes

One of the most exciting discoveries of the last few years is that RNA molecules, without the help of proteins, can catalyze a number of chemical reactions. RNAase P is an enzyme that contains both a protein and an RNA moiety. In the presence of a sufficient concentration of Mg^{2+}, the RNA component alone will hydrolyze its substrate—a highly specific RNA sequence [16]. The cleavage occurs at a unique point in the substrate RNA.

The self-splicing of RNA is even more remarkable. The first system discovered involves the elimination of an intron from Tetrahymena ribosomal RNA. A central segment of the precursor RNA is eliminated, and the two end segments join together spontaneously. The mechanism is understood in some detail and depends on a series of trans-esterification reactions [17].

These examples of catalysis by RNA, together with laboratory experiments on template-directed synthesis, show that RNA catalysts (Ribozymes) and templates can bring about a number of interesting sequence-specific transformations of other RNA molecules and of the monomeric nucleotides. The key and as yet unanswered question is, "Can RNA molecules either alone or with the help of associated small molecules such as co-enzymes, catalyze reactions which do not depend on the direct base-pairing interaction of the catalytic RNA with a sequence of another RNA molecule?" If Ribozymes could catalyze either the replication of an arbitrary polynucleotide sequence or the reactions of intermediary metabolism, for example, biological organization could have evolved considerably before the invention of protein synthesis. This clearly should be a key issue in discussions of the origins of life.

6. The origin of the genetic code and of protein synthesis

Experiments designed to reveal the physico–chemical basis of the genetic code have proved inconclusive. Theoretical models seem to me to ascribe to short polypeptides properties that the real molecules are unlikely to possess. I suspect that the genetic code did not evolve as directly as most mathematical models propose.

The evolution of complex adaptations can only be explained by Natural Selection if there are many intermediate forms, each of which is already at a selective advantage. The selective mechanisms in the intermediate stages may or may not be related to the "purpose" of the final adaptation. In the case of the evolution of the genetic code it seems to me essential that the attachment of amino acids or short peptides to the 3'-termini of RNA molecules must have favored the replication of the RNA molecules prior to the evolution of protein synthesis. Polypeptides might, for example, have directed incoming activated nucleotides to the 3'-terminus of the template.

The most interesting recent work relevant to the origin of protein synthesis concerns the stage immediately after the appearance of the genetic code. Gilbert has argued persuasively [18] that contemporary RNA molecules contain in their sequences clues to the nature of the very earliest catalytic polypeptides. It appears likely that modern proteins have been formed by combining and recombining very primitive polypeptides usually made up of 30–40 amino acids. There are already clues that these original polypeptides are based on a few structural "themes". This is a rapidly developing field in which important discoveries can be expected. Hopefully they will serve to define the nature of the most primitive coded polypeptides.

References

[1] F. Wöhler, Ann. Phys. (Leipzig) 12 (1828) 253.
[2] A.I. Oparin, Proiskhozhdenie zhizny (Moscovsky Robotchii, Moscow, 1924), translated in: J.D. Bernal, The Origins of Life (The World Publishing Co., Cleveland, New York, 1967).
[3] S.L. Miller, Science 117 (1953) 528.
[4] S.L. Miller and L.E. Orgel, The Origins of Life on the Earth (Prentice–Hall, Englewood Cliffs, NJ 1974).
[5] R.D. Brown, in: Origin of Life, Proc. 3rd ISSOL Meeting and 6th ICOL Meeting, Jerusalem, June 22–27, 1980, ed. Y. Wolman (Reidel, Dordrecht, Boston, London, 1981) pp. 1–10.
[6] P.G. Stoks and A.W. Schwartz, in: Origin of Life-Proc. 3rd ISSOL Meeting and 6th ICOL Meeting, Jerusalem, June 22–27, 1980, ed. Y. Wolman (Reidel, Dordrecht, Boston, London, 1981) pp. 59–64.
[7] A.G. Cairns-Smith, Genetic Takeover and the Mineral Origins of Life (Cambridge University Press, Cambridge, New York, 1982).
[8] T. Mizuno and A.H. Weiss, in: Advances in Carbohydrate Chemistry and Biochemistry, Vol. 29, eds R.S. Tipson and D. Horton (Academic Press, New York, London, 1974) pp. 173–227.
[9] J.D. Watson, Molecular Biology of the Gene, 3rd Ed. (Benjamin, Menlo Park, CA, 1976).
[10] L.E. Orgel and R. Lohrmann, Accounts of Chemical Research 7 (1974) 368.
[11] T. Inoue and L.E. Orgel, J. Mol. Biol. 12 (1982) 201.
[12] T. Inoue and L.E. Orgel, Science 219 (1983) 859.
[13] T. Inoue, G.F. Joyce, K. Grzeskowiak, L.E. Orgel, J.M. Brown and C.B. Reese J. Mol. Biol. 178 (1984) 669.
[14] T. Haertle, unpublished work.
[15] R. Naylor and P.T. Gilham, Biochemistry 5 (1966) 2722.
[16] C. Guerrier-Takada and S. Altman, Science 223 (1984) 285;
C. Guerrier-Takada, K. Gardiner, T. Marsh, N. Pace and S. Altman, Cell 35 (1983) 849.
[17] T.R. Cech, Int. Rev. Cytol. 93 (1985) 3.
[18] N. Lonberg and W. Gilbert, Cell 40 (1985) 81.

Discussion, session chairman N.K. Jerne

Weisskopf: In both examples you showed, you have used RNA, which is already a very complicated construction. I have two questions. First, how did Nature come to such a complicated thing, and are there other molecules thinkable that would do the same job?

Orgel: The experiments described would not necessarily show what happened historically on earth, but they would show a great deal about the various ways in which matter organizes itself. RNA seems an economical choice, and it is my guess that the original molecules were something similar.

Bjørnholm: You have shown so nicely how polynucleotides may replicate spontaneously. Do you think of the origin of life as a matter of nucleotides only, with the polypeptides (enzymes) at a later stage? After all, one may also view the cooperative functions of polynucleotides and polypeptides, forming feedback loops, to be the essence of life on the molecular level.

Orgel: I am inclined to think that the first thing to happen was the evolution of the self-replicating molecule, and that the manipulation of amino acids came as a subsequent development.

Fowler: I got the impression that in the molecular replication, when a mutation occurs it is always successful. Why can the opposite not occur?

Orgel: The impression that I wanted to give was that the successful mutations were those that were *retained*; neutral or disadvantageous mutations were eliminated.

Casimir: If you are able to inhibit the replication by means of added druggs, and mutations show up to overcome the effect of the inhibitive drug, how sure are you that the mutations are there anyway, unaffected by the presence of the drugs?

Orgel: This is a central point, best described as "anti-Weissmanism": both the molecular "phenotype" and "genotype" are indeed affected.

Casimir: I may be influenced by my industrial past, but it seems to me the DNA can be regarded as a blueprint of a product to be manufactured. Customers may complain about the product and they may themselves make certain modifications, or service mechanics may do so. That is changing the phenotype. But if pressures are very large and the products do no longer sell, then the blueprint will be changed. So I am willing to believe that if a species is under extreme pressure this may stimulate the occurrence of mutations, although I know this is against the generally accepted dogma.

Broglia: Is there any way of getting an estimate, rough as it may be, of the timescales needed to produce a simple viable organism?

Orgel: I'm afraid the answer is "No".

Kristensen: Is it so that a condition for the evolution of new species is an environment that changes, but not too fast?

Orgel: That is probably correct if "environment" is taken to mean not only the abiotic, physical surroundings, but also, for instance, competition from other species.

The Lesson of Quantum Theory, edited by J. de Boer, E. Dal and O. Ulfbeck
© Elsevier Science Publishers B.V., 1986

Physics, Biological Computation and Complementarity

John J. Hopfield

California Institute of Technology
Pasadena, California, USA

Contents

1. The domain of physics in biology

Biology as we know it lies within a restricted domain of physics. The laws of elementary particle physics and cosmology and the history of the universe serve merely to determine the nature of a planetary environment. The dynamical equations of quantum mechanics and quantum electrodynamics (and their classical equivalents when appropriate) are the essential elemental laws of physics which lead to biology. Some physicists make claims that "we shall never understand life until we understand the origins of the elementary particles". But the real mysteries of biology lie in the way in which these dynamical laws of physics, and the substrate of electrons, photons and nuclei on which they operate, produce the complex set of counter-intuitive phenomena labeled with the term biology.

Biology is a problem in dynamics—an organism functions by irreversibly preying on the available free energy which it finds in its environment in order to maintain its dynamic state. Driven (or non-equilibrium) physical systems of simple components already show complex and almost unpredictable behaviors. Turbulence in fluid flow, deterministic chaos and fractal forms in snowflakes are a few of the complex phenomena which arise from the same simple physical laws and substrates that rule biology. Some of the basic unsolved problems of biology, such as the generation of complex forms in large systems, can already be seen in these simple systems. Our understanding of such problems in physics is far from complete. The physics of large systems in equilibrium is unified and simplified through our understanding of the derived or secondary laws of statistical mechanics and thermodynamics. There is no comparably general theory of strongly non-equilibrium systems. The major conceptual problems in biology have the additional complication that the biological matter is itself very complex due to a long and selective evolutionary history.

Biology is not a quantum-mechanical problem. Because the masses of nuclei are large compared to the masses of electrons, the adiabatic separation of electron and nuclear motion in the Schrödinger equation is usually adequate. The electrons can then be effectively removed from the problem, leaving a problem of only nuclear motion with effective interactions of some complexity between nuclei. This remaining problem of nuclear motion is partly in the classical regime, and partly a quantum-mechanical problem. The hydrogen stretching vibrations are of large quantum energy compared with kT at room temperature, and are rigid in the molecular dynamics. The rotational quantum numbers are large at room temperature, and the rotational motion is thus essentially classical, as is much of the translational motion. The molecular dynamics of liquid water at room temperature can be described in classical terms as the motion of rigid molecules having a complex set of two- and three-body forces between them. Accurate descriptions of the viscosity, rotational relaxation and dynamic neutron diffraction of liquid water have been obtained from the computed dynamics of such a model of water. Non-equilibrium problems such as the turbulence of water are in essence problems of classical physics, as is biology.

This is not to deny that there are intrinsic limitations in the knowledge with which we can know positions and momenta of nuclei. In this regard, the classical approximation ignores a quantum-mechanical limitation. But the essential mysteries, phenomena and complexity of biology are not a problem of Planck's constant. They are a problem of the large size of Avagadro's number combined with the non-equilibrium nature of the system. There do not seem to be any important larger quantum coherent aspects to living matter, contrary to the romantic hopes of some physicists of the 1930s.

2. Logical, physical and biological computers

Biology can be seen as a hierarchy of computations and computational devices. The translation of DNA into protein structure is a kind of computation. The construction of a complex organism from the instructions in DNA is also the following of an algorithm. The recognition of a familiar object is a neural computation. These diverse computations share some common characteristics because they share an evolutionary background. The species which exist in biology today are those which have survived the competition with other organisms and the cataclysms of weather and geology. At the other end of the size scale, the proteins which exist within a given species have survived a competition with different molecules which might have performed the same functions, and with the alternative of simply not having this function performed at all. The fundamental survival and competition problem for the organism or for the enzyme molecule is to develop an algorithm for predicting the future, or more accurately, to develop a behavior such that physical processes (or actions) taken now will be likely to be appropriate to the future environment in which the organism or molecule finds itself, and promote the survival or reproduction of the organism. The organism which survives best will be that which most clearly "sees" the consequences of its possible present actions. The prediction of the future from present information is a kind of computation. To understand complex aspects of biology beyond the descriptive level, it becomes necessary to think about the computational aspects—what is computation and how does biological computation differ from that which we think about in conventional computers. Before delving into neurobiological computation, we will illustrate some of the issues.

Computation has three conceptual elements: an input, an output and a "device", which reads the input and produces the output. The particular output, or range of outputs, is the consequence both of the particular input and the nature of the computing "device". The classic conceptual device of computational theory is the Turing machine. This machine has several internal states. It can read an input–output tape, shift the tape, write output on the tape and change the internal state of the machine. Universal computation can be performed by such machines. Turing machines are intrinsically digital (or logical) machines, having a small number of reading and writing symbols and a finite number of internal states. When computation is a physical process, as in a real computer or in biology, each of these elements has a physical manifestation.

The description of computers as logical devices has been thoroughly developed in the past 50 years. A computer is also a physical dynamical system, which follows the laws and limitations of physics. The object of a computer designer is to develop a real dynamic system which behaves as similarly as possible to some given logical design. Physical systems have noise and imperfections not envisioned in the simple logical view of computer function. As a result, considerable effort in hardware design is invested in a problem of no logical concern, namely trying to get reliable computation in a noisy and fault-filled world. Biology compounds this problem by adding a unique view of the nature of errors and of logic itself. We will examine these points by an example from protein synthesis.

A growing cell is constantly producing more proteins [see, for example, Watson (1976)]. In this process, the information in a strand of messenger RNA is used as an instruction for making a protein. The input tape consists of a single strand of RNA. The output is a protein, a linear polymer of amino acids which then folds into a functional three-dimensional structure. The mRNA is a polymer made up of four kind of units, A, U, G and C. The protein is a polymer made up of twenty different kinds of amino acids, glycine, alanine, tyrosine, ...

AUGGGUCCAAAGAGCCUGUGG.......UGA......mRNA
Met Gly Pro Lys Ser Leu Trp   stop   protein

Protein synthesis is carried out at a polymolecular assembly of proteins and RNA called a ribosome, which is the "Turing machine" for the process. Many different molecules participate in the protein synthesis on the ribosome, including tRNA, GTP and a host of co-factors. The ribosome reads the mRNA input tape and inserts the appropriate amino acids into the protein in sequence. The logical operation performed might be described as the following instruction set:

(1) Read the next three bases on the mRNA molecule "tape".
(2) Look up the corresponding amino acid in the genetic code "dictionary".
(3) Add that amino acid to the protein.
(4) Shift the input tape by three bases.

The program contains a possible "stop" instruction in the code dictionary. The signal for starting is more complex.

The logical operation of reading a particular next codon in the mRNA is actually carried out in a chemical reaction which knows nothing about logic. The reaction might schematically be written:

$$\text{peptide-}n + \text{correct amino acid} \rightarrow \text{peptide-}n + 1.$$

Correct refers to the amino acid in correspondence with the next mRNA triplet. There is, however, also a competing reaction

$$\text{peptide-}n + \text{incorrect amino acid} \rightarrow \text{incorrect peptide-}n + 1.$$

Within the logical view of computation this competing reaction simply "doesn't happen". From the point of view of chemistry, the competing reaction can happen, but has a different energy barrier for taking place. The reaction is rather similar to the one desired, and any enzymatic system which allows the correct reaction to take place must also allow the incorrect reaction, albeit with rather slower rates. The equilibrium constant for adding the correct amino acid and for adding the incorrect one are essentially identical. Equilibrium processes—physical processes which take place sufficiently slowly—result in useless proteins being formed because a particular incorrect amino acid is as likely to be added as a correct one. The logic which we would like the biological system to display—correct amino acid only—is not possible without errors, and is possible with low but finite errors only by deliberately running the system out of equilibrium. In non-equilibrium processes, the choice between products can be made on the basis of rates. The biological process must be designed to make use of kinetics to obtain accurate logical calculation. The qualitative description of biochemistry—that reactions take place because molecules fit together, and other reactions do not because the corresponding enzymatic binding does not happen—consists of logical statements, and by being only logical overlooks the essential non-equilibrium element to real physical computation.

Ordinary computers also make use of dissipation to obtain their accuracy. The course of a computation might be described as a motion in computer state space, or in a physical phase space. The initial data define a starting point in that space, and the computer is supposed to follow an appropriate and determinate path to the solution, represented by another point in that space. This is illustrated by the solid line with arrows in fig. 1. The effect of noise and imperfections is to cause the flow to deviate from the desired direction, and an accumulation of noise or statistical imperfections will result in a wrong answer.

COMPUTER STATE SPACE

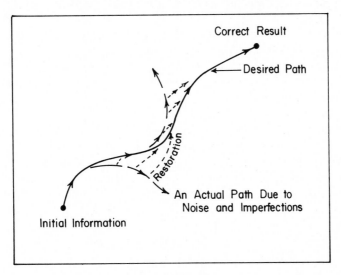

Fig. 1. The dynamical trajectory of a computing machine from initial information to the appropriate answer. In a noiseless perfect machine, the desired path would be directly followed. Noise and imperfections cause increasing divergence from this path, while restoration returns the trajectory toward the ideal one.

The way to avoid this problem is to introduce a physical process which squeezes the system back down onto the proper track in state space. If we can keep this compression going on, it will result in finding the correct answer in spite of noise. Compressing a bunch of trajectories in phase space down into a smaller volume is easily done by a dissipative process. This squeezing down is essential to avoiding errors in a system which has the physical possibility of making them. The reason that protein synthesis must be run out of equilibrium is exactly this necessity for compacting the occupied state space in a dissipative fashion to avoid errors. (In electronic digital computers, the electrical power necessary to do a computation is in principle bounded by this aspect, though in fact the power dissipation is far larger for reasons which have nothing to do with fundamental physics.) In digital computers, this idea is called restoration (von Neuman 1952, Mead and Conway 1980), and the fact that a nominal digital "one" (really a particular voltage level) will recover to the appropriate level after a transient perturbation is given to the circuit is an illustration of the presence of restoration. In thinking about the computations which are performed in neurobiology, we must expect that the system will show such restoration in an obvious fashion, or else the system would be unable to compute. Restoration—which gives a robustness against noise and computer imperfections–is a universal necessity to physical computers, while an irrelevancy in logical computers.

Even with restoration, biological computers are inaccurate. The intrinsic accuracy of protein synthesis of the simplest sort based on elementary chemical recognition, is about 1 error in 100 in difficult cases (Pauling 1957, Hopfield and Yamane 1980). The protein synthesis system also has a proofreading system (Hopfield 1974), which consumes more free energy and adds another layer of restoration, resulting in error levels on the scale of $1/3000$, a drastic level from the point of view of electronic machines, but apparently good enough for biology.

The one additional oddity about biological computers is that they do not perform as well as possible. Streptomycin resistant mutants of bacteria proofread better (Yates 1979) and are more accurate than the normal wild-type bacteria, but are an artifact of an evolutionary history with streptomycin present in the growth environment. In the absence of streptomycin, the bacterium reverts to the wild type—which is less accurate. The same general kind of less-than-best accuracy has been also demonstrated in DNA synthesis in T4 bacteriophage by Muzyczka et al. (1972).

In biology, being as accurate as possible in computations may be a loosing proposition! Biology is not interested in perfect logic. The possibility of making progress through random accidents—called creative thought when they occur in neurobiology, or evolution and speciation when they occur in molecular biology—seems to be an essential part of biological computation.

3. Neural computation

We turn next to computation in neurobiology. The object is to understand how a set of neurons makes decisions, generates actions, generalizes and learns and profits from past experiences. The emphasis here must be on our understanding. To merely know the input–output relation of a set of neurons by exhaustive study is not satisfying. Worse, a clump of neurons with 100 input neurons and 100 output neurons might easily require more than 10^{40} bits of information to characterize the input–output relation, making exhaustive study impossible. It would be equally unsatisfying to know the neural hardware in sufficient detail to be able to simulate the hardware on a monsterous digital computer and predict correctly the behavior of a neural system. This would correspond to being able to simulate the behavior of a classical gas of complex molecules, without the conceptual understandings brought to such gases by statistical mechanics and thermodynamics.

Perhaps the largest computational burden placed on our brains is involved in visual perception. The end result of this computation is our decisions about what we have seen. (It is appropriate to emphasize decisions, for decisions are the essence of computation. Strictly linear systems do not truly compute, although they can be very useful elements in a computational system.) Given a flash exposure to a typical visual scene, we note the presence of a few familiar objects, some rough characteristics of each, such as color, general size, etc. The immense supply of almost non-meaningful information—more than 10^9 bits of information were processed by the retinal cells which begin this calculation—is compressed into significant perceptual information, of which there are probably only a few thousand bits. In a digital machine such a computation would be done by making a very large number of sequential decisions, but somehow the essence of biological decisions seems to be rather more holistic, collective, or Gestalt. We want to understand how such decisions are made.

Aspects of neural computations in higher animals which are particularly puzzling from a physics viewpoint include:

(1) The system makes very effective use of its computational resources, whether measured by speed of calculation or volume or energy considerations.

(2) In such systems, the computations done by the system are very resistant to damage to the neural system (fail soft).

(3) Emergent properties such as self-awareness seem to be present.

(4) In higher animals, neurobiology manages to function without a determined circuit diagram.

There are two reasons to think that progress might be made in understanding neural function. First, the system is large, the connectivity between neurons is large, the behavior is somewhat insensitive to the destruction or misfunction of components, and the calculation seems to have a somewhat holistic character. This suggests that collective effects might be involved, and that a search for collective effects in neural networks [Little (1974); Little and Shaw (1978); Hopfield (1982)] and neural computation might be fruitful. (Note that this is not the way that most chip designers work—they do not make use of collective effects, and would attempt to suppress them if they ever were to be noticed.) Second, the system cannot be as complex as it might appear. It is true that a general module of neurons with 100 inputs and 100 outputs could require 10^{40} bits to specify its behavior, and would require such a number if all we think of is to list them. But a general module would require also 10^{40} bits of information to describe how to build it. A simple module, which can be described in perhaps 10 000 bits cannot produce a general input–output relation. It must produce a very special kind of input–output relation, which can be only apparently complex, not truly complex. (In this same fashion, the random-number generators which are used in computers appear to provide highly random numbers, but in fact generate very special sequences because the generators can be described by short programs.)

4. Classical neurodynamics

We must understand what computational or circuit facilities a nervous system has at its disposal in order to see what is a mystery and what is trivial. The anatomy of a "typical" cortical neuron is sketched in fig. 2. The morphological and functional diversity of such cells is very large. We will briefly review the electrophysiology of such cells, which is covered in detail in textbooks (see, for example, Kandel and Schwarz 1981).

A small electrode can be inserted into a cell body, and the potential difference between the inside and the outside of the cell can be studied as a function of time. A typical result of such a recording is a baseline potential of about -90 millivolts, on which is superimposed a set of more or less stereotyped voltage "spikes" called action potentials, rising to a potential of about $+50$ millivolts, and being of about 1 millisecond duration. An action potential propagates from a cell body down an axon by an active regenerative process, and can thus propagate long distances without attenuation.

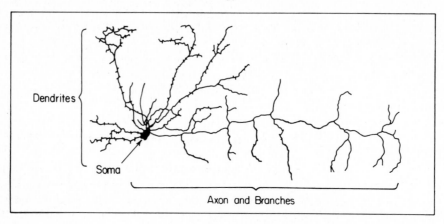

Fig. 2. A "typical" neuron from brain, showing the dendritic arbor (chiefly inputs), the cell body or soma, and the axon or chief output path.

The rate of generation of action potentials depends on the input to the dendritic arborization. Inputs to the dendrites produce current flows which depolarize the cell body, and if the cell body is depolarized enough, an action potential will be generated. This process can be readily simulated by passing a positive DC current into the cell body. The rate of firing (generation of action potentials) as a function of that positive current is sketched in fig. 3. Such curves have generally a sigmoid form, going from zero for large negative currents to a saturating maximal value of 100 to 1000 per second (depending on the neuron) for large positive currents.

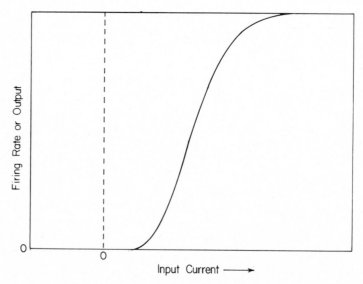

Fig. 3. The mean firing rate of a typical neuron as a function of the input current to the cell body. A sigmoid curve of this general shape is taken as the input–output relation $V = g(u)$ for the model neurons.

The ends of branches of axons of a cell are connected by synapses to the dendritic arborization of other cells. When an action potential reaches a synapse, a small amount of a chemical neurotransmitter is released at the axon terminal. Some of that transmitter becomes bound to receptors on the adjacent dendrite, and a current leak across the dendritic cell membrane is briefly produced. The detailed time history of this process depends on the type of cell, and on the particular neurotransmitter and receptor molecules involved, but is typically on the millisecond scale in simple motor neurons. This net current introduced into the cell body by such a process contains positive terms from synapses which excite the cell, and negative terms from synapses which inhibit (or suppress) the activity of that cell.

The electrical capacitance of a neuron is not negligible—cell membranes are only about 50 Å thick. The integrative time constant of a cell is appreciable, on the time scale of milliseconds. An individual action potential arriving at a synapse is generally not sufficient to result in the production of an action potential by the next cell. More typically, many action potentials must arrive (generally at many different synapses) on the dendritic arborization of a single cell within an integrative time constant before the post synaptic cell will fire.

A minimal set of properties which might be used to describe the functioning of a neural network might:

(a) Regard the individual action potentials as irrelevant quantization, and replace the true output of a cell i (a set of action potentials occurring at particular times) by a smooth, continuous property V_i describing the short-term average of the number of action potentials per unit time being sent out by that neuron. (Similarly, the discrete electrons are replaced by continuous charges and currents when one is dealing with electronic circuit behavior, and the quantization of electrical charge is generally neglected except when a noise analysis is done.)

(b) Describe the effective output V_i of a cell by a monotonic increasing sigmoid function $V_i = g_i(u_i)$ (where u_i is the input voltage of cell i) with $g(-\infty) = 0$ and $g(+\infty) = 1$. (The choice of 1 is only a convenient scaling.)

(c) Linearize the input dendritic arborization. The connection from neuron j to neuron i can then be described by a transconductance T_{ij}. The input to cell i can then be described as $\sum_j T_{ij} V_j + I_i$. I_i represents the input to a cell coming from sensory input (e.g. light falling on a light-sensitive neuron, resulting in a transduction from external light to an internal neural signal), external electrodes placed by a neurobiologist, or other neurons not being described at the moment.

(d) Represent the effect of the transmembrane resistance and capacitance of cell i by constants ρ_i and C_i. These approximations result in the following equations of motion for the state of the system of neurons:

$$C_i \frac{du_i}{dt} = \sum_j T_{ij} V_j - \frac{u_i}{R_i} + I_i,$$

$$u_i = g_i^{-1}(V_i).$$

These equations (Sejnowskii 1981, Hopfield 1984), describe a classical neurodynamics—classical in the sense of classical physics. They omit the effect of the quantization of the action potentials, the "pulse coding" by which many neurons communicate. They are also classical in treating the communication speed as infinite, neglecting the time necessary to propagate an action potential.

These are equations both of extreme complexity and also of drastic oversimplification. The complexities of these equations can be demonstrated by the properties of simple circuits of these elements. For example, two neurons can be connected together to make a bistable flip–flop circuit. Three or five neurons with negative couplings can be connected in a ring, forming a ring oscillator or clock. By choosing the correct offsets in the gain functions and the correct T_{ij}, a neuron can be set up so that its output is very nearly zero unless at least two other neurons which are driving it have outputs near one. This makes a logical circuit which performs the "and" operation. Continuing in this fashion, we can construct all the elements necessary to build the sequential logical operations and memory necessary for a universal computer. Any digital computer which can be built can be duplicated (as far as its logical operation is concerned) by appropriate choices of the set of g, T, I, R and C. The complexity of the behaviors which can be shown by these equations is beyond description, for they comprise all possible computations and computers.

In real brains, the synapses themselves change with time due to the activity of the network. Learning is believed to chiefly involve the modification of the synaptic strengths T_{ij}. A complete set of neurodynamical equations must also describe the rate of change of the synaptic strengths, an equation of motion for the T_{ij} themselves. Considerable experimental work and modeling of this procedure has been done, but cannot be reviewed here. The dynamic equations described above might describe the response of a brain to a new situation in the light of what the brain already knows, but does not include learning new facts or relations.

These equations are, at the same time, a gross oversimplification of the realities known from experimental neurophysiology and neuroanatomy. Individual action potentials are known to be important in some parts of neurocomputation, especially in early sensory processing in the auditory system and in the visual analysis of motion. Cells in invertebrates often display bursting behavior even when driven by a constant input, generating a train of equally spaced spikes followed by a period of no action potentials. The description of inputs from axons to dendrites is an idealization of motor neurons, while neurons in brains often show connections from dendrites of one cell to dendrites of another cell, and some connections where three processes (e.g. two dendrites and an axon) make mutual close contact. Signals in the dendritic arborization do not simply add—for there are rather more complicated interactions between inputs, and connections made close to the cell body may "veto" the inputs coming from more distant parts of that dendritic arbor by forming a shunting impedance path. Some general neurotransmitters have relatively longer-lasting or modulatory effects on neurons, and several time scales of integration are present.

The above list of emissions and oversimplifications could be easily extended to great length, making many neurobiologists feel that such a set of equations must be inadequate. But for looking at collective behaviors, it is essential to model a neuron as simply as possible, realizing that if any interesting computational properties occur in a simple system, the real system will certainly have these properties, and perhaps more. An excessively complex model of the capabilities of a cell would preclude the possibility of understanding the collective aspects of the system. So we are led to ask whether these very simple networks will do in a natural, elegant, and collective fashion some of the kinds of computation which biology seems to do uniquely well.

In conventional computer hardware, memory is located by address. Information is stored somewhere, at a particular address, and can be retrieved by an instruction to read the contents at that address. This is most obvious in the case of a magnetic recording medium, where the information is stored as magnetization at a physical location on the recording disc, and the address is a description of that location. It is equally true of the chips used in a simple semiconductor memory like a Read Only Memory (ROM), where there are two separate sets of signal wires to the chip. One is a set of address leads, into which information is sent about the locations to be read. The others are information leads, on which the information from the desired location is sent out. If we were in possession of part of the information stored at some address and wanted the rest of that information, but did not know the address, there would be no alternative but to examine each memory in turn. Partial information is not a key to more information in ordinary machine memory hardware.

Memory in animals seems to be of an entirely different nature. Each of us knows many individuals, and a particular acquaintance can be described by a large number of characteristics (eye color, height, face shape, accent, vocation, spouse, children, shared experiences, education, nationality, political views...). All these characteristics belonging to a single individual are somehow associated or linked in a memory. The entire set of features connected to one individual can be retrieved from an incomplete and partially incorrect set of information about that individual. "The short jovial tennis-playing physicist from MOT" is sufficient information to identify Prof. Feshbach, in spite of the fact that MIT has been misspelled and has hundreds of physicists and many tennis players. In this retrieval we have no notion of address. Any substantial subset of the information can be used to access the memory. Psychologists refer to such linkages as associative memory. Associative memory is a general and important feature of memory in all animals which learn, and the classical association paradigms of Pavlovian condition, often thought of in terms of higher animals, have been demonstrated by Gelperin (1983) and Sahley et al. (1981) on animals as simple as a garden slug (*Limax maximus*).

Although associative memories seem natural to biology, they are foreign to conventional computer hardware. They turn out to be natural for collective biological hardware, because the neurodynamics we have given describes a dynamic system, not a logical calculation, and the idea of associative memory is a natural construct in a dynamical system.

Consider an "information space" about people, with many dimensions. Each axis could be labeled with a characteristic such as height, weight, name of spouse, or age. In such a space, each person you know is represented by a point, whose location describes all characteristics of the individual in question. You know very few of all possible people, so the set of points which represent your memories is very sparse in this space. Suppose now that you are given partial information about some

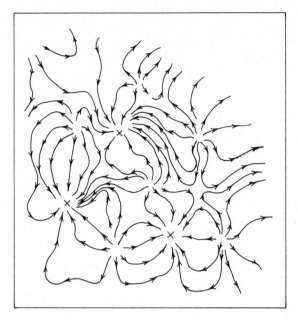

Fig. 4. A simple motion map of the change of the state of a dynamic system, characterized by a set of stable states. The stable states represent memories (in the case of associative memory), or more generally, possible answers in a computation.

particular friend. This information represents an approximate location in the information space. Presumably this approximate location is closer to the position of this friend in the information space than to the positions of other individuals. An associative memory functions by using this closeness to locate or move to the complete information. If a dynamical system is described by a flow characterized by stable points, as in fig. 4, the initial information could start the dynamical system at a location close to one of the stored memories. If the state-space motion is such that the motion goes to the most nearby stable point, the dynamical system would be functioning as an associative memory (Hopfield 1982). Many dynamical systems display state-space motions with multiple stable states. Any such system is a candidate for an associative memory if in addition the stable states can be placed wherever they are desired. Thus associative memory is not an esoteric property of computational systems. It is natural behavior of many simple dynamical systems.

In the case of the neuronal dynamics model, when the matrix of connections is symmetric, Hopfield (1984) showed that the flow is simple, and has no cyclic or strange attractors. Somewhat less stringent conditions will suffice. The symmetry of the connections is natural to the associative memory problem because the idea of simple association is itself symmetric. The essential step in this proof is the construction of a Lyapunov function, the "computational energy", which decreases monotonically during the neurodynamical motion. This energy is minimized (at least locally) by the state-space motion. Of course, the system of neurons is itself unaware that it is doing anything so grand as minimizing some global function—it merely obeys its own neurodynamical equation. This computational energy is simply our way of describing a global or collective property which enables us to understand the nature of the behavior of a large and complex system.

The idea of restoration is intrinsic to the notion of computation in a world of noise and error. The entire computation of this neural dynamics is intrinsically restorative, for the motion of the system is downhill on an "energy" terrain. Minor noise, or perturbations in the shape of the energy surface will not change the valley to which the state-space trajectory is descending, except when it starts at or traverses a logically ambiguous location such as a saddle point in the terrain.

Associative memories based on these principles of collective behavior and neurodynamics have been simulated on digital computers. The T_{ij} construction is itself amazingly simple. The T_{ij} can be built up by adding one memory at a time, as biology would require. The particular connections between i and j do not require any global information about the memory to adjust themselves, but only make use of information locally available from the "experience" of the network being held in a state to be remembered by the external input to it. Hebb (1949) postulated biological learning rules for synaptic connections which are of the nature that this associative memory would require.

Because the neuronal equations also correspond to elementary (but highly connected) electrical circuits of novel form, they also suggest the utility of making devices of such a structure. John Lambe first constructed such circuits, and there is a growing effort to investigate such devices in both electrical and optical formats (Lambe et al. 1985, Sivilotti et al. 1985, Jackel et al. 1986, Psaltis and Farhat 1985).

The secret to the simple flow pattern to stable states is the "computational energy" function which is always being decreased by the equations of motion. The operation of the collective circuit decision could be anthropomorphically described as the attempt of the circuit to find a minimum of this function. Computational problems can often be stated as minimization or optimization problems. When a mapping can be found which creates a correspondence between an energy function and the quantity which is to be optimized, then this kind of network may be effective in solving the problem. Many perceptual and reasoning problems from biology are of this minimization nature. What is the best route home, or the best pathway to move your hand in order to reach out and grasp the sandwich? Human speech is normally a continuous sound stream. Given such a continuous stream, where is the best place to interpret the breaks between words? (If you think this is not a problem, try to remember the first time you heard French spoken if your native language is not French). In the scene at which you are now looking, what is the best way of associating features together so that they will make reasonable objects? There is an immense field of optimization problems which biology must solve, and the fact that the neural networks are spontaneously capable of making a collective optimization decision suggests the way that a nervous system may go about doing such computations.

In an effort to understand the power of such networks in difficult computations, Hopfield and Tank (1985) took a non-biological optimization problem which is computationally very difficult (NP-complete) and examined the kinds of solutions which a neural network that we designed would generate. The object was not to see how a brain might solve this problem, but rather to study the computing power which an assembly of 100–1000 richly interconnected neurons could possess. The Traveling Salesman Problem investigated is defined as follows. Given a particular set of cities—Copenhagen, Stockholm, Oslo, Bergen, Malmö, Aarhus,..., in what order should one fly between the cities such that each city is visited once, one returns to the starting point, and the total distance flown is as short as possible? This problem is connected to biological computation of problems like word boundaries because in each case, there is a combinatorially larger number of possibilities which must be considered.

The simulation showed that a simple network of 900 "neurons" could find a good solution to a problem on 30 cities in a single convergence. The settling time for a set of neurons involved in such a computation would be 0.1–0.2 seconds. The network did not find the best solution. But of the more than 10^{30} possible solutions, the network found one of the best 10^7, rejecting poor solutions by a factor of 10^{23}. This is an immense amount of computation for such a small set of elements. A microcomputer, with an intrinsic speed about 10^5 times faster and containing 100,000 times more transistors, would achieve a comparable result in a comparable time. The fact that the neuronal network found a very good solution but not the best one is typical of biological computation. As was evident from the earlier molecular examples, biology does not insist on perfect solutions to computational problems.

The immense computing power of this small network comes from features which are explicitly those of neurobiology. First, large connectivity between neurons (or in engineering, amplifiers) is required. Brains have connectivities on the scale of 1000, while typical transistors in integrated circuits get inputs from only a few other transistors, and send outputs to only a few others. Second, the system operates in an analog mode, using the smooth, graded and nonlinear response of a neuron to its input. By contrast, the conventional digital machine emphasizes logical operations, ("on" or "off", 1 or 0), and suppresses as much as possible the fact that the fundamental hardware is itself an analog circuit. Third, a single collective decision is smoothly made by many neurons at once, rather than a sequence of minute logical decisions. The operation is as a dynamical physical system rather than as a logical one.

5. Beyond neurodynamics: complementarity

I turn finally to topics which I do not usually write about. They are subjects about which Niels Bohr would have asked, and on this centenary I am obliged to try to give answers—such as they are—to some queries which clearly come forth from the legacy of Bohr's essays.

Bohr's view of complementarity extended rather more broadly than the narrow confines of the wave-particle duality or the quantum-mechanical uncertainty relations. His view came from the understanding of quantum mechanics and its relation to classical systems, but did not fundamentally rely on quantum ideas. He asked whether there were other contexts in which the conflicting demands of observer and of normal unperturbed system behavior produced an essential paradox, and inability to know all about a particular system in the same fashion that we cannot ask simultaneously wave and particles questions. Such limitations on complete knowledge need not even necessarily arise from observer-experiment problems, but might arise in other situations. Bohr particularly raised this question with respect to the operation of biology, where he remarked (Bohr 1958, 1963):

> "The basis for the complementary mode of description in biology is not connected with the problems of controlling the interaction between object and measuring tool, already taken into account in chemical kinetics, but with the practically inexhaustible complexity of the organism."

This clearly expresses his view that issues of complementarity and the limits of knowledge in biology are of the nature that I described for turbulence, and not fundamentally involved with quantum mechanics in a profound way. Bohr's (1958) view toward complementarity in biology evolved during his lifetime, as biology and the understanding of quantum physics also rapidly evolved. His later essays themselves seem somewhat complementary. The next paragraph attempts to combine the views expressed chiefly in several assays written in the last five years of his life (Bohr 1963).

Bohr viewed the action of the human brain as perhaps the most likely case of the occurrence of a complementarity limitation to the precision of knowledge of biological behavior. There appeared to Bohr to be intrinsic complementary aspects to the detailed study of the anatomy of a particular brain and the thought processes in that brain. Since the thought processes themselves change the microanatomy of the brain at the molecular level—how else could we remember our thoughts—observation of the function of the brain in a whole animal inevitably changes the anatomy of that brain. But an observation of the anatomy of the brain in sufficient detail to predict the functioning of that brain is so destructive as to preclude making the observations which would test the predictions. In this description, microanatomy and psychology become complementary views of a brain, to be united by a description which like the Schrödinger equation and its interpretation, should show why this limitation must forever remain.

What does a generalization of the classical neurodynamic equations in order to take into account thermal noise, and more realistic anatomy and physiology, suggest about such questions? If, with such generalizations, an adequate number of precise measurements were made (including studies of the properties of the modification processes) we could with sufficiently large computers predict the state of the neuronal system for a short time, and thus be able to simulate its behavior. Adequately precise information about a mammalian brain for such purposes might well require 10^{20} bits of information about a single brain, clearly a hopeless task. It might be too complex to ever simulate. This might be a fundamental limitation to knowledge of such systems. But while Bohr considered this aspect, it is not the complementarity which he viewed as most important.

His essential complementarity problem is whether even a complete set of electrophysiological measurements for simulation of neurons in complete detail can be connected to particular higher mental processes which are normally described in macroscopic or cognitive terms. There must be something observably different in the microscopic behavior of my neurons which occurs between the time at which I do not understand a mathematical theorem and the time slightly later when I do understand it. That difference can be studied by physiologists. That difference can also be pursued by psychologists asking questions about the nature of my under-standing. But even if we can observe all the physiological microdetails, these microdetails may not necessarily lead to any knowledge of what "to understand the theorem" means. There may be no way to deduce the information available to the psychologist from the data obtained by the physiologist. In such a case, there would necessarily be parallel or complementary descriptions of mental phenomena. The analog in physics might be a phase transition so subtle that even though we can see macroscopically that the phase transition has occurred, the phase itself is so complex that it cannot be described microscopically by a list of symbols or words short enough to be read during a lifetime. In that case, a complete Hamiltonian, the equivalent to the equations of neural dynamics, would still need to be augmented by the complementary description of macroscopic physics in order to describe observa-ble phenomena.

The complementarity issues just discussed have to do with the issue of computa-tional complexity and complexity measures, a topic which has barely begun to enter physics. There is, in addition, one peculiar aspect of the mathematics of collective neuronal computation which has the feeling of complementarity. This involves the question as to whether we will be able to understand the logical process by which the brain computes.

Computation is, in the usual description, a logical operation. But when we look at the way problems like the one of the traveling salesman are solved on a neural network (Hopfield and Tank, 1985), it does not turn out to be a logical procedure. Figure 5a shows a representation of the state of the neural network at an inter-mediate time during the settling to a solution to such a problem. Each point in the array of squares represents the output of a particular neuron. When the square is as large as the one at the lower right, that neuron has an output of 1. When the system has found a good solution to the problem, each of the outputs will be either 1 or 0. Thus, in the final state of the system, the output will appear as in fig. 5b, with one 1 in each row and each column.

In this computational process, each of the neurons can be thought of as standing for a proposition. An output of 1 means that the proposition is considered to be true, and an output of 0 means that the proposition is not true. In fig. 5, the propositions (which are of the general form "the salesman should go to city D as the 7th city in the tour") are laid out in such a fashion that each proposition contradicts the other propositions in each row and column. An appropriate final state must have no more than one 1 in each row and each column.

What are we then to make of the intermediate state of the form shown in fig. 5a? At this time, the proposition that city D should be in position 8 has a weight of 0.25 while at the same time, the contradictory propositions that city D should be in position 7 and that city C should be in position 8 also have weights of about 0.25. The system might be characterized as simultaneously having partial belief in several mutually contradictory propositions. There is no obvious logical characterization of this state of partial belief in contradictory propositions. We cannot view this partial

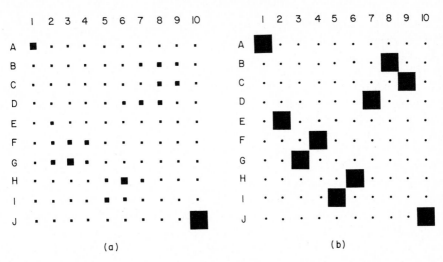

Fig. 5. The linear dimension of each square represents the activity of that neuron. The neuron in square A5 represents the proposition that city A should be visited 5th. (a) An intermediate state of the calculation. City J has been arbitrarily placed in location 10, so that the output of the corresponding neuron is of maximum size. (b) The final stable state of the network, represent the path AEGFIHDBCJ (and return to A).

belief in terms of probabilities, for the entire system is deterministic. There is a logical space in which a proposition is true or false, a space which for N propositions can be described as the vertices on an N-dimensional cube. The neural network computes by moving on a trajectory in the *interior* of this cube, while only the corners have logical meaning. Thus an insistence on finding correct logic during the decision process of the network destroys the computation as done by the network. The ability to use the non-logical interior of this space seems to be an important part of the power of these collective "neuronal" networks.

Is this a sensible way to go—to try to understand a brain by giving up simple logic and moving toward analog, analytic, and collective behavior? John von Neumann (1948) thought about the relation of the brain to the digital computers which he had been so influential in helping develop, and concluded:

> "All this will lead to theories (of computation) which are much less rigidly of an all-or-none nature... than formal logic. They will be of a much less combinatorial, and much more analytical, character."

He continued this theme in later lectures (von Neumann 1952), but did not find what he regarded as a satisfactory solution to the problem of how a brain computes.

6. Summary

We have learned a great deal about the chemical and electrophysiological basis of brain since Bohr wrote the cited essays. The way in which important aspects of brain function and neural computation could be of a collective nature is beginning to emerge. We see some simplified governing equations which are relatable to collective behaviors. These collective behaviors even in rudimentary form seem to show how associative memory and combinatorially difficult decision-making of the kind necessary in visual or auditory perception can be carried out with great effectiveness by a modest collection of neurons.

It is not clear where the limits of such pursuits will lie. Will we be able to explain some of the phenomena of cognitive psychology on the basis of details of microscopic neuronal behavior? Or will the macroscopics never be microscopically understandable? If Bohr's argument that a form of complementarity may be involved in brain turns out to be correct, the mathematical explanation for it will lie in the nature of computational complexity and the meaning of computation. Will there be a mathematical complementarity in brain? My own guess is that the complications of the system are not intrinsic, and that we will be able eventually to go astonishingly far toward understanding the language of psychology from a microscopic viewpoint. While weather prediction on the basis of molecular modeling is impossible, we do understand why there is weather and what storms are on the basis of an atomic approach to atmospheric phenomena. In this same sense, new approaches to theory in neurobiology should lead us at least to an understanding of the *existence* of psychological constructs on a microscopic basis, and potentially lead much further.

Postscript

The editors have expressed the hope that someone might read this book in connection to the second Bohr Centenary. The science described here will, 100 years from now, be amusingly naive. How science is done—who influenced whom, and how—may be of greater interest. I close on a personal note, which would otherwise be lost. I did not know Niels Bohr. John Wheeler was the strongest influence in my taking a position at Princeton, and Max Delbrück strongly influenced the Caltech scene and my interest in moving there. Both of these men were in turn involved with Bohr and Copenhagen. I have deep admiration for the imaginative scientific spirit of each of them. Reading the essays of Niels Bohr and listening to this symposium, I became increasingly aware that many elements I find admirable in these scientists are related to the spirit in which Bohr worked. But whether this similarity is due to influence or due to mutual affinity is not obvious. Francis O. Schmitt and the Neurosciences Research Program did much to stimulate my interest in neurobiology. The significance and liveliness of the many discussions about biological and network computation with David W. Tank is gratefully acknowledged.

This research was supported in part by NSF grant PCM-8406049.

References

Bohr, N., 1958, Atomic Physics and Human Knowledge (Wiley, New York).
Bohr, N., 1963, Essays (1958–1962) on Atomic Physics and Human Knowledge (Interscience, New York) p. 21.

Gelperin, A., 1983, in: Neuroethology and Behavioral Physiology, eds F. Huber and H. Markl (Springer-Verlag, Berlin) p. 189.

Hebb, D.O., 1949, The Organization of Behavior (Wiley, New York).

Hopfield, J.J., 1974, Proc. Natl. Acad. Sci. USA **71**, 4135.

Hopfield, J.J., 1982, Proc. Natl. Acad. Sci. USA **79**, 2554.

Hopfield, J.J., 1984, Proc. Natl. Acad. Sci. USA **81**, 3088.

Hopfield, J.J., and D.W. Tank, 1985, Biol. Cybern. **52**, 141.

Hopfield, J.J., and T. Yamane, 1980, in: Ribosomes, eds G. Chamblis, G.R. Craven, J. Davies, K. Davies, L. Kahan and M. Nomura (University Park Press, Baltimore) p. 585.

Jackel, L.D., R.E. Howard, H.P. Graf, B. Straughn and J.S. Denker, 1986, J. Vac. Sci. Tech. **B4**, 61–63.

Kandel, E.R., and J.H. Schwartz, 1981, Principles of Neuroscience (Elsevier, New York).

Lambe, J., A. Moopenn and A.P. Thakoor, 1985, Jet Propulsion Lab Publication, **November**, 85.

Little, W.A., 1974, Math. Biosci. **19**, 101.

Little, W.A., and G.L. Shaw, 1978, Math. Biosci. **39**, 281.

Mead, C.A., and L. Conway, 1980, Introduction to VLSI Systems (Addison-Wesley, Menlo Park, CA) p. 335.

Muzyczka, N., R.L. Poland and M.J. Bessman, 1972, J. Biol. Chem. **247**, 7116.

Pauling, L., 1957, in: Festschrift Arthur Stoll (Birkhäuser, Basel) p. 597.

Psaltis, D., and N. Farhat, 1985, Opt. Lett. **10**, 98.

Sahley, C., A. Gelperin and J.W. Rudy, 1981, Proc. Nat. Acad. Sci. USA **78**, 640.

Sejnowskii, T.J., 1981, in: Parallel Models of Associative Memory, eds G.E. Hinton and J.A. Anderson (Lawrence Erlbaum Associates, Hillside, NJ) p. 189.

Sivilotti, M., M.R. Emmerling and C.A. Mead, 1985, in: Conf. on Very Large Scale Integration, ed. H. Fuchs (Computer Science Press, Rockville, MD) p. 329.

von Neumann, J., 1948, in: Collected Works, Vol. 5, ed. A.H. Taub (Pergamon Press, New York, published in 1963) p. 304.

von Neuman, J., 1952, ibid, p. 354.

Watson, J.D., 1976, Molecular Biology of the Gene (Benjamin, Menlo Park, CA).

Yates, J.L., 1979, J. Biol. Chem. **254**, 11550.

Discussion, session chairman N.K. Jerne

Zimanyi: You have used a symmetric connectivity matrix T_{ij}. But in a neuronal network we have about as many inhibitory neurons as excitatory ones. The part of the connectivity matrix representing these types of neurons should be antisymmetric instead of symmetric. Howe can one overcome that problem?

Hopfield: Even networks which do not in detail look symmetric may be constructed equivalently to symmetric networks and possess the stability properties of symmetric networks. The symmetry assumed here serves to ensure viability of the mathematics. Asymmetric systems open a Pandora's box of totally uncontrollable mathematical behavior, but have a much richer computational structure.

von Weizsäcker: I have no quarrel with reductionism. Bohr used to explain complementarity by the difficulty of language. Can you explain the complementarity structure in language by your network mode? An example: My uncle, Victor Weizsäcker, a medical psychologist, who reminded me in many respects of Niels Bohr, once said: "When I ask you: 'what are you thinking now?', and you start to answer, you are already lying."

Hopfield: You are absolutely right in referring to Bohr's pointing out the language as an intrinsically complementary system; complementarity must be left out from microscopics when deriving collective results in a reductionist attitude.

Anderson: Going back to one of your first sentences about the complications in biology arising not from the size of Planck's constant, I should like to point to the size of Avogadro's number, a biological constant in itself, as it represents a piece of matter approximately our size. This may be the amount of matter necessary to produce the required complexity.

Berry: Let me return to your point concerning how thinking causes microscopic changes in physiology and your skepticism whether microphysiology could tell us when a person has solved a problem. Strictly, I should say "when the person *thinks* the problem is solved." I suspect you don't really believe that, because you and everybody in this room knows how solving a problem makes you feel better, quite literally. A problem must be awfully easy and awfully trivial to generate no such feeling at its solution. Consequently, do you not think the microphysiology should display some chemical signal when the person thinks he or she has found the solution?

Hopfield: Emotional states are relatively easy to recognize by chemical signals. I think it could probably be done now. But you would still not be able to tell what the person's understanding of the answer was or whether it was solved from your point of view.

The Lesson of Quantum Theory, edited by J. de Boer, E. Dal and O. Ulfbeck
© Elsevier Science Publishers B.V., 1986

Complementarity in Linguistic Observation, Description and Explanation

Wolfgang U. Dressler

University of Vienna
Vienna, Austria

Contents

1. Complementarity

In choosing Niels Bohr's notion of complementarity as the organizing principle of my contribution I do not intend to speak about complementarity in the fallacious meaning exposed by Niels Bohr (1958, p. 81) when he writes "the relation between national cultures has sometimes been described as complementary". Also within linguistics it makes no sense to speak of complementary relations between languages, and in fact at least theoretical linguistics is not so much concerned with any of the more than 6000 languages that either are spoken today or have been sufficiently described before they died, as it is concerned with what is common to all languages or at least available to the linguistic capacity of all human beings.

Another, rather fallacious way of using the notion of complementarity would be taking descriptions in terms of classical physics to be complementary to descriptions in terms of quantum theory because classical explanation (and therefore also description) has been proven to be a special case of quantum-theory explanation. Relations between the macro-level and the micro-level, however, are quite different in linguistics from such relations in physics. As an illustration let me mention a false claim: Two linguists (M. Wandruszka cited by R. Anttila 1977, p. 221) cited the

Austrian biologist Wolfgang Wieser's description of the brain as not working exactly, often blundering and correcting itself, not proceeding logically, but according to similarities, being extremely redundant etc. so that a grammatical description of language by means of precise rules would be unrealistic. Now precise rules and rigorous statements, as proposed by some linguists, refer to the macro-level of human linguistic competence and not to the micro-level of neurological activities of the brain which constitute the material substratum of linguistic competence. This situation may remind you of physics where Werner Heisenberg's uncertainty relation on the micro-level does not vitiate the precise working of laws of classical physics on the macro-level. And this situation may be compared to the contrast between the working of the brain as studied by neurophysiologists and linguistic operations as studied by linguists.

There is, however, a crucial difference between physics and linguistics: Whereas physicists have bridged the gap between the micro-level and the macro-level by an explanatory chain of arguments, there is an apparently unbridgable abyss between the neurological or biological micro-level and the linguistic macro-level with no possibility of reducing one to the other, although the interdisciplinary study of language pathology sheds some light on the question. Therefore we may speak of complementarity here: Niels Bohr (1958, p. 76, cf. p. 92) distinguished two complementary ways of description for biological facts when he wrote:

> "Actually, we must recognize that the requirements of objective description, in tendency at least, are fulfilled by the characteristic complementary way in which arguments based on the full resources of physical and chemical science, and concepts directly referring to the integrity of the organism transcending the scope of these sciences, are practically used in biological research."

To these two complementary levels we must now add a third level, the level of linguistic description. We will come back to important consequences of these differences. Notice that this level of observation, description and explanation, typical for the analytic disciplines within human sciences, does not coincide with the world of the arts as differentiated by Niels Bohr (1958, pp. 79ff.) from the world of sciences. And this distinction of levels implies that we may not reduce the level of linguistics to the level of biology nor, via a transitive reduction, to the level of physics (cf. Putnam 1981). With this I negate only the possibility of a total reduction of linguistic regularity to biological regularity, whereas I am in favor of partial reductionism, e.g. of trying to find biological, physical (or psychological) foundations of linguistic regularities. Thus, in the models of Natural Phonology and Natural Morphology (cf. Dressler 1984a, 1985a, b, Mayerthaler 1981) we try to find the extralinguistic bases for universal linguistic preferences, e.g. why at the end of a word many languages prefer a voiceless [t] to a voiced [d], but never the other way round (cf. below).

If I am correct, a main feature of Niels Bohr's thinking on complementarity is that mutually exclusive theoretical concepts may be indispensable for an exhaustive account of the physical phenomena (Bohr 1958, pp. 74, 76ff.). One example would be the particle (or corpuscular) theory and the wave theory of the atom (cf. Heisenberg 1971, pp. 67ff., 134, 171ff.; Hutten 1956, pp. 170, 189).

2. *Word formation*

Let us see whether we can find something similar in linguistics and let us examine for this purpose the area of word formation. The rules of word formation of a language tell us how to combine words to compound words such as *quantum* and *mechanics* to the compound word *quantum mechanics*. Or in order to take our reunion, more than a year ago we were first invited to a *Jubilee Conference to celebrate the centennial of Niels Bohr*; but soon a single, but complex compound word was used: the *Niels Bohr Centenary Symposium*. Other word-formation rules tell us how to derive one word from another by means of endings, e.g. *convene* → *convention* → *conven-tion-al* → (and if we want) → to *conven-tion-al-ize* → the *conven-tion-al-iz-ation*. Even if we have never heard a word like *conven-tionalization* and even if we strongly disagree of this word for stylistic or esthetic reasons, all of us would agree that *conventionalization* is a well-formed word according to the rules of English word formation that speakers of English share, albeit in a subconscious way (cf. Bauer 1983). However, if we read or hear the new word formed by James Joyce to characterize a singer, i.e. his creation *endlessnessnessness*, then we might enjoy this word, but we would agree that it is not well-formed or that it violates the rules of English word formation. Various linguistic schools have elaborated descriptive and also explanatory models or rules of word formation both for English and many other languages.

However, any model of word-formation rules has troubles in accounting for how *blackmail* is formed from *black* and *mail*, *gooseberry* from *goose* and *berry*; and what should be made of *cran* in *cranberry*? And what about *heliocentric*, *heliograph*, *heliotype* which have no English word *helio* as a common basis. Of course, it would be anachronistic to assume the presence of the Greek word *hélios* ("sun") in the minds of native speakers of English. And if we take another compound with the same historical origin, *heliotrope*, already the different pronunciation of its first part tells us that it cannot be derived by an English rule of word formation from a word for "sun", i.e. *heliotrope* is a thoroughly fossilized compound word but still a compound.

In order to describe such compound or derived words we obviously need another model, namely a lexical model of lexical description, which tells us which words exist in English, i.e. which words are stored in the mental lexicon, what they mean and whether and how they are related to each other. In this way we have two complementary models, a rule model and a lexical model. They complement, they supplement each other insofar as they refer to mutually exclusive observational situations, e.g. *conventionalization* is not an actually existing English word, but it can be well described as being formed by rules of English word formation, whereas *heliotrope* is an existing word of English but can hardly be formed by rule from the primitive elements *helio* and *trope*.

On the other hand, there are many compound or derived English words where it is very difficult to decide whether they should be described by the rule model because they seem to be processed in a rule-like way by our brain, or whether they are simply lexically stored, e.g. the noun *convention*. Here, both models compete with each other. The question is a case of undecidability which may be somewhat comparable to the technical problems involved in Werner Heisenberg's uncertainty relation, or even more to the ultimate undecidability between the particle and the wave model and their complementary use in different situations (cf. Hutten 1956, p. 200). Or is it rather the case that the brain operates according to both models at the same time? This is a question of the psychological reality of linguistic constructs.

This is one situation where the aforementioned abyss between the neurological micro-level and the linguistic macro-level can be partially illuminated by interdisciplinary work in language pathology, especially through the investigation of aphasia, a central cerebral syndrome, by the concerted efforts of neurophysiologists, neuropsychologists and neurolinguists, to mention the new and very exciting linguistic subfield, neurolinguistics. For our discussion I want to cite results of a team working in Moscow (especially Glozman 1974). They found that some of their aphasic patients showed rather good access to lexical storage, but had extreme difficulties to form words and to evaluate their relationships, whereas other patients freely used word formation rules compensating thus for great deficiencies in lexical storage. In a similar vein, normal children have been observed of freely using word formation rules when their lexicon was still very deficient (see e.g. Clark and Berman 1984).

3. An intuitive approach

We now need an interactive approach in order to describe and explain word formation, an approach which assumes interaction between autonomous parts of the language system, as it has become fashionable in contemporary linguistics (cf. Chomsky 1980a, b, Beaugrande and Dressler 1981).

Notice also that such an interactive approach can hardly be reconciled with a deterministic science theory or with deterministic explanations. Therefore, as a linguist, I agree with Niels Bohr (1958, p. 72) (cf. Heisenberg 1971, p. 29; Prigogine 1979, p. 66) when he draws attention to "the renunciation of the very idea of determinism" (cf. Hutten 1956, pp. 256ff., 249ff.) and to the ensuing "radical revision of the fundamentals to the description and comprehension of physical experience".

In other words the widening of knowledge, in linguistics as well as in physics (according to Niels Bohr 1958, pp. 67ff., 82), has led to a widening of the conceptual framework and to revisions in the standards of explanations which in its turn has had important consequences for the ways of describing and even of observing the phenomena we are interested in.

4. Linguistics and Galilei

Physics is very often seen as the very model of a scientific discipline by representatives of other disciplines. This is true for many linguists as well, especially adherents of generative grammar as founded by Noam Chomsky (1957, 1980a, b). [For a history see Newmeyer (1980), for a recent critique Boas (1984).] Taking over ideas as first practised in physics, Noam Chomsky (1980a, pp. 8ff., 218ff.) and other linguists [e.g. Botha (1982); here I follow Botha (1984, pp. 4ff., 161ff.)] have propounded the so-called Galilean style of inquiry in linguistics. Three aspects of this approach (both within and beyond generative grammar) are given below.

The first aspect is "an attitude of epistemological tolerance towards promising theories that are threatened by still unexplained or apparently negative data" (Botha 1984, p. 5; cf. Chomsky 1980b, p. 10). This attitude (amply documented for Niels Bohr) is highly commendable, but how long should the period of grace last? How long should one refrain from criticizing a rival school of thought which doggedly refuses to tackle problems which one considers crucial, and which have been considered to be crucial before?

The second aspect of the Galilean style is said to be aiming at "depth of understanding in restricted areas—and not gross coverage of data". Combined with the first aspect this means that coverage of data will be broadened in the long run.

Before I critically assess the extent of progress (cf. Dressler 1984b, 1985c) achieved within generative grammar, let me add the third aspect of the Galilean style, i.e. to "make radical abstractions and idealizations in defining the initial scope of inquiry" (Botha 1984, p. 5; cf. Chomsky 1980a, b, pp. 218ff.). This is an aspect underlined by Niels Bohr (1958, pp. 68, 70) (cf. Heisenberg 1971, pp. 187ff.; Hutten 1956, p. 248) as well.

5. Generative grammar

Let us now inspect three problematic idealizations and restrictions which generative grammar has practiced for nearly 30 years:

(1) the theoretical study of language has been reduced to the study of grammar, thus eliminating e.g. all questions about what human language is used for. We will return to this reduction below.

(2) The grammars to be described and explained are the grammars of ideal native speakers/listeners of their respective languages.

(3) The primary and most direct source of data in linguistic research are the linguistic intuitions of native speakers.

The idealization made in (2) eliminates the social properties of language and relegates them, at best, to the status of secondary, intervening variables studied by the linguistic subdiscipline of socio-linguistics. Quite apart from the question of whether the social character of language is not one of its fundamental properties, we must pass beyond questions of explanation and description to a fundamental question of observation: Niels Bohr (1958, pp. 69ff., 72, 74) has underlined a problematic aspect of observation in quantum theory, i.e. "the influence exerted by an observation on the object to be observed" (Klein 1967, p. 92ff.).

This is much more so in linguistic observation, and has been called the observer's paradox (Labov 1970, pp. 46ff.): the most regular and systematic type of speech is unobserved speech. Thus, linguists should concentrate on studying unobserved speech. However, in doing so they usually become observers noticed by their subjects, who can therefore not produce "unobserved speech" any more. This confirms social constraints on language performance, and it is language performance which gives us the raw data at the level of linguistic observation. It also raises the question of how to validate linguistic observations in the face of the ideal speaker/listener who is a speaker/listener in abstracto without any considerations for problems in speech performance.

The problem of data validation becomes even more critical in view of restriction (3) strongly upheld within generative grammar. If we want to know whether the aforementioned words *conventionalization* and *endlessnessnessness* are well-formed or grammatically possible English words, we should rely on the grammatical intuitions of native speakers of English. However, whenever a linguist asks an informant whether a word or sentence is correct or grammatical, the informant evaluating his own intuitions of grammaticality, is neither an ideal speaker/listener nor free (or freer) of performance problems than any actual speaker or listener in his performance of speech production or receptive speech processing, respectively (cf. Ringen 1975). This problem becomes even more acute, if the linguist is his own informant and observer at the same time, i.e. if he evaluates his own grammatical intuitions.

Therefore, at the level of observation, we must not rely on one particular source of data, but take oral or written speech production and perception as seriously as speech evaluation. Moreover, why should we exclude or regard as secondary other data sources, such as the aforementioned areas of child language, language pathology or poetic language? The non-necessity of "gross coverage of data" must not be confused with the in-depth study of a restricted problem area within a large range of different types of observation.

Now let us return to the Galilean style of linguistic inquiry in order to underline another aspect, the need for "unifying, principled theories deductively removed [...] from the primary problematic data" (Botha 1984, p. 5). This, I think, is in perfect agreement with Niels Bohr's views on quantum theory and scientific knowledge in general. Regarding this aspect Noam Chomsky and linguists directly and indirectly inspired by him have made much headway and have proposed unifying principles which have aroused much interest even beyond linguistics, such as in philosophy and psychology. For example the grammaticality or ungrammaticality of a great variety of very complex interrogative, comparative, relative etc. sentences has been shown by Chomsky (1977) to be accountable by a single syntactic rule and a small set of principles restraining it.

6. Phonology

When native speakers of German, Russian or Polish learn English, they frequently mispronounce English words such as *Ted, five, lab, peas* as *Tet, fife, lap, peace*, i.e. they devoice the word-final voiced consonants *b, d, g, v, z*, [ʒ, dʒ, ð] and pronounce them as *p, t, k, f, s, sh, ch, th*. This is called final devoicing, and has been explained as simple transfer from the native pronunciation of German, Russian, Polish etc. which have final devoicing to English which does not have it in standard pronunciation. For example *Te*[t] is the obligatory German pronunciation of *Ted*, and this is said to be carried over to English spoken by native speakers of German (such as Henry Kissinger). This is a very simple and common-sensical description and explanation apparently corroborated by our daily observations of "foreign accents": people carry over the articulatory habits of their native language to foreign languages.

We may also extend the coverage of data to other observations: what about native speakers of languages which lack final consonants of the type d, t, z, s,..., altogether? Such speakers have been observed (Donegan and Stampe 1979, p. 132; cf. Dressler 1984a, p. 47, §5.1.4) to apply final devoicing to languages such as English and French. I once worked with a native speaker of Yoruba, an African language lacking such final consonants; although he spoke English and French quite fluently, his "foreign accent" in both languages comprised final consonant devoicing, although his native language Yoruba has no final consonants to devoice. He could therefore not have learned in his own language the articulatory habit of unvoicing final consonants which he could then have carried over into his pronunciation of English and French.

Let us add a third source of data, first language acquisition by small children. Children all over the world (cf. Locke 1983) have been observed to devoice final consonants, including British and North-American children. Such observations have induced the American linguist David Stampe (1969) (cf. Donegan and Stampe 1979), the founder of the model of Natural Phonology, to posit the following more abstract unifying hypothesis (in my modified version; see Dressler 1984a, 1985a): All small children have the process of final devoicing at their disposal. If they acquire German, Russian or Polish as their first language they can happily continue devoicing final consonants. But if they acquire English or French, they must learn to inhibit final devoicing. Now Yoruba children never meet final consonants in their language; thus they do not inhibit final devoicing but may retain this process in a latent form; and when they afterwards learn English or French words which have such final consonants, this latent process of final devoicing surfaces.

Final consonant devoicing has a physical and a psychological basis: It is easier to pronounce and to perceive a voiceless consonant (of the *t*-type and the *s*-type) than a voiced consonant (of the *d*-type and the *z*-type) at the end of a word. And it is the psychological principle of least effort which explains why human beings prefer sounds and sound constellations which are easier to pronounce and to perceive [cf. Lindner's (1975) articulatory theory]. But this holds only *ceteris paribus*. The physical and psychological explanation can only be a partial one in phonology, the universal conditions and hierarchies of conditions must be explained within linguistics, as well as the factors favoring the transformation of a universal preference into an obligatory rule within a specific language.

A fourth source of data for final devoicing is a study on alcoholism, i.e. on the influence of whiskey on final devoicing. Alcohol is known to disinhibit inhibitions, and so it may also disinhibit the inhibition of consonant devoicing: the more whiskey the American subjects of Leland and Skousen (1974) drank, the more frequently (and consistently) they devoiced word-final English consonants.

In a fifth source of data, the language disturbance of aphasia, (cf. Dressler 1982) English aphasics have been observed to devoice final consonants. This again can be explained by our unifying, abstract hypothesis: we assume that they have lost control of inhibiting final devoicing. That is, if they intend to pronounce final voiced consonants, they often cannot inhibit devoicing them. The respective sound intentions of children, foreigners and aphasics are thus modified by the process of final consonant devoicing.

Linguistics can thus not be reduced to its physical and psychological basis.

7. On the language of physics

Let me add a very brief appendix on the language of physics because there seems to exist some curiosity about this subject among physicists. Niels Bohr and other physicists have expressed the belief that quantum mechanics has brought about or would need an entirely new "language". I have to disappoint those physicists who think that the situation in physics is very different from the situation in other rapidly developing sciences.

Either I have misunderstood the problem or I can largely follow Heisenberg's (1971, pp. 160–181) essay on "Sprache und Wirklichkeit in der modernen Physik" and Hutten's (1956) book (cf. also von Weizsäcker 1974a, b). The language of physics pertains to so-called languages for special purposes. If we may formulate it in a semiotic way, i.e. by means of sign theory, physicists use words and sentences with meanings they agree upon in reference to their objects of study. If these objects of study change such as it happened in quantum mechanics, then also the signs used must change (cf. Mittelstaedt 1972, p. 86). Either words or argumentations extend or otherwise change their meanings, or new words or argumentations are introduced or at least proposed. Innovations of signs are of course difficult to diffuse and they easily get into conflict with inertia in general sign usage. The Austrian writer Robert Musil remarked some 50 years ago that the expression "swift as an arrow" (German: *pfeilschnell*) is still used, although for centuries objects have existed which move much swifter than arrows. No wonder then that Eddington's proposed word *wavicle* met with little success, although it was a grammatically well-formed new word (i.e. Eddington did follow the rules of English word formation).

Physics like any other science strives towards unambiguous precision of scientific language use (cf. Weizsäcker 1974a, b), and this has clear repercussions on the levels of the word, of the sentence, and of the text, i.e. in the ways sentences are combined to larger entities such as paragraphs. Here the form of physical language follows its function.

More precision implies more clarity, and there is—according to an acute insight of Niels Bohr—some sort of "complementarity" between "clarity and truth" (German: *Klarheit und Wahrheit*). This originates in the restrictedness of any language system insofar as in everyday language a linguistic sign usually refers to a large and vaguely delimited collection of objects or sensations. Clarity in scientific language means then that the meaning of words and sentences must be "defined" (i.e. restricted) to a homogeneous class of referents (e.g. objects referred to). However, if the objects referred to are not a simple structure, such "definitions" may become very difficult because language—in order to be viable—must under-differentiate reality. Therefore a "truthful" description of complex structures necessarily becomes itself very complex and thus less clear. In this way clarity of description entails a simplified (sometimes superficial) depiction of reality.

8. Summary

Niels Bohr (1958, pp. 76ff.) (cf. Klein 1967, p. 76) acknowledged for biology and human sciences complementary "between a mechanistic and a finalistic approach". If we want to describe and explain linguistic intentions or the functions of language, not just of grammar (cf. Seiler 1978), we need such a complementary "finalistic" approach to linguistic explanation and description. This is in accord with the ways important philosophers of language such as Ludwig Wittgenstein have tackled language (cf. also Campbell and Ringen 1981). Moreover such a "finalistic" approach can be linked to a philosophical meta-level which is applicable to all human sciences, i.e. the semiotic model of Charles S. Peirce (1965). I hope to have been able to show, first, that "unity of knowledge" can be demonstrated by parallels between physics and linguistics on the level of observation, description and explanation, and second, that Niels Bohr's notion of complementarity fits linguistics quite well.

References

Anttila, R., 1977, Lingua **42**, 219–222.
Bauer, L., 1983, English Word-formation (Cambridge University Press, Cambridge, MA).
Boas, H.U., 1984, Formal versus Explanatory Generalizations in Generative Transformational Grammar (Niemeyer, Tübingen).
Bohr, N., 1958, Atomic Physics and Human Knowledge (Wiley, New York).
Botha, R.P., 1982, Lingua **58**, 1–50.
Botha, R.P., 1984, Stellenbosch Papers in Linguistics **13** (University of Stellenbosch).
Campbell, L., and C. Ringen, 1981, Phonologica **1980**, 57–68.
Chomsky, N., 1957, Syntactic Structures (Mouton, The Hague).
Chomsky, N., 1977, in: Formal Syntax, ed. P.W. Culicover (Academic Press, New York) pp. 71–132.
Chomsky, N., 1980a, Rules and Representations (Colombia University, New York).
Chomsky, N., 1980b, Lectures on Government and Binding (Foris, Dordrecht).
Clark, E., and R. Berman, 1984, Language **60**, 542–590.
de Beaugrande, R., and W.U. Dressler, 1981, Introduction to Text Linguistics (Longmans, London).
Donegan, P., and D. Stampe, 1979, in: Current Approaches to Phonological Theory, ed. D. Dinnsen (Indiana University Press, Bloomington) pp. 126–173.
Dressler, W.U., 1982, Wiener Linguistische Gazette **29**, 3–16.
Dressler, W.U., 1983, Fachsprache **5**, 51–57.
Dressler, W.U., 1984a, Phonology Yearbook **1**, 29–51.
Dressler, W.U., 1984b, Folia Linguistica **18**, 571–573.
Dressler, W.U., 1985a, Morphonology (Karoma Press, An Arbor).
Dressler, W.U., 1985b, Journal of Linguistics **12**, 2.
Dressler, W.U., 1985c, in: The Identification of Progress in Learning, ed. T. Hägerstrand (Cambridge University Press, Cambridge) pp. 139–142.
Glozman, Z., 12974, Voprosy Psihologii **5**, 81–87.
Heisenberg, W., 1971, Schritte über Grenzen (Gesammelte Aufsätze und Reden) (Zausen, Leck).
Hutten, E.H., 1956, The Language of Modern Physics (Azzen & Unwin, London).

Klein, O., 1967, Glimpses of Niels Bohr as scientist and thinker, in: Niels Bohr, His Life and Work as seen by his Friends and Colleagues, ed. S. Rozental (North-Holland, Amsterdam) pp. 74–93.

Labov, W., 1970, Studium Generale **23**, 30–87.

Leland, L. and R. Skousen, 1974, Papers of the Parasession on Natural Phonology (Chicago Linguistic Society, Chicago) pp. 233–239.

Lindner, G., 1975, Der Sprechbewegungsablauf (Akademie-Verlag, Berlin, GDR).

Locke, JK., 1983, Phonological Acquisition and Change (Academic Press, New York).

Mayerthaler, W., 1981 Morphologische Natürlichkeit (Athenaion, Wiesbaden).

Mittelstaedt, P., 1972, Die Sprache der Physik (Bibliographisches Institut, Wiesbaden).

Newmeyer, F.J., 1980, Linguistic Theory in America: the first quartercentury of transformational generative grammar (Academic Press, New York).

Peirce, Ch. S., 1965, Collected Papers, eds C. Hartshorne and P. Weiss (Harvard University Press, Cambridge).

Prigogine, I., 1979, Vom Sein und Werden (Piper, Munich).

Putnam, H., 1981, in: Mind Design. ed. J. Haugeland (MIT Press, Cambridge) pp. 205–219.

Ringen, J., 1975, in: Testing Linguistic Hypotheses. eds D. Cohen and J. Wirth (Hemisphere Publ. Corp., Washington, DC) pp. 1–42.

Rozental, S. (ed.), 1967, Niels Bohr, his Life and Work as seen by his Friends and Colleagues (North-Holland, Amsterdam).

Seiler, H.J., 1978, Leuvense Bijdragen **67**, 257–265.

Stampe, D., Papers from the 5th Regional Meeting (Chicago Linguistic Society, Chicago) pp. 443–454.

Stegmüller, W., 1979, Hauptströmungen der Gegenwartsphilosophie II (Körner, Stuttgart).

von Weizsäcker, C.F., 1974a, in: Die Einheit der Natur (dtv, Munich) pp. 61–83.

von Weizsäcker, C.F., 1974b, in: Die Einheit der Natur (dtv, Munich) pp. 39–60.

Wüster, E., 1979, Einführung in die Allgemeine Terminologielehre und Terminologische Lexikographie (Springer, New York, Vienna).

Discussion, session chairman N.K. Jerne

Jerne: Although you do not intend to discuss the origin of language, do you know of any record proving that oral language preceded the written one, e.g. a written record from old times describing a spoken language of even older times?

Dressler: Written language is derived from oral language like a Morse code. No child starts communicating by Morse code. Language has two functions, a communicative one, and a cognitive one. Both are needed in a primitive society.

Jones: Even recent developments in physics can be described by means of language that existed earlier. It thus seems that we have not reached the limit of our capability of employing a linguistic description of nature. My question: is physics more limited than language in this sense?

Dressler: There seems to be some difference of applicability between the existing languages. If for instance, like in some Oceanic languages, all numbers above three are expressed by "numerous", mathematics would be quite difficult but not impossible to describe. It might be that the problem in physics is smaller than in other sciences due to the use of mathematics as a meta-language.

Rüdinger: There is little doubt as to Bohr's opinion about the latter question: there is no way in which we can avoid using ordinary language as the means of communicating scientific results.

The Lesson of Quantum Theory, edited by J. de Boer, E. Dal and O. Ulfbeck
© Elsevier Science Publishers B.V., 1986

Great Connections Come Alive:
Bohr, Ehrenfest and Einstein

Martin J. Klein

Yale University
New Haven, Connecticut, USA

On an occasion like this an appropriate scriptural text cannot be out of place, and which text could be more appropriate than this one from Ecclesiasticus: "Let us now praise famous men, and our fathers that begat us. The Lord manifested in them great glory, even his mighty powers from the beginning." Praising our great predecessors fulfills a deep human need, but one might question whether it is a historian's task. Perhaps one measure of the greatness of those we would praise is the degree to which such praise is indeed compatible with the professional obligations of the historian. I have no doubts on that score today.

Commemorating a great scientist of the past normally means celebrating his work —exploring what it has meant to the development of his subject and how it has opened new aspects of the world to our understanding. That, very properly, is what we have been doing at this symposium to celebrate Niels Bohr's centennial. We have been acting in the spirit of what Einstein wrote when Isaac Newton's tercentenary was being celebrated in 1942: "To think of him is to think of his work." [1] I would not want to take issue with the truth of Einstein's remark, but when we apply it to Bohr, or to Einstein himself, this truth becomes a "deep truth." Let me remind you that "deep truth" is a term that acquired a very specific meaning in Bohr's Copenhagen. In contrast to a truth, which is a statement whose opposite is clearly false, a "deep truth" is a statement whose opposite also contains deep truth. Surely when we think of Bohr's work, or of Einstein's, we must think of the men themselves. Their stature as human beings is comparable to the stature of their contributions to science. Their greatness as men is realized and embodied in the way they lived their lives in science. Of how many distinguished scientists could it be said: "However great a scientist he was, he was even a rarer phenomenon as a noble character"? [2] How many could be described as men "whom it is good to have known and consoling to contemplate"? [3] These words were used about Bohr and Einstein, respectively, but could have been applied to either.

These two lives, remarkable enough individually, were linked in an equally remarkable way. Bohr and Einstein knew each other for thirty-five years, but met only infrequently. What we know about the ways in which they shared their concerns for the fundamental issues of physics comes in large part from the wonderful essay Bohr wrote for Einstein's 70th birthday, his "Discussion with Einstein on Epistemological Problems in Atomic Physics" [4]. Written after a quarter of a century of their profound searching into the depths of quantum physics, Bohr's essay necessarily laid stress on the nature of their disagreement on fundamental issues. By doing so he wanted to emphasize how strongly he had been influenced by Einstein's critical attitude, and how positive an influence his occasional meetings with Einstein had been on the development and clarification of Bohr's own position. He did not, perhaps unfortunately, sufficiently convey to his readers the warm personal ties that bound the two men.

It would be natural to assume that we could supplement Bohr's essay and learn more about the nature of that friendship by studying the Bohr–Einstein correspondence. We find, however, only a handful of letters exchanged over the years. These are impressive and interesting, to be sure, but they do not give us the detailed record of developing ideas and personal interactions we might have hoped to find. The communication between Bohr and Einstein, indeed the close intellectual and emotional ties that bound them, involved a third man, Paul Ehrenfest, who played an essential role in their relationship.

It may not be too much to say that Ehrenfest himself was one of the strongest links between his two great friends. His rich and extensive correspondence with them and with others presents us with a fresh and lively picture of Bohr and Einstein as they appeared to an unusually perceptive contemporary of their own generation. Ehrenfest did not change physics as his two close friends did—and no one knew that better or felt it more keenly than he—but he was in his own way as extraordinary a human being. "Passionately preoccupied with the development and destiny of men," as Einstein wrote of him, "his relations with his friends played a far greater role in Ehrenfest's life than is the case with most men." [5] Fortunately for us, as it was for him, Bohr and Einstein were two of these friends.

Bohr came to know Ehrenfest before he met Einstein. Perhaps surprisingly, it was Bohr who took the first step toward Ehrenfest by writing to him in the spring of 1918. To see why it happened this way and at this particular time we must have a look at the background of their first encounter. Bohr was thirty-two at the time he wrote, and had been professor of theoretical physics at Copenhagen for two years. His name was already known to physicists everywhere for his brilliantly successful theory of the hydrogen atom and its spectrum [6]. This novel combination of the quantum ideas of Planck and Einstein with Rutherford's nuclear atom depended crucially for its success on Bohr's bold departure from all past theories of the emission of light. According to Bohr, the frequency of the light emitted by an atom was not the frequency with which anything in the atom actually moved. The audacity of this step was, however, balanced by Bohr's careful concern to show that his new way of determining the frequencies emitted did correspond to the classical results in the limit of long wave lengths. In undertaking to write "On the Constitution of Atoms and Molecules", as Bohr did in 1913, he knew that he would be exploring a domain new to theoretical physics, a domain dominated by the remarkable stability of these small structures [7]. He knew too that the novelty of this domain demanded constant attention to the guides provided by successful past explorations of more familiar regions of experience, together with a mastery of the phenomena continually being found along the way.

The famous trilogy of papers of 1913 was followed by several more during the next couple of years, but since 1915 Bohr had not published anything. Not that he had been idle. It was in the first years after his return from Manchester to take up his professorship at Copenhagen that J. Rud Nielsen was struck by

> "... the speed with which he moved. He would come into the yard, pushing his bicycle, faster than anybody else. He was an incessant worker and seemed always to be in a hurry. Serenity and pipe-smoking came much later." [8]

Bohr had much to keep him busy then, including his teaching and the early planning for the new institute of theoretical physics whose construction was approved by the university faculty in 1917 [9]. But his greatest efforts went into his own research.

During the war much progress was made on the quantum theory and its application to atomic spectra. Bohr followed this closely, and continually reviewed and revised his own thinking in the light of what he learned. But despite his admiration for the elegant new mathematical methods introduced in Arnold Sommerfeld's papers—of which Bohr said "I do not think that I have ever enjoyed the reading of anything more than I enjoyed the study of them" [10]—Bohr went on searching for something deeper. He recognized, as others did, that "many difficulties of fundamental nature" remained unsolved, but he was especially aware of the basis reason for this fact: "it has not been possible hitherto to replace these [classical] ideas by others forming an equally consistent and developed structure." And that structure was Bohr's goal.

He worked on a long paper that would try to unify what had been done so far, would bring out the assumptions underlying the existing theory, and perhaps "throw light on the outstanding difficulties by trying to trace the analogy between the quantum theory and the ordinary theory of radiation as closely as possible" [11]. In this work where the correspondence principle was developed, Bohr made important use of Einstein's new quantum theory of radiation and of Ehrenfest's principle of adiabatic invariance. He was one of the first to grasp the significance of Ehrenfest's work, and the adiabatic principle had been part of Bohr's set of working concepts for several years before 1918 [12]. Since this principle provided a criterion for finding some results of classical physics that would "stand unshaken in the midst of the world of radiation phenomena whose anti-classical quantum character stood out ever more inexorably," as Ehrenfest would write later [13], Bohr found it especially attractive. He renamed it the "principle of mechanical transformability" to emphasize its connection with the necessary stability of the stationary states of a quantum system.

When Bohr finally had copies of the first part of his long paper in hand—he now saw it as the first of four parts—what could be more natural than to send one to Ehrenfest. He had not met Ehrenfest, but he did have some sense of what he was like through his conversations with Hans Kramers. Kramers had left Holland at the age of 22 after several years of study at Leiden, made his way to Copenhagen in 1916, and became Bohr's valued assistant, the first research student of the new professor. Bohr's letter to Ehrenfest reported his pleasure in Kramers' progress, as well as discussing the value he placed on Ehrenfest's own work and explaining his changed terminology [14].

Ehrenfest received Bohr's letter and paper at a peculiar moment in his life [15]. He had moved to Leiden in 1912 to follow Lorentz in the chair of theoretical physics at its university. This prized position came to Ehrenfest at the early age of thirty-two, but he had already lived and worked in Göttingen and St. Petersburg, as well as in his native Vienna, which he detested. From the moment of his arrival in Leiden Ehrenfest had thrown himself into the creation of a genuine scientific community. His efforts succeeded quickly, and Lorentz noted with warm approval that Ehrenfest "had gotten the students talking." He also pursued his own ideas on a variety of physical problems, and strove valiantly to keep up with his close friend Einstein's work on the general theory of relativity, discussing it often with Lorentz. Some months before Bohr wrote to him, however, Ehrenfest had become fascinated by the problem of economic equilibrium [16]. He had put aside his work in physics and devoted all his energy to trying to develop the parallels between economics and thermodynamics. About the time he received Bohr's letter, Ehrenfest wrote Einstein that he was still "miles away from all of physics." He knew he would return eventually "full of remorse and ready to mend [his] ways," but for the moment physics had lost its charm [17]. His return to physics came much sooner than Ehrenfest expected.

Bohr's letter and his paper, "On the Quantum Theory of Line-Spectra" apparently succeeded in breaking the spell that mathematical economics had cast over Ehrenfest. There is some irony in this. Ehrenfest had not been one of those who instantly grasped the significance of Bohr's 1913 papers, who saw as Einstein had that it was an "enormous achievement" and "one of the greatest discoveries" [18]. On the contrary, in a letter to Lorentz written in August 1913 Ehrenfest commented: "Bohr's work on the quantum theory of the Balmer formula (in the *Phil. Mag.*) has driven me to despair. If this is the way to reach the goal, I must give up doing physics." [19] He had not changed his mind three years later, when he wrote to Sommerfeld to congratulate him on his theory of the hydrogen fine structure:

> "Even though I consider it horrible that this success will help the preliminary, but still completely monstrous, Bohr model on to new triumphs, I nevertheless heartily wish physics at Munich further successes along this path." [20]

Ehrenfest had begun to change his mind when he saw the connections between the adiabatic principle and Bohr's theory, and encouraged his student Jan Burgers to write his dissertation on "the Rutherford-Bohr model of the atom" [21]. By the time Bohr's letter reached him, Ehrenfest was ready to receive it.

His reply indicated how very pleased he was to find that Bohr had given the adiabatic principle such a major place in his work, and that Bohr even appreciated Ehrenfest's point about the adiabatic invariance of the statistical weights assigned to the stationary states. "Oh, if you only knew how much aggravation I had until I managed to convince anyone that there was a problem here," he wrote. Ehrenfest was also happy to learn that Kramers was working so effectively. He liked Kramers very much and appreciated his ability, but Ehrenfest had been concerned about one thing: "He always does only what comes easily to him at the time—nothing ever seems to seize him in such a way that he has even a little feeling of real urgency about it." [22] Ehrenfest always had to feel that his students were possessed by their work before he was convinced they had a genuine vocation for physics.

Bohr had closed his letter with the hope that he would be able to come to Holland and meet Ehrenfest when the war was over. Ehrenfest responded in kind, in his own lively way. "I impatiently await the day when you enter our house as our guest." And then he added a remark that showed how highly he had come to think of Bohr: "I hope that Einstein can also be with us then." Within a few days Ehrenfest was back to physics, jotting down notes and questions on the quantum theory, ideas for further development prompted by his study of Bohr's paper.

Ehrenfest wasted no time in arranging for Bohr to visit Holland. There was to be a meeting of Dutch scientists in April 1919, and in January Ehrenfest invited Bohr to take part, urging him to spend weeks rather than days in Holland, to stay with the Ehrenfests "for as long as you can stand it" in their chaotic household, and promising once again to try to get Einstein there at the same time [23]. Bohr probably had no way of knowing that Ehrenfest was hoping to provide him with the greatest gift he could imagine by trying to arrange for him to meet Einstein. Ehrenfest had compared his friend's last visit to Leiden in 1916 with a marvelous concert, composed entirely of the finest music, a concert whose "echoes, that go on resounding inside you afterwards" offer as much satisfaction as the event itself. For Ehrenfest, Einstein was not only one of the "wonders of Nature" for his intellectual gifts, but also

> "a marvelous interweaving of simplicity and subtlety, of strength and tenderness, of honesty and humor, or profundity and serenity (a somewhat melancholy serenity, to be sure)." [24]

While Ehrenfest, like Yeats, loved to "have the new friend, meet the old," he did not offer to share Einstein unless he felt his gift would be properly appreciated.

Bohr's visit to the Netherlands in the spring of 1919 was a great success, even though Einstein could not be there, regretfully declining despite his "intense wish" to meet Bohr, "that man of magnificent intuitive gifts" [25]. Bohr lectured at Leiden in English, the language in which he had published almost all his scientific work up to then. Ehrenfest had assured him that this would be no problem since all Dutch scientists spoke the language well. Ehrenfest did not, however, as he made clear when he introduced Bohr:

> "Please allow me to welcome our honored guest, Professor Bohr, in the German language. I could, of course, do this in English, but then, to my great regret, it would be totally unintelligible, and this might be considered unsuitable." [26]

His introductory remarks also showed that he had already learned something essential about Bohr. Ehrenfest expressed his conviction that this personal encounter with Bohr would give Dutch physicists "a much deeper sympathetic understanding of his ideas" than could be obtained by the mere study of his publications. He had now decided that Bohr's ideas were not only well worth understanding, but could be understood, despite the almost impenetrable form in which they sometimes appeared in the journals. Perhaps Kramers had told him of how one learned by working with Bohr day after day; perhaps Ehrenfest had now seen this for himself after only a few days in Bohr's presence. In any case he knew that Bohr in person was quite different from Bohr in print.

He devoted himself to making sure that Bohr's visit was a happy and fruitful one, introducing Bohr to the physicists of the Netherlands and making sure there was time for their unhurried talk, but also seeing to it that Bohr did not become exhausted by rushing from one conversation to another [27]. Ehrenfest knew how essential it was for a man like Bohr to have tranquil hours and days when he could be alone with his thoughts. He and Bohr had much to talk about together—from the current problems of the quantum theory to the Icelandic sagas, from the stages of a child's development to the difference between genuine physicists and the others. Their exchanges ranged over heaven and earth as Ehrenfest showed his new friend the treasures of the Dutch museums and the brilliant colors of the bulb fields [28].

Bohr was deeply moved by this encounter, calling it "a most wonderful time of the greatest intellectual enrichment." He tried to say how much this friendship meant but found that difficult. He then added,

> "Dear Ehrenfest, you do not know how miserable and stupid I feel when writing this letter. I am sitting and thinking of all what you have told me about so very many different things, and whatever I think of I feel that I have learned so much from you which will be of great importance for me; but, at the same time, I miss so much to express my feeling of happiness over your friendship and of thankfulness for the confidence and sympathy you have shown me. I find myself so utterly incapable of finding words for it." [29]

The many-layered emotional and intellectual response evoked in Bohr by this meeting with Ehrenfest did not fade away after his return to Copenhagen. In a long letter written at the end of October 1919, Bohr returned to some of the matters they had discussed, and especially to some of the non-scientific issues. His comments in the middle of his letter on the problems of expressing certain kinds of ideas are worth quoting:

> "Dear Ehrenfest, writing to you I feel so poor, not only in what I am thinking, but especially in means of expression. Although of course in general I express myself freer in English than in German I feel, when I want to enter on a question of finer sentiments, if possible a still greater poverty in English than in German, because in such matters it is the choice of the words and not their grammatical use which is essential, and in this respect the German words can almost always be got by direct transcription of the corresponding Danish word, while it is not so in English. Therefore, although I am happy being allowed to write to you in English, you must take what I write not as an image of what I like to say, but as a picture in which I am bound to restrict myself to use the limited number of colors given by my poor collection of English words." [30]

At another point in this letter Bohr used another metaphor to illustrate the "utter deficiency of our means of realizing and discussing ideas of 'irrational character'".

> "You will see how right I am in trying beforehand to excuse my poor means of

expression, and you may take the foolishness I write, and the inadequate words I use, as disconnected 'Anschläge' on a musical instrument, where the only contents of the words are the chains of ideas and sentiments they automatically produce by reflex, just as you listen to the tunes produced by the instrument in harmonious connection with the original tone."

During that same month of October 1919 Einstein paid his first post-war visit to Ehrenfest in Leiden. This visit came at a high moment in Einstein's life, only a few weeks after he learned that the English eclipse expedition had confirmed the correctness of his predicted value for the gravitational deflection of light by the sun. To Einstein these two weeks spent with Ehrenfest were "really a beautiful and tranquil time", a time that confirmed the importance of their friendship. "I know that it is good for both of us," he wrote "and that each of us feels less of a stranger in this world because of the other." [31]

Just as Ehrenfest had wanted to share Einstein with Bohr, he now was keen on sharing Bohr with Einstein and evidently sang the praises of his new friend. On his return to Berlin Einstein wrote that he was now going "to bury himself" in Bohr's papers. "You have shown me that there is a man of profound vision behind them, one in whom great connections come alive." [31] Since we know that Bohr remembered how, even as a child, he had liked "to dream of great interrelation-ships," we can imagine how pleased he would have been to know that he was being characterized in this way [32].

Some six months later Bohr lectured to the German Physical Society in Berlin, speaking "On the Series Spectra of the Elements" [33]. It was on this occasion that Bohr finally met Einstein. For well over a decade, ever since the days when he was working on his doctoral thesis, Einstein had been a major presence in his intellectual life. From Einstein's papers he had learned the indispensable role of the energy quantum in individual atomic events, the inevitability of energy quantization in accounting for thermal radiation, and the close connections between the essential features of his own atomic theory and the equilibrium of matter and radiation. No matter what Einstein was like as a person, meeting him would have been a major event in Bohr's life. Nine years earlier, when he met J.J. Thomson soon after his arrival in Cambridge, Bohr had written to his brother Harald: "You should know what it was for me to talk to such a man." [34] The prospect of a meeting with Einstein would have seemed even more momentous in Bohr's mind, particularly after all he must have heard from Ehrenfest. Though Einstein was only six years older than he, Bohr would always see him as one of the grand masters, as if he were of another generation.

Bohr and Einstein hit it off as well as Ehrenfest hoped they would. In his first letter to Einstein, Bohr wrote:

"To meet you and talk with you was one of the greatest experiences I have ever had, and I cannot say how grateful I am for all the kindness with which you met me on my visit to Berlin... You don't know how very stimulating it was for me to have the long awaited opportunity to hear directly your views on the questions that I have been working on. I will never forget our conversations on the way from Dahlem to your house...." [35]

Einstein, not usually given to emotional expression, was even warmer:

> "Rarely in life has a person given me such joy by his mere presence as you have. Now I understand why Ehrenfest is so fond of you. I am now studying your great papers, and while doing this I have the pleasure—whenever I get stuck somewhere—of seeing your friendly youthful face before me, smiling and explaining. I have learned much from you, especially how you confront scientific matters." [36]

Two weeks later Einstein was in Leiden telling Ehrenfest "with extraordinary warmth" about his meeting with Bohr. Ehrenfest's reaction came in a card to Bohr saying: "When will I have both of you here together some time." [37] That dream of Ehrenfest's would not be realized for several years.

The meetings of Bohr, Ehrenfest and Einstein did continue, but as pairwise encounters. Einstein made a brief lecture tour in June 1920 speaking in Oslo and then in Copenhagen. His visits to both cities were major news events, drawing admiring crowds, and extensive coverage by the press [38]. For Einstein the best part of his trip was the hours he spent with Bohr, as he reported to Lorentz [39].

In April 1921 the first of the post-war Solvay Conferences was held in Brussels. All three men had been invited and planned to go. "I am looking forward to the new year more than I can say," Bohr wrote to Ehrenfest in November 1920, "especially because it will bring me into personal contact with you and Einstein again; we shall all really meet together for the first time in Brussels...." [40] It did not happen that way. Einstein changed his plans, withdrawing from the Solvay meeting in order to go to the United States with Chaim Weizmann on a fund-raising expedition for the new Hebrew University in Jerusalem [41]. And at the last minute Bohr, too, found it impossible to go to Brussels. The years of intense scientific work, together with the great strain of planning and supervising every detail of his new Institute for Theoretical Physics, officially opened in March 1921, finally caught up with him [42]. About a week before the Solvay meeting was to begin Bohr had to withdraw; his doctor had ordered several weeks of complete rest [43].

His friends had been concerned for months about the extent to which Bohr was overworked. Since the previous August Ehrenfest had been urging Bohr to give up everything that was not absolutely essential, and in particular to cancel his plans to attend the Solvay meeting and give the review he had promised of the quantum theory of atomic structure:

> "You understand that it is sad for me to lose this opportunity to see you and Einstein together, but I *still* advise you to do it. In any event decline to lecture—just come for the discussions!" [44]

At the end of December, Ehrenfest offered to help Bohr with the review he persisted in trying to write. This had been Lorentz's suggestion [45] and Ehrenfest put it forward diffidently, emphasizing (by triply underlining the phrase) that he would do it only if Bohr wanted him to [46]. There was one real difficulty that had to be faced: the two men wrote in utterly different styles. The more Bohr worked over his manuscripts, the more complicated they became in their successive versions. As he understood a subject better and saw more deeply into its details and difficulties, he tried harder to qualify his remarks so that no shade of meaning, no subtle difference between situations, could be overlooked by the reader. Bohr seemed to fear oversimplification more than anything else, and exerted all his efforts to avoid giving an illusion of clarity at the expense of suppressing some feature of the situation that did not fit into a simple pattern. As Pauli once wrote: "He knew well what he wished *not* to say when he strove in long sentences to express himself in his scientific papers." [47] Ehrenfest's way of dealing with physics was completely different. He tried always to seek out the one essential feature of a situation, sharpening the statement of a problem as much as possible—even pushing it into paradoxical form if he could—so that the solution could then appear with devastating clarity. He worked like a caricaturist, capturing the defining traits at the expense of all else.

Under these circumstances Ehrenfest proposed that all he could do to help Bohr would be to prepare a short manuscript composed of a series of propositions distilled from Bohr's papers. He feared, of course, that Bohr would not find anyone else's words an adequate expression of his ideas. "I would gladly adapt myself to your genetic way of thinking," Ehrenfest wrote "but I would still ultimately have to assert (or at least inquire into) a thesis, a proposition; it is inevitable that a few dogmatically hardened bones get stuck into the genetic mollusk." [46]

The printed proceedings of the Third Solvay Conference included both Bohr's draft of the first part of his promised report and Ehrenfest's brief distillation of the propositions that seemed to him to constitute Bohr's correspondence principle [48] that principle which appeared to some like a "somewhat mystical magic wand, which did not act outside Copenhagen", as Kramers once put it [49]. Although Ehrenfest made no attempt to deal with the complexities as Bohr would have, his article ends with remarks that express the spirit of Bohr's approach. Ehrenfest thought the deepest significance of the correspondence principle was that it seemed to come closer than anything else to that future theory, whose very outline was not yet apparent, which would be capable of handling the problems of radiation. For that reason it would be premature, he felt, to try to codify the current version of the principle just so that it could be applied more conveniently. One should accept its tentative nature, leaving it open to change. Ehrenfest had realized that Bohr was struggling to formulate a language for describing the new domain of experience provided by atomic physics, and he accepted Bohr's view that this struggle demanded its own methods. When he ended the text for the published version of his Solvay report this way, Ehrenfest might have been remembering some words that Bohr had recently written to him:

> "Throughout this year, ever since the moment when I came upon the first traces of this conceptual structure, I haven't really been in a position until now to try to think things through in an orderly way. 'Think through' is a phrase that applies very badly in this domain where nothing is fixed and where everything really depends only on a feeling for harmony. But you don't know what great joy I have had recently, apart from some hours of doubt, and I should like so much to give you an impression of it, since you have had so much complaining and bother from me." [50]

By the end of June 1921 Bohr had recovered his strength and was making plans for the future. These included making an occasion to see Ehrenfest again, an occasion that would help to make up for the missed opportunity in Brussels. Bohr invited Ehrenfest to come to Copenhagen at his convenience, to lecture or not, if he preferred just to lead some discussions. Money was available to cover travel costs, and both Ehrenfests would be made welcome at the Bohr home in the Institute. "I don't need to tell you what an event it would be for all the Copenhagen physicists, and what a joy it would be for me and my wife." [51]

Ehrenfest had been looking forward to such a visit for a long time, but Bohr's letter stirred up mixed emotions. After days of staring at the letter sitting there at his desk and chewing on his pen, Ehrenfest finally poured out his feelings. First his "jubilation" at the opportunity to see Bohr and especially on his own home ground. "It was *splendid* to get your letter. But almost immediately thereafter gray depression and doubt crept in again." All of Ehrenfest's old self-doubt was revived. "What can *I* lecture on in Copenhagen? What would I know that Kramers, for example, doesn't know much better and couldn't say much better than I?" He also felt guilty because the same money could be better used to invite one of the many deserving physicists from the conquered nations (he listed no fewer than 17 of them!). But after going through a long litany of reasons why he should not accept, Ehrenfest did recognize what he could contribute by his visit:

> "You see, I might really be useful for discussing subjects you propose, because it really isn't totally useless for you to have to present your thoughts and conjectures to a listener who is tolerably critical, quick to understand, and anxious to learn. And I can, finally, also give lectures for other people (not for you!) if you make clear to me what you want from me. ... Dear, dear Bohr, I really would like so terribly much to be with you again." [52]

Bohr was, of course, really anxious to have Ehrenfest and not some more or less worthy substitute. He knew his friend well enough to see past the self-doubt that tormented Ehrenfest, that peerless lecturer whom Einstein would call "the best teacher in our profession whom I have ever known" [53]. After many reassurances to this effect [54], Ehrenfest did finally make his trip to Copenhagen, the first of many, in December 1921. We know his reactions from a postcard he sent to Einstein after three weeks in Denmark:

> "It is marvelously beautiful here—everything! He is a prodigious physicist. I sound to myself like a badly told joke when I open my mouth in his presence. A lecture of his will appear shortly in the *Zeitschrift für Physik* (the manuscript goes off tomorrow), in which he explains how he has now deciphered the structure of *all* the atoms. It is something tremendous, both methodologically and for its results. ... I felt so happy in Bohr's house, happier than I have been in a long time. ... If only my poor wife weren't so miserably stuck at home, I would be completely joyful now and would wish for nothing more than to be Bohr's assistant." [55]

Ehrenfest described his impressions of Bohr in much more detail in a long letter he wrote to Lorentz after his return to Leiden. Best of all had been the possibility of talking to Bohr again and again about his work:

> "To be sure, this is certainly not easy to do. In the first place he was overloaded with proof-reading, which he does in a completely grotesque way, because he incessantly changes everything, making fundamental changes again and again—even in the galley proofs! In the second place it's hopeless to put one individual question to him. He

reacts to that like a very rapidly spinning top, 'completely transversely.' His brother Harald consoled me by saying... 'If Niels tells me something, I absolutely don't understand for 59 minutes what he is talking about and what he is driving at, but in the 60th minute light suddenly dawns, and then I see that everything he had said before was really necessary.' I eventually chose this procedure too: ask and then listen patiently and attentively to whatever follows, but just don't come back to the question. In this way I learned an enormous amount but had hardly a tenth of my questions answered. Nothing is more impossible for him than to say 'yes, yes' or 'no, no.' If one grasps him with the yes-or-no tongs, he becomes very *wretched* like a caged bird.

He is a very, very great scientist. His way of developing the quantum theory is really very different from that represented by Sommerfeld's book, for example. Thank God! His work is permeated by the conviction that we have just touched the *beginning* of an essentially *new* physics. His model of the atom is an extraordinarily symbolic symbol for him. At the same time he has a very powerful feeling for which features of the symbol should be taken inexorably seriously and which he may plainly ignore or at least treat very lightheartedly. (He once said in conversation, quite naively and not trying to be 'witty': 'Oh no, I can't believe that. That is much too concrete for it to be real; that is only formal.')"

Ehrenfest summed up his very lengthy report to Lorentz on Bohr and his work by writing:

"What I find so liberating in Bohr is that one can *think* again instead of just calculating." [56]

On November 10, 1922 the Royal Swedish Academy of Sciences announced its choices for the Nobel prizes in physics. Since the Academy had deferred awarding the physics prize in 1921, it could now announce two winners: the prize for 1921 went to Einstein, and that for 1922 to Bohr. The very next day Bohr wrote to Einstein. His letter contained the following striking passage:

"To me it was the greatest honor and joy ... that I should be considered at the awarding of the prizes at the same time as you. I know how little I have deserved this, but I should like to say that I have felt it as the greatest good fortune that—quite apart from your great contribution to the world of human thought—the fundamental contribution that you have made to the more special field in which I work should be recognized and also quite publicly, just as the contributions of Rutherford and Planck were, before I was considered for such an honor." [57]

Bohr's letter finally caught up with Einstein in Japan, and Einstein's answer was just as remarkable:

"I can say without exaggeration, that it pleased me as much as the Nobel prize. I find especially delightful your fears that you might have received the prize before me—that is genuinely 'Bohrish'. Your new investigations of the atom have accompanied me on the trip, and they have even increased my love for your intellect." [58]

This brief exchange tells us a great deal about the ways in which these two thought and felt about each other. Einstein had already put his view of Bohr as a physicist on the record earlier that year when he proposed to the Academy of

Sciences at Berlin that Bohr be elected as a corresponding member. His recommendation referred to "the rare blend of boldness and careful consideration" in Bohr as a scientific thinker, to his "intuitive grasp of hidden things and his keen critical sense." "With all his knowledge of details," Einstein wrote "his gaze is steadily directed to the underlying principle. He is unquestionably one of the greatest discoverers of our age in the field of science." [59]

In December 1925 Ehrenfest's dream of having Bohr and Einstein in Leiden together finally came true. The occasion was the golden anniversary of Lorentz's doctorate, an event that the Dutch scientific community wanted to celebrate in proper style [60]. Einstein, who loved and revered Lorentz as he did no other human being, had of course agreed to be present [61]. Bohr, who had received an official invitation in April, apparently did not decide to attend until he learned from Ehrenfest early in September that Einstein would be there. To be able to spend a week with Einstein was a treat that Bohr could not pass by. Ehrenfest was delighted with Bohr's sudden decision, and wrote at once to Einstein:

> "Bohr is now struggling mightily with the problems of quanta and he needs to talk about his ideas with you more than with anyone else. It is so important to him to know to what extent you have run into the same deep difficulties as he has. I know that no man alive has seen so deeply as you two into the real abysses of the quantum theory, and that no one else but you two really sees what completely radical new concepts are needed." [62]

Ehrenfest asked both his friends to give him unconditional authority to ensure their isolation. He continued,

> "I believe that you two could not remain so isolated in either Berlin or Copenhagen as in my house, in case I obtain the necessary authority from you to arrange it. ... You realize what a great experience it would be for me to hear the two of you discussing the quantum riddles, but you can rely on me to leave you by yourselves almost all the time. I am especially interested in what will come of the discussion between you and Bohr concerning the experiments that you are always thinking up on the boundary between waves and particles. While I expect that you two will feel quite similarly about the general puzzles, some fruitful conflict might develop between you concerning this special area." [62]

Einstein answered at once, "with what was for him a completely unheared of speed," as Ehrenfest described it to Bohr [63]. He granted Ehrenfest the requested power of attorney for isolation, looked forward with great anticipation to seeing Bohr, but also commented on one substantial aspect of Ehrenfest's letter:

> "I am no longer thinking up experiments about the boundary between waves and particles; I believe that this was a mistaken effort. One will probably never arrive at a reasonable theory by an inductive route, though I also believe that quite basic experiments like those of Stern and Gerlach and Geiger and Bothe can seriously help." [64]

When he wrote to Bohr about his forthcoming visit, Ehrenfest was full of advice. He wanted Bohr to stay in his home; he and Einstein could have the two little rooms upstairs, where they would have the privilege of smoking, a privilege limited to those guests rooms and perhaps to those special guests.

"I urgently request that you bring no sort of writing work with you. It would really be a sin to spoil this opportunity—this rare opportunity to peer for once with Einstein into the furthest depths of physics that are accessible at present to anyone's gaze. And you would certainly, and at the same time uselessly, spoil it for yourself, if you were to bring such work with you." [63]

Ehrenfest painted an appealing picture of what Bohr could expect—quiet chats while strolling along the canals of Leiden, or on the nearby beaches, free from all interruption, on calm, sunny days. And with Einstein, that "wonderfully deep and good man", who has "suffered much and who feels the sorrow in the world" [63].

This meeting took place at one of the crucial times in the turbulent history of the quantum theory. Earlier in 1925 several experiments had quickly ruled out the Bohr–Kramers–Slater theory, a theory which had tried to eliminate the light quantum from physics, even at the expense of abandoning the exact validity of the laws of conservation of energy and momentum [65]. Although Kramers successfully used some features of the abandoned theory in his work on dispersion, Bohr recognized that an even more fundamental change in the conceptual basis of physics had become necessary. In April he had written to Rutherford referring to "our present theoretical troubles" which he found to be "of an alarming character indeed" [66]. He was now prepared to consider "a radical departure from an ordinary space–time description" in discussing atomic events [67]. "We must take recourse to symbolic analogies of a still higher degree than before," Bohr wrote to Max Born at the beginning of May. "Just lately I have been racking my brains trying to imagine such analogies." [68] It was a time when Bohr and his co-workers were occasionally "close to despair" [69]. He was indeed anxious to talk to Einstein about the paradoxical and tormenting problems of the wave-particle duality of radiation, the duality that Einstein had been wrestling with for two decades.

Einstein's concern with duality had deepened during the past year. Struck by S.N. Bose's new and intriguing derivation of the Planck law for black-body radiation, Einstein had applied the same statistical arguments to a gas of material particles. When he analyzed the statistical fluctuations implied by the distribution law for this gas (the Bose–Einstein distribution), Einstein found the same dual structure that he had first traced for radiation at least as early as 1909. Just at this time he had been reading an interesting thesis sent him from Paris by Paul Langevin. The thesis was by Louis de Broglie, and Einstein saw new support for de Broglie's matter waves in the fluctuation properties of the Bose–Einstein gas. Perhaps matter showed the same kind of duality as radiation. It was an idea that could be put to experimental test, and Einstein was eager to have this done [70].

There were other weighty matters to discuss, foremost among them the new matrix mechanics begun by Heisenberg, and just applied to the problem of the hydrogen atom with brilliant success by Pauli [71]. Only days before departing for Leiden Bohr had learned of Pauli's "wonderful results" [72]. "I do not know whether to congratulate him or you the most," he wrote to Heisenberg [73]. Bohr, like Pauli, was convinced that Heisenberg's work marked a new stage of "wonderful progress as regards the development of the rational quantum mechanics." [74]

Also brand new and crying out for discussion was the proposal of the spinning electron, just published by two of Ehrenfest's students, George Uhlenbeck and

Samuel Goudsmit. This was strongly opposed by Pauli, who had already dissuaded Ralph Kronig from publishing similar ideas almost a year earlier. Bohr, too, was skeptical about spin at first since he saw no reason why this spin should be coupled to the orbital angular momentum of the electron [75].

The question of the electron spin was the first subject discussed by Bohr and Einstein. A few months later Bohr wrote:

> "Einstein asked the very first moment I saw him what I believed about the spinning electron. Upon my question about the cause of the necessary mutual coupling between the spin axis and the orbital motion, he explained that this coupling was an immediate consequence of the theory of relativity. This remark acted as a complete revelation to me, and I have never since faltered in my conviction that we at last were at the end of our sorrows." [76]

Most of the conversation must have dealt with the problems of radiation. Bohr reported to Slater that as a result of his long discussion with Einstein in Leiden he now believed there was agreement on the general ideas, and that Einstein was no longer searching for contradictions between the wave theory and light quanta [77]. Perhaps the "fruitful conflict" on this issue that Ehrenfest had looked forward to did not take place at Leiden. It would certainly take place at later meetings.

For Bohr the time in Leiden was "a wonderful experience", and his conversations with Einstein were "a greater pleasure and more instructive than I can say." [78]

There would be several more occasions when all three met together, most notably at the 1927 Solvay Conference. By this time it was clear that Bohr and Einstein did not "feel quite similarly about the general puzzles." [79] In a long letter to his students Ehrenfest described the nature of their discussions at the meeting in Brussels:

> "Bohr towering completely over everybody. At first not understood at all... then step by step defeating everybody. Naturally, once again the awful Bohrish conjuring terminology. Impossible for anyone else to summarize. (Poor Lorentz as interpreter between the English and the French who were absolutely unable to understand each other. Summarizing Bohr, and Bohr reacting with polite despair.) Every night Bohr came to my room at 1 a.m. to say 'just one single word' to me, until 3 a.m. It was splendid for me to be present at the dialogues between Bohr and Einstein. Like a game of chess. Einstein always with new examples. Something like perpetual motion devices of the second kind, but to violate the uncertainty relations. Bohr always searching, out of a dark cloud of philosophical smoke, to find the tools to shatter example after example. Einstein like a jack-in-the-box: jumping out again fresh every morning. Oh, that was priceless. But I am almost without reservation pro-Bohr and contra-Einstein. He is now behaving toward Bohr exactly the way the defenders of absolute simultaneity behaved toward him." [80]

Despite his definite statements in the last two sentences, Ehrenfest did not find it possible to be simply "pro-Bohr and contra-Einstein." Bohr's own account of these conversations at the Solvay meeting contains a comment on this point:

> "I remember also how at the peak of the discussion Ehrenfest, in his affectionate manner of teasing his friends, jokingly hinted at the apparent similarity between

Einstein's attitude and that of the opponents of relativity theory; but instantly Ehrenfest added that he would not be able to find relief in his own mind before concord with Einstein was reached." [81]

The deep differences in the views of his two close friends were painful for Ehrenfest to bear, and he did what he could to promote mutual understanding between them. In September 1931 he tried to arrange another meeting in a letter addressed to both men:

"Now I am naturally very, very anxious to have you, Einstein, here when Bohr is here. I cannot begin to tell you both how important it would be to me to hear you two discussing the current state of physics in a *tranquil* conversation. I have already confessed to you that I oscillate like a pith ball between the plates of a condenser when I go from one of you two to the other. What I want to accomplish most of all is that Bohr gets to see quite clearly how completely you, Einstein, know and understand his ideas and efforts and nevertheless consider it fully justified to continue searching for the 'genuine' nonlinear microscopic differential equations. I know very well, Einstein, that you have no trace of the urge to propagandize in your soul, and therefore can feel little incentive for such a discussion. For me, however, it is enormously important to see as sharply as possible up to what point you are both forced to think alike, and where your freedom for a parting of the ways begins. I promise not to interrupt you at all. I do hope, though, that I would be able to help a little bit now and then because I am so very familiar with your two extremely different ways of speaking, and Bohr's terrible clouds of politeness are a major hindrance to communication if they aren't vigorously blown away from time to time." [82]

I have deliberately limited my discussion to the first years of the Bohr–Ehrenfest–Einstein relationships, the years in which those remarkable friendships were formed. The warm understanding and strong bonds of sympathy established then between Bohr and Einstein sustained them through another thirty years of occasional meetings and deepening differences in their views on physics. Without this solid base of friendship the Bohr–Einstein dialogue might not have continued to be a model of civilized disagreement over fundamental scientific issues. Without Ehrenfest's catalytic role that friendship of Bohr and Einstein might never have achieved its full strength. Each of the three men felt revived, cheered and refreshed after contact with any of the others. Each felt, as Einstein had once written to Bohr, that the mere presence of the others was a source of joy. Perhaps after all, as Lionel Trilling once suggested, "truth is the expression, not of intellect, nor even ... of will, but of love." [83]

Acknowledgement

Unpublished letters by Bohr are quoted by permission of Professor Aage Bohr. Unpublished letters by Ehrenfest are quoted by permission of the Ehrenfest family. Unpublished letters by Einstein are quoted by permission of the Hebrew University of Jerusalem.

References

[1] A. Einstein, Isaac Newton, in: Out of My Later Years (New York, 1950) p. 201.

[2] F. Frankfurter, Niels Bohr, in: Of Law and Life and Other Things That Matter. Papers and Addresses of Felix Frankfurter 1956–1963, ed. B. Kurland (New York, 1969) p. 251.

[3] B. Russell, Preface, in: Einstein on Peace, eds O. Nathan and H. Norden (New York, 1960) p. xvi.

[4] N. Bohr, Discussion with Einstein on epistemological problems in atomic physics, in: Albert Einstein: Philosopher–Scientist, ed. P.A. Schilpp (Evanston, Illinois, 1949) pp. 201–241.

[5] A. Einstein, Paul Ehrenfest: in memoriam, in: Out of My Later Years (New York, 1950) pp. 214–217.

[6] N. Bohr, On the constitution of atoms and molecules, in: Philos. Mag. 26 (1913) 1–25, 476–502, 857–875. Reprinted with a valuable historical introduction by Léon Rosenfeld: N. Bohr, On the Constitution of Atoms and Molecules (Copenhagen, New York, 1963). Reprinted again in: Niels Bohr, Collected Works, Vol. 2, Work on Atomic Physics (1912–1917), ed. Ulrich Hoyer (North-Holland, Amsterdam, 1981) pp. 159–233. *

[7] W. Heisenberg, Physics and Beyond. Encounters and Conversations, transl. Arnold J. Pomerans (New York, 1971) p. 39.

[8] J.R. Nielsen, Memories of Niels Bohr, in: Physics Today 16 No. 10 (October 1963) 22–30. Quote from p. 23.

[9] P. Robertson, The Early Years: The Niels Bohr Institute 1921–1930 (Copenhagen, 1979).

[10] N. Bohr to A. Sommerfeld, 19 March 1916, in: Works 2, pp. 603–604.

[11] This quotation and the two immediately preceding it come from N. Bohr, On the quantum theory of line-spectra, Part I, in: Kgl. Danske Vidensk. Selsk. Skrifter, Naturvidensk. og Mathem. Afd. 8 Raekke, IV. 1 (1918), pp. 1–36. Reprinted in: Works 3, The Correspondence Principle (1918–1923), ed. J. Nielsen, pp. 67–102. Quotes from p. 70.

[12] See the letter cited in ref. [10].
Also see Bohr's paper, On the application of the quantum theory to periodic systems, scheduled to appear in the April 1916 issue of the Philosophical Magazine but withdrawn by Bohr after the appearance of Sommerfeld's papers. The text appears in Works 2, pp. 431–461.

[13] P. Ehrenfest, Adiabatische Transformationen in der Quantentheorie und ihre Behandlung durch Niels Bohr, in: Naturwissenschaften 11 (1923) 543–550. Reprinted in Paul Ehrenfest, Collected Scientific Papers, ed. Martin J. Klein (Amsterdam, 1959) pp. 463–470. Quote on p. 463.

[14] N. Bohr to P. Ehrenfest, 5 May 1918, in: Works 3, pp. 11–12.

[15] M.J. Klein, Paul Ehrenfest. The Making of a Theoretical Physicist (Amsterdam, 1970).

[16] Ibid., pp. 305–306.

[17] P. Ehrenfest to A. Einstein, 8 May 1918.

[18] Einstein as quoted in: G. von Hevesy to N. Bohr, 23 September 1913 and G. von Hevesy to E. Rutherford, 14 October 1913.
See L. Rosenfeld's introduction, ref. [6], pp. xli–xlii.

[19] P. Ehrenfest to H.A. Lorentz, 25 August 1913.

[20] P. Ehrenfest to A. Sommerfeld, May 1916.

[21] J.M. Burgers, Het Atoommodel van Rutherford–Bohr, (Haarlem, 1918).

[22] P. Ehrenfest to N. Bohr, 10 May 1918.

[23] P. Ehrenfest to N. Bohr, 13 January 1919.

[24] P. Ehrenfest to H.A. Kramers, October 1916.

[25] A. Einstein to P. Ehrenfest, 22 March 1919.

[26] P. Ehrenfest, Research Notebook XXV, 25 April 1919.

[27] P. Ehrenfest to N. Bohr, 28 February 1919.

[28] See the notebook of ref. [26], 29 April–1 May 1919.

[29] N. Bohr to P. Ehrenfest, 10 May 1919.

* Bohr's Collected Works are under the General Editorship of Léon Rosenfeld (Volumes 1–3) and Erik Rüdinger (starting with Volume 5). Publication of this series of volumes began in 1972. I shall refer to the series simply as Works.

[30] N. Bohr to P. Ehrenfest, 22 October 1919.

[31] A. Einstein to P. Ehrenfest, Received 9 November 1919. See also M.J. Klein, Ehrenfest, pp. 310–314.

[32] J. Kalckar, A glimpse of the young Niels Bohr and his world of thought, in: Works 6, Foundations of Quantum Physics I (1926–1932), ed. J. Kalckar, pp. xvii–xxvi. Quote on p. xix.

[33] N. Bohr, Über die Serienspektren der Elemente, in: Z. Phys. 2 (1920) 423. English translation by A.D. Udden in: Works 3, pp. 242–282.

[34] N. Bohr to H. Bohr, 29 September 1911. Quoted in John L. Heilbron and Thomas S. Kuhn, The Genesis of the Bohr Atom, in: Historical Studies in the Physical Sciences 1 (1969) 211–290. Quote on p. 223.

[35] N. Bohr to A. Einstein, 24 June 1920, in: Works 3, pp. 634–635.

[36] A. Einstein to N. Bohr, 2 May 1920, in: Works 3, p. 634.

[37] P. Ehrenfest to N. Bohr, 14 May 1920, in: Works 3, p. 609.

[38] See the reports from the German Embassies in Oslo and Copenhagen to the German Foreign Office dated, respectively, 22 June and 26 June 1920, Albert Einstein in Berlin 1913–1933, Teil I. Darstellung und Dokumente, eds C. Kirsten and H.-J. Treder (Berlin, 1979) pp. 226–227.

[39] A. Einstein to H.A. Lorentz, 4 August 1920.

[40] N. Bohr to P. Ehrenfest, 22 November 1920, in: Works 3, pp. 611–612.

[41] A. Einstein to P. Ehrenfest, 1 March 1921.
See also R.W. Clark, Einstein. The Life and Times (New York, 1971) pp. 382–391.

[42] J. R. Nielsen, Introduction to Works 3, pp. 24–26, 28–32.

[43] N. Bohr to P. Ehrenfest, 23 March 1921, in: Works 3, p. 614.

[44] P. Ehrenfest to N. Bohr, 19 August 1920, in: Works 3, p. 609.

[45] H.A. Lorentz to P. Ehrenfest, 17 December 1920.

[46] P. Ehrenfest to N. Bohr, 27 December 1920, in: Works 3, pp. 612–613.

[47] W. Pauli, Niels Bohr on his 60th birthday, Rev. Mod. Phys. 17 (1945) 97–101. Quote on p. 99. On one occasion Pauli agreed to criticize a manuscript of Bohr's, and then commented: "The result of this will be that you are going to prolong your sentences still further." W. Pauli to N. Bohr 13 January 1928. Works 6, pp. 41–42, 435–436.

[48] N. Bohr, in: Atomes et Électrons (Inst. Int. Phys. Solvay, Paris, 1923) pp. 228–247.
P. Ehrenfest, ibid., pp. 248–254.
Discussion, ibid., pp. 255–262.
The revised English text of Bohr's report, and the French translation of Ehrenfest's appear in Works 3, pp. 359–395.

[49] H.A. Kramers, quoted in Helge Kragh, Niels Bohr's second atomic theory, in: Historical Studies in the Physical Sciences 10, 123–186. Quote on p. 156.

[50] N. Bohr to P. Ehrenfest, 16 September 1921, in: Works 3, pp. 626–627.

[51] N. Bohr to P. Ehrenfest, 28 June 1921.

[52] P. Ehrenfest to N. Bohr, 3 July 1921.

[53] A. Einstein, Out of My Later Years, p. 216.

[54] N. Bohr to P. Ehrenfest, 10 July 1921, 24 October 1921, 11 November 1921
Margrethe Bohr to P. Ehrenfest, 18 November 1921.

[55] P. Ehrenfest to A. Einstein, 27 December 1921.

[56] P. Ehrenfest to H.A. Lorentz, 4 February 1922.

[57] N. Bohr to A. Einstein, 11 November 1922, in: Works 4, The Periodic System (1920–1923), ed. J.R. Nielsen, pp. 28, 685.

[58] A. Einstein to N. Bohr, 10 January 1923, in: Works 4, pp. 28, 686.

[59] A. Einstein, Wahlvorschlag für N. Bohr zur Aufnahme als korrespondierendes Mitglied in die Akademie der Wissenschaften in: Albert Einstein in Berlin, Teil I, p. 130.

[60] G.L. De Haas-Lorentz (ed.), Reminiscences, in: H.A. Lorentz. Impressions of His Life and Work, (Amsterdam, 1957), pp. 146–148.

[61] A. Einstein to P. Ehrenfest, 18 August 1925.

[62] P. Ehrenfest to A. Einstein, 16 September 1925.

[63] P. Ehrenfest to N. Bohr, 19 September 1925, in: Works 5, The Emergence of Quantum Mechanics (Mainly 1924–1926), ed. K. Stolzenburg, pp. 326–328.

[64] A. Einstein to P. Ehrenfest, 18 September 1925.
[65] See Works 5, pp. 3–216.
     Also see M.J. Klein, The first phase of the Bohr–Einstein dialogue, in: Historical Studies in the Physical Sciences 2 (1970) 1–39.
[66] N. Bohr to E. Rutherford, 18 April 1925, in: Works 5, p. 488.
[67] N. Bohr to R.H. Fowler, 21 April 1925, in: Works 5, pp. 81–82.
[68] N. Bohr to M. Born, 1 May 1925, in: Works 5, pp. 85, 310–311.
[69] N. Bohr to S. Rosseland, 6 January 1926, in: Works 5, pp. 484–486.
[70] See M.J. Klein, Einstein and the wave-particle duality, in: The Natural Philosopher 3 (1964) 1–49. See also A. Pais, Subtle is the Lord... The Science and Life of Albert Einstein (Oxford, New York, 1982) pp. 423–439.
[71] See Works 5, pp. 219–226.
[72] N. Bohr to W. Pauli, 25 November 1925, in: Works 5, pp. 455–456.
[73] N. Bohr to W. Heisenberg, 26 November 1925, in: Works 5, pp. 370–371.
[74] N. Bohr to R.H. Fowler, 26 November 1925, in: Works 5, pp. 225, 337–338.
[75] See Works, 5, pp. 226–240.
[76] N. Bohr to R. Kronig, 26 March 1926, in: Works 5, pp. 234–236.
[77] N. Bohr to J.C. Slater, 28 January 1926, in: Works 5, pp. 497–498.
[78] N. Bohr to P. Ehrenfest, 22 December 1925, in: Works 5, p. 329.
[79] See especially Bohr's essay listed in ref. [4]. See also Works 6, pp. 35–41, 99–103.
[80] P. Ehrenfest to S.A. Goudsmit, G.E. Uhlenbeck and G.H. Dieke, 3 November 1927, in: Works 6, pp. 37–41, 415–418.
[81] See ref. [4]. Quote from p. 218.
[82] P. Ehrenfest to N. Bohr and A. Einstein, 12 September 1931.
[83] L. Trilling, The Liberal Imagination (New York, 1948) p. 86.

The Lesson of Quantum Theory, edited by J. de Boer, E. Dal and O. Ulfbeck
© Elsevier Science Publishers B.V., 1986

Niels Bohr and Nuclear Weapons

Margaret Gowing

Modern History Faculty, Broad Street
Oxford, UK

Abstract

The chapter shows that, in the light of the Bohr–Wheeler paper of September 1939 which interpreted the fission process, Bohr believed that an atomic bomb was impossible. It relates how he was brought out of occupied Denmark by the British to join the Anglo–American wartime bomb project, how he immediately saw the political implications of the bomb, and advocated unsuccessfully that in the interests of postwar international control, Russia should be told about the bomb before it was used. The chapter also discusses Bohr's postwar Letter to the United Nations. It assesses the value of both these arms control initiatives by Bohr.

It was one of the most fateful coincidences of history that the discovery of uranium fission by Hahn and Strassman and the theoretical explanation of the phenomenon by Meitner and Frisch were published at the beginning of the year of the outbreak of World War II. In April 1939 Joliot–Curie's team in Paris was the first to announce experimental evidence that in fission spare neutrons were released; this opened the possibility of a nuclear chain reaction and an atomic bomb. The widespread and agitated discussion of this project diminished when, on September 1st, the day Germany invaded Poland, an article by Bohr and Wheeler was published in *Physical Review* [1], giving the classic interpretation of the fission process. This included the important deduction (already foreshadowed by Bohr in a letter of February 7, 1939 to the same journal [2]) that it was the rare uranium U 235 nuclei, not the uranium U 238 nuclei, that fissioned: a deduction consistent with the observation that fission was much more likely with moderated, slow neutrons than with fast ones. It seemed that these slow-neutron chain-reactions might produce power, but not the fantastically fast reaction necessary for a bomb.

This basic knowledge of fission was available to the whole world and in the subsequent two years, scientists in the belligerent European countries—Britain, France and Germany—and the non-belligerent United States worked on bomb possibilities. The most effective work was done in Britain by the famous Maud Committee. The strange name of this committee was derived from a telegram sent by Lise Meitner from Sweden in May 1940, just after Denmark was invaded by the Germans, to the physicist O.W.R. Richardson: "Met Niels and Margarethe recently. Both well but unhappy about events. Please inform Cockcroft and Maud Ray Kent." Cockcroft believed the last words were an anagram for "radiumtaken" and the words seemed a good code name. When Bohr arrived in England in 1943 he asked whether the message ever reached his old governess Maud Ray who lived in Kent.

The Maud Committee was composed of British and refugee scientists. A paper of April 1940 by Otto Frisch and Rudolf Peierls—both refugees at Birmingham University, England—had instigated the Committee by showing that a small lump of U 235 would give the fast reaction necessary for a bomb and by proposing an industrial method for separating the U 235. When France fell, two members of Joliot–Curie's team fled to Britain and a slow-neutron team developed round them. Two members of the team soon suggested that in a slow neutron reaction the element 94, foreshadowed by Macmillan and Abelson in the United States in May 1940, would be produced and that it would also be an efficient super-explosive with a small critical mass. Unknown to the British, Berkeley scientists in March 1941 demonstrated experimentally that this was so.

Several groups of American scientists were working on many aspects of "the uranium problem" but in a diffuse, leisurely way. It did not become urgent to them until in July 1941 they were shown the British Maud Report, which showed most coherently and cogently why and how an atomic bomb was possible. Their government set up, even before the attack on Pearl Harbour in December 1941 ended the United States' neutrality, what was soon to become the huge Manhattan Project.

The great fear was that Germany might make a bomb first, but, most mercifully, their project floundered in its science and its organisation.

After the invasion of Denmark in 1940, Bohr was preoccupied with the grave problems of his country and his Institute, including its refugees from Nazism; he faced them with dignity, courage and deep patriotism. Atomic bombs were not a main concern for him. He had explained in a lecture at the outbreak of the war [3] that an explosion could indeed be achieved with a sufficiently large amount of U 235 but he did not think that it would be technically possible to separate enough U 235. He did not at that time consider the possible slow-neutron route to a bomb. He was therefore deeply disturbed by a visit from the great German physicist Heisenberg in October 1941. Robert Jungk in his well-known book *Brighter than a Thousand Suns* calls this visit

> "a little known peace feeler. By the expedient of a silent agreement between German and Allied atomic experts, the production of a morally objectionable weapon was to be prevented."

This suggestion of German moral scruples is supported in the book by a letter to Jungk from Heisenberg about this visit to Bohr. Aage Bohr, who was so close to his father in these nuclear events, has written that Heisenberg put no proposal to Niels for a physicists' agreement not to develop nuclear weapons, but that he left the strong impression that the Germans attributed great military importance to atomic energy.

Early in 1943 a message reached Niels Bohr in a micro-dot in a key handle from James Chadwick, the British physicist who had discovered the neutron and who was the informal scientific leader of the British atomic project. Chadwick wrote that he had heard,

> "you have considered coming to this country if the opportunity should offer. I need not tell you how delighted I myself should be to see you again ... There is no scientist in the world who would be more acceptable both to our university people and to the general public ... I have in mind a particular problem in which your assistance would be of the greatest help..."

Bohr still felt that it was his duty to remain in Denmark, but he replied to Chadwick that he might leave if he felt he could be of real help. He said that he did not think this probable, adding:

> "I have to the best of my judgment convinced myself that, in spite of all future prospects, any immediate use of the latest marvellous discoveries of atomic physics is impracticable."

Two months later he reported to Chadwick rumours of German preparations for producing metallic uranium and heavy water in order to make atomic bombs. However, he was still sceptical about such bombs. The Bohr–Chadwick messages were buried in the garden at Carlsberg to be found after the war.

In September 1943 Bohr and his family, who were now in danger of arrest, fled to Sweden and the British Atomic Directorate arranged for Niels and Aage to go to England. Niels departed on October 6, 1943, in an unarmed bomber which flew at great height. The earphones did not fit his large head and, not hearing the order to turn on the oxygen, he became unconscious but recovered as the plane lost height. Aage arrived a week later.

On his arrival in England, Niels was immediately told everything about the British and American projects: it now seemed almost certain that the Americans would produce nuclear weapons within a year or two. Father and son were received most warmly by the scientists, by the administrators of the code-named Tube Alloys project and by the Minister in charge—Sir John Anderson, later Lord Waverley, who was Chancellor of the Exchequer and who was henceforth to be a very warm friend of the Bohr family. The Bohrs had arrived at an important moment for the British project. Their Maud Report had pushed the scattered, ill-organised American project off the ground, but the over-confident British had preferred an independent atomic project in co-operation with the Americans rather than the full integration between the projects which Roosevelt had suggested when they received the Maud Report. However, the American project soon far outstripped the British and neither needed nor wanted British help. The British, already highly mobilised and unable to build huge atomic plants, became desperate: they could not proceed on their own and were cut off from American knowledge. It was only after a great struggle that Churchill persuaded Roosevelt to sign the Quebec Agreement in August 1943 which enabled British scientists to participate in some parts of the American project, notably at Los Alamos where the bombs were to be fabricated.

Bohr, so welcome to the British for his own sake, was also, as a member of their team in America, a trump card for them in implementing the Quebec Agreement.

Bohr promised that he would not allow himself to be drawn into the American orbit, that he would assist the common effort and also do everything he could to make the association between America and Britain a real partnership. He and Aage arrived in the United States early in December 1943 under the cover names Nicholas and James Baker, or affectionately to colleagues Uncle Nick and Jim. They were not attached to any specific team but Bohr's main scientific contribution was to the work at Los Alamos, where he found many of his former students. There he stimulated and liberated scientific ideas which gave rise to theoretical and experimental activities which cleared up unanswered questions—for example on the velocity selector, bomb assembly and the design of the initiator. Old and new friendships flourished here.

Bohr was fascinated by the vast Manhattan Project, built as it was on theoretical foundations he had laid. However, he was infinitely—and immediately—more impressed with the implications of this weapon of unparalleled power for the future of the world. He had the reputation for being the most unworldly of scientists but the unworldliness was purely behavioural. His knowledge of philosophy, history and politics was profound and had been deepened by the experience of the refugees from Nazism at his Institute in the 1930s and by the German occupation. His exceptionally imaginative intuition marked not only his science but also his view of world politics. He immediately realised when he saw the Manhattan Project that it was only a beginning; at Los Alamos scientists already foresaw a hydrogen bomb.

Bohr was concerned privately with the question of how soon the weapon would be ready for use and what role it might play in the Second World War but he took no part in discussions about whether the bomb should be dropped on Japan. He looked rather to the years after the war and the terrifying prospect of an atomic arms race. After his very first visit to Los Alamos he wrote to London that future effective control would involve not only the most intricate technical and administrative problems, but also concessions over exchange of information and openness about industrial efforts and military preparations that were hardly conceivable in terms of prewar international relationships. Bohr felt that the invention of atomic bombs was so climacteric that it would facilitate a whole new approach to these relationships.

Before long his thoughts crystallised into a precise proposition. Despite the wartime alliance with Russia, after she entered the war in June 1941, Bohr believed that there would be tension between the West and Russia after the war and that confidence might be promoted by telling Russia about the bomb before it was used. Conversely, he believed that it would be disastrous if Russia should learn of it on her own. Knowing very well the competence of the Russian physicists, Bohr felt certain that the margin of time before the Russians made a bomb themselves would be very small. This conviction was strengthened when, in London, in April 1944, he received a letter from Peter Kapitza, written six months earlier when Bohr escaped to Sweden, inviting him to settle in Russia. This reinforced Bohr's belief that the Russians were aware of the American project. He sent back a warm, innocuous reply to Kapitza and showed the correspondence to the British authorities.

The political implications of the bomb had become Bohr's prime concern and he spent much of his time writing "political" memoranda and in haunting the offices and anterooms of those who had political power or access to it. His discursive talk and his low, indistinct voice were not easy to follow but he made important converts among British Ministers and officials: Lord Halifax and Sir Ronald Campbell, respectively Ambassador and Minister at the British Embassy in Washington and, most significantly, Sir John Anderson and Lord Cherwell (the scientist who was Churchill's personal adviser) and Field Marshal Smuts (Prime Minister of South Africa). Halifax told Bohr, however, that because of America's preponderant share in the project, any initiative would have to come from President Roosevelt. It seemed fortunate therefore that Bohr was able to resume a prewar friendship with Mr Justice Frankfurter, a Supreme Court Judge, and a friend of Roosevelt, who already knew about the bomb. He communicated Bohr's ideas and hopes to the President, who said the whole thing "worried him to death" and that he was most eager to explore it with Churchill.

In March 1944 Sir John Anderson wrote to Churchill that the Americans would almost certainly get a bomb first but that Russia would most probably put forward a great effort once they had expelled the Germans. Moreover, the project would come within the capacity of other countries. There were two alternatives: a particularly vicious armaments race in which at best America and Britain would for a time enjoy a precarious and uneasy advantage, or a form of international control must be devised. If it was decided to work for international control there was much to be said for communicating to Russia in the near future the bare fact that the Americans expected by a given date to have this devastating weapon and for inviting them to collaborate in preparing a scheme for international control. If the Russians were told nothing they would learn sooner or later what was afoot and might then be less disposed to co-operate. There was little risk that Russia, if she chose to be unco-operative, would be much helped by such a communication. Cherwell added his plea:

> "I must confess that I think plans and preparations for the postwar world and even the peace conference are utterly illusory, so long as this crucial factor is left out of account."

Churchill disagreed profoundly and constantly repeated his conviction that the project must be kept absolutely as secret as possible.

Pressed by Smuts, Cherwell and Sir Henry Dale, President of the Royal Society, Churchill saw Bohr on May 16, 1944. This was only two weeks before the Allied invasion of France, and, perhaps partly for that reason, the meeting was a tragic failure. His friends had feared that Bohr's "mild, philosophical vagueness of expression and his inarticulate whisper" might prevent a "desperately pre-occupied Prime Minister" from understanding him and so it proved. The main point was never reached. "We did not speak the same language," said Bohr afterwards. Later Churchill told Cherwell: "I did not like the man when you showed him to me, with his hair all over his head."

However, Churchill realised that he must discuss the long-term problem of the atomic bomb with the President when they next met in September 1944. Before then Roosevelt had received a memorandum by Bohr which outlined the scientific basis of the project, his own feelings on seeing the project after his escape from Denmark,

Kapitza's approach to him, his belief that the project offered an opportunity for a new spirit and new hope in international relations, his fears of a nuclear arms race between Russia and the West. On August 26, Roosevelt had an interview of $1\frac{1}{2}$ hours with Bohr in complete privacy.

Bohr reiterated his belief that there was a great opportunity for better world relations provided it was seized now rather than later. He expanded on his reasons for urging an approach to Russia and on his arguments against those who said that the West would lose thereby. He said it must be assumed that the Russians knew great efforts were being made in the United States to make a bomb; that the Russians themselves were studying the matter and would be free to develop a full effort at the end of the German war; that the Russians would probably obtain the German secrets at the end of the war. If America and Britain said nothing before a bomb was used they would, urged Bohr, arouse Russian suspicions and create a greater risk of fateful competition in atomic weapons. They would lose the opportunity of using an approach to Russia in order to establish confidence. Bohr emphasised that it was not necessary to begin by giving the Russians detailed information about the bomb. The approach should be general and if the Russians responded in a co-operative spirit the way would be open for frank discussions. If not, the West would know where they stood. Bohr believed that an approach might be possible through preliminary and noncommital contact between scientists.

The President was most friendly to Bohr and open and frank in his discussions of the political problems raised by the bomb. He said that an approach to Russia must be tried and that it would open a new era of human history. Stalin, he believed, was a sufficient realist to understand the implications of this scientific and technological revolution. Encouraged by his talk with Roosevelt, Bohr drafted a letter to Kapitza on the lines discussed and held himself ready to go to Russia. Bohr's high hopes were soon destroyed. In September 1944 when Churchill and Roosevelt met and discussed the atomic bomb, the results were very different from those foreshadowed during Bohr's interview with Roosevelt. They signed an agreement which not only said that no other country was to be told about the bomb but also included a paragraph saying that enquiries were to be made about Professor Bohr and steps taken to ensure that he leaked no information, particularly to the Russians.

This agreement, besides turning down Bohr's proposal for an approach to Russia, put his good faith and honour in question. Churchill wrote forcefully to Lord Cherwell:

> "The President and I are much worried about Professor Bohr. How did he come into the business? He is a great advocate of publicity. He made an unauthorised disclosure to Chief Justice Frankfurter who startled the President by telling him he knew all the details. He said he is in close correspondence with a Russian professor, an old friend of his in Russia, to whom he has written about the matter and may be writing still. The Russian professor has urged him to go to Russia in order to discuss matters. What is all this about? It seems to me Bohr ought to be confined or at any rate made to see that he is very near the edge of mortal crimes."

Bohr's ministerial friends rushed to defend him and to say that Churchill was talking nonsense. Cherwell sent a strong reply to Churchill telling him how Bohr had come into the business, about the Bohr–Frankfurter talks, the story of the approach by Kapitza and the reply that had been agreed by British Intelligence:

> "I have always found Bohr most discreet and conscious of his obligations to England to which he owes a great deal and only the very strongest evidence would induce me to believe that he had done anything improper in this matter. I do not know whether you realise that the possibilities of this super weapon have been publicly discussed for at least six or seven years. The things that matter are which processes are proving successful, what the main stages are and what stage has been reached. Most of the rest is published every silly season in most newspapers."

Cherwell repeated these views to Roosevelt in the presence of Vannevar Bush, the eminent American scientist, who agreed with them. Churchill accepted Cherwell's opinion about Bohr and the matter was dropped. Bohr, when he heard of the misunderstanding, was distressed; he might have been deeply offended but his sense of humour was always stronger than his pride.

We do not know the reasons for Roosevelt's *volte face*. As for Churchill, he believed passionately in the desirability and possibility of keeping atomic weapons secret. At home he kept the matter secret from most of the War Cabinet (including the Labour leader, Mr Attlee, who in July 1945 became Prime Minister) and from his Defence advisors, and he refused to impart any information to the French, to whom the British had atomic obligations. He wrote:

> "You may be quite sure that any power that gets hold of the secret will try to make the article and that this touches the existence of human society. The matter is one out of all relation to anything else that exists in the world and I could not think of participating in any disclosure to third or fourth parties at the present time. I do not believe there is anyone in the world who can possibly have reached the position now occupied by us and the United States."

Meanwhile, Bohr found himself exercising a restraining hand on Einstein, who in December 1944 sent him a cri de coeur about the prospect of a postwar arms race. The politicians, he said, did not appreciate the threat. In all the principal countries influential scientists had the ear of political leaders—Bohr himself, Compton, Cherwell, Kapitza and Joffe. These men should come together to bring pressure to bear on their political leaders to strive for the internationalisation of military power. "Don't say impossible," wrote Einstein to Bohr, "but wait a few days until you have accustomed yourself to these strange thoughts." Bohr went to see Einstein and explained to him that it would be quite illegitimate and might have the most deplorable consequences if anyone who knew about the bomb should take the initiative into his own hands. Bohr assured Einstein that the attention of responsible statesmen in England and America had been called to the implications of the bomb. Einstein thereupon agreed to abstain from action and to impress on his friends the undesirability of doing anything that might complicate the statesmen's task.

Bohr, conscious that time was running out, became increasingly convinced that postponement of any discussion with Russia until a bomb was demonstrated might give the appearance of an attempt at coercion in which no great nation could be expected to acquiesce. He emphasised yet again that Russia would soon learn, at the least, about the German work. In April 1945, Lord Halifax and Frankfurter walked through Rock Creek Park in Washington discussing how to get Bohr's proposals properly considered. As they ended their walk, they heard all the bells in Washington tolling. Roosevelt was dead.

In May 1945 in Washington, the Secretary of State for War chaired a committee of scientists which inter alia discussed disclosure to Russia and possible forms of international control. Members of the committee were torn between a desire for scientific openness and a conviction that the business could not remain secret for long on the one hand, and by anxieties over deteriorating Russian behaviour on the other hand. The anxieties won and the committee decided early in June 1945 that no information should be revealed to Russia or anyone else until the first bomb had been dropped on Japan [4].

On July 24, eight days after the atomic bomb test at Alamogordo and thirteen days before a bomb was dropped on Hiroshima, President Truman told Stalin simply that the United States had a new weapon of unusual destructive force. Bohr's wartime pleas had failed. As books told about them from the 1960s onwards, they were seen as the remarkable intuition of a remarkable scientist. More recently, a leading historian of international relations, however, attacked them [5]. He wrote:

> "the concept of 'international control' in the minds of Bohr and others was essentially a cop-out, a flight into higher mysticism away from the unpleasant and unacceptable world of politics."

Such strictures were inappropriate to Bohr's essentially practical proposal. He knew that Russian physicists were extremely good and that once a bomb was dropped there could be no secret. To inform Russia officially would therefore carry little risk and might conceivably bring benefits. *Not* to inform Russia would bring little benefit and would intensify suspicions. Bohr's idealism, that is, was set in a very practical framework of cost–benefit analysis as he looked to a future when all civilised life might be destroyed in a flash.

If Russia had been told about the bomb during the war it might have made no difference. But she had already begun her own project in 1942 and also knew a great deal about the Manhattan Project from spies. The fact that she was told virtually nothing by the Allies guaranteed that attempts made just after the war to establish international control, which might have failed anyway, were doomed.

Bohr, as I have noted, made no representation in advance about the use of the atomic bomb against Japan, and he did not argue about past events once the war was over. He privately deplored the spirit in which the bomb was used and the opportunities that were lost but he neither made nor joined any written protestations. His thoughts were on the future and the postwar world. With his inbred and unquenchable optimism he was convinced that while atomic bombs introduced unprecedented threats to the world, they also gave a unique opportunity for a new approach to international relationships.

In the spring of 1945 Bohr had written another memorandum looking beyond the question of informing the Russians about the bomb during the war. Bohr warned that the American–British effort, immense though it was, had proved far smaller than might have been anticipated and that any information, however scanty, that might have leaked from it would have greatly stimulated efforts elsewhere. Probably within the very near future means would be found to

> "simplify the methods of production of the active substances and intensify their effects to an extent which may permit any nation possessing great industrial resources to command powers of destruction surpassing all previous imagination. Humanity will therefore be confronted with dangers of unprecedented character unless in due time measures can be taken to forestall a disastrous competition in such formidable armaments and to establish an international control of the manufacture and use of the powerful materials."

Extraordinary measures would be necessary to counter secret preparations for the mastery of the new means of destruction. Not only must there be universal access to full information about scientific discoveries but every major technical enterprise, industrial as well as military, must be open to international control. The special character of the production of the active materials, and the peculiar conditions governing their use as dangerous explosives, would, said Bohr, greatly facilitate such control and should ensure its efficiency, provided the right of supervision was guaranteed. Detailed proposals for the establishment of an effective control would have to be worked out with the assistance of scientists and technologists appointed by governments and a standing expert committee of an international security organisation might be charged with keeping account of new scientific and technical developments and with recommending appropriate adjustments of the control measures.

On recommendations from the technical committee, the organisation would be able to judge the conditions under which industrial exploitation of atomic energy sources could be permitted, with adequate safeguards to prevent any assembly of active material for an explosive. All material prepared for armaments might ultimately be entrusted to the security organisation to be held in readiness for eventual policing purposes. The prewar bonds between scientists of different nations would be especially valuable in creating controls.

Bohr had foreseen proliferation—that it would be possible for any nation with large industrial resources to command these unimaginable powers of destruction. But he also saw that the special character of the production of fissile material would greatly facilitate efficient control provided that an effective organisation with the right of supervision was established. In all this, his key belief was that there must be openness about scientific discoveries and about industrial and military enterprises.

Elements of Bohr's ideas were to be found in the early postwar proposals for atomic energy control discussed at the United Nations Commission on the subject and, later, in the non-proliferation safeguards to be operated by the International Atomic Energy Agency. However, the United Nations proposals came to nothing and the non-proliferation arrangements did not apply to the existing atomic powers.

On-site inspection, which Bohr regarded as essential to "openness" and which has been an issue in all attempts to control nuclear weapons and installations, has generally been unacceptable to the Soviet Union. However, Bertrand Goldschmidt, who has been continuously involved with atomic energy and with international control since 1940, reminded the Niels Bohr Symposium on Nuclear Armaments of events at the United Nations Atomic Energy Commission in 1946 and 1947. The Soviet Union had made a serious proposal for the establishment of an International

Control Commission to inspect atomic facilities which would, however, still be in national hands. The United States and the Western Powers on the other hand wanted a supra-national authority. There were other problems in the negotiations but in retrospect Goldschmidt believes a unique chance may have been missed since the Soviet Union proposed the maximum opening of their territory which they were ever to put to the international community.

Goldschmidt's paper makes Bohr's proposals for openness less impracticable than they have since appeared. As it was, coming at this early stage of atomic development, the Commission was the first and possibly the last real opportunity for international control. After the funeral of the United Nations Atomic Energy Commission in 1948, openness became *more*, rather than *less*, unthinkable on both sides. Nevertheless, Bohr continued his campaign on every possible occasion. The darker the international outlook grew, the more he was convinced that a great issue, "suited to invoke the highest aspirations of mankind" must be raised. To him this issue was *openness*, with free access to information about all aspects of life in every country. He pleaded that the initiative should be taken—even if the chances of getting agreement were thin—because an offer of openness would strengthen the moral position of the supporters of international co-operation. The opposition of those who refused to join would amount to a confession of lack of confidence in their own cause.

In 1948, Bohr had written in these terms to General Marshall, the United States Secretary of State, urging that America should take the initiative in openness and stressing that this would not entail an a priori commitment to disarmament. His efforts culminated in June 1950 in his Open Letter to the United Nations pleading for

> "an open world with common knowledge about social conditions and technical enterprises, including military preparations, in every country."

Like all Bohr's other memoranda, the Open Letter was written with the same examination and re-examination of every word and every nuance that marked his scientific papers. The opaqueness of Bohr's prose may have obscured the message of the Open Letter. However, as it turned out, the prose style probably made little difference. The letter could not have appeared at a worse moment. The Cold War was rapidly intensifying and the Korean War broke out at much the same time. Fear was rampant. Oppenheimer, for one, was deeply pessimistic that anyone in a position of political responsibility would take openness as a basis for action. The Letter brought little public reaction outside Scandinavia and in Britain even the liberal *Manchester Guardian* newspaper wrote unsympathetically that we "must keep our feet on the ground". Rudolf Peierls replied eloquently: "let us also try to keep our heads out of the sand."

Bohr himself remained dedicated to his main theme of openness so much so that he would not weaken it by joining the other peace moves and appeals from men such as Einstein and Bertrand Russell. In 1956 Bohr wrote a further letter to the Secretary of the United Nations, Dag Hammarskjøld.

Looking back on the Open Letter thirty-five years after it was written, I do not think it has ever been considered very seriously except as a moving, albeit unrealistic, expression of idealism. I suggest that on the contrary, the proposal was realistically farsighted. In the war, Bohr had been among the very first to realise that atomic weapons would change the world, that their significance was far greater than their simple but terrifying arithmetical equivalents of thousands of tons of TNT. He quickly appreciated the potential of thermonuclear weapons. He realised that horrific weapons could develop from new advances in biology and chemistry as well as in physics. He also foresaw developments in communications and electronics which would revolutionise information-gathering. Indeed he lived to see Sputnik and missiles.

Above all, he was among the very first to realise that a nuclear arms race has *no logic*. Since a small number of existing atomic weapons are enough to cause unimaginable destruction so that their only rational function is to deter rather than to fight, it is highly undesirable that either superpower should attempt to acquire a quantitative or qualitative lead over the other. Is there not more logic, as well as a bias to peace, in deliberate openness between the nuclear states about their weaponry and its scientific and industrial infrastructure, than in the depressing tales of espionage? It was a very senior intelligence expert who recently said that the greatest danger to the world today is misperceptions caused by lack of proper knowledge.

Experience as well as logic suggests that Bohr was right. The United States McMahon Act of 1946 forbade the transmission of almost all atomic information to any country, including her closest ally Britain, with penalties including death or life imprisonment. Yet ten years later, under the Atoms for Peace programme, much information hitherto considered top secret was positively thrust upon the world and the heavens did not fall. Co-operation has become particularly strong in some areas such as thermonuclear fusion which were once particularly secret. There has been besides the paradox that surveillance through satellite has probably enhanced rather than diminished security.

So, I suggest, Bohr's wartime and postwar view of the nuclear future was hard-headedly realistic as well as clear-mindedly visionary. He himself realised that his open world was a remote possibility in the world of 1950. But who can say that he was wrong when he believed that amidst the stiff technicalities of arms control, which are wellnigh incomprehensible except to the expert, mankind also needs *hope* —as he said, an issue to invoke its highest aspirations.

In short, Bohr showed in his reflections on nuclear weapons the wisdom, imaginative intuition and optimism which informed his science and his whole being.

References

General sources

A. Bohr, The war years and the prospects raised by the atomic weapons, in: Niels Bohr: his Life and Work as seen by his Friends and Colleagues, ed. S. Rozental (North-Holland Amsterdam, 1967) p. 191.

M. Gowing, Britain and Atomic Energy 1939–1945 (Macmillan, London, 1964).

Journal articles

[1] N. Bohr and J.A. Wheeler, Phys. Rev. 56 (1939) 426.
[2] Letter from N. Bohr, 7.2.39, Phys. Rev. 55 (1939) 418–419.

[3] N. Bohr, Recent investigations on the transmutation of atomic nuclei, translated from Fysik Tidskrift for 1941.
[4] R.G. Hewlett and O.E. Anderson, The New World 1939–1946 (Pennsylvania State University Press, 1962).
[5] D.C. Watt, The historiography of nuclear diplomacy, in: Science 194 (1976) 174–175.

The Lesson of Quantum Theory, edited by J. de Boer, E. Dal and O. Ulfbeck
© Elsevier Science Publishers B.V., 1986

Niels Bohr: The Man and his Legacy

John Archibald Wheeler

The University of Texas
Austin, Texas, USA

Contents

1. Introduction

Commitment, diligence, openness. The year 1985 marks the 100th anniversary of the birth of Niels Bohr—and of seventy million other human beings born in the same year. What is special about this one man of blue eyes and average height, always lighting a match to his pipe but hardly ever managing to smoke it, that we should single him out for special attention? Are we here to put him on a pedestal? No! Niels Bohr was against making a hero of anybody because he was so ready to see the heroic in everybody.

Do we make Niels Bohr the center of this inspiring conference because he elucidated the structure of the atom and the structure of the nucleus, because he masterminded our understanding of the quantum, because of his *genius*? Please, no, he says to us again, and quotes to us with a smile and a wave of his pipe his favorite words from Theodor Fontane:

> Gaben, wer hatte sie nicht?
> Talent, Spielzeug für Kinder!
> Erst der Ernst macht den Mann,
> Erst der Fleiss das Genie;

or, translated:

> Gifts? Who hasn't?
> Talent? Toy for children!
> Commitment only makes the man;
> Only diligence the genius.

Diligence? Sense of commitment? Thousands upon thousands of the seventy million born a hundred years ago lived lives of conscientiousness and drive. What additional trait made Bohr special? Openness! Openness to the most wildly different ways of looking at any issue.

If openness, commitment and diligence were the central features of Bohr the man, and of the way of doing science that he taught the world, how can we be blamed for wanting to recall the blooming of that openness?

Openness manifest in dialog. One of the schoolmates of Niels and his younger brother Harald speaks of the amazement it brought to everyone around the two boys to see their warm-hearted, lively, never ceasing to-and-fro:

> "Their way of thinking seemed to be co-ordinated; one improved on the other's or his own expressions, or defended in a heated yet good-humored manner his choice of words. Ideas changed their tone and became polished; there was no defense of preconceived opinions, but the whole of the argument was spontaneous. This way of thinking *à deux* was so deeply ingrained in the brothers that nobody else could join it."

They formed their own discussion group in university years. They contributed to it in the same lively way. As the evening went on, the others fell silent in delight and admiration as the back-and-forth of the two brothers went on to produce a new position, a new conclusion or a new outlook to which all could subscribe. That was openness in action.

As to the child and the university student, so to the later Bohr dialog was central to openness, and humor the magic element of dialog. Invited to speak to his classmates on their twenty-fifth jubilee, he asked at one point in his talk how had he and his comrades managed to avoid one-sidedness—despite the "strength which it may offer us"? And how had they succeeded in acquiring a more balanced outlook on life? Out of a happy combination, Bohr suggested, of "our academic tradition" and "our Danish popular sense of humor." Like Abraham Lincoln, with his famous method to persuade a reluctant Civil War cabinet to consider a new proposal in a new light, Bohr knew that nothing has more power than a joke to jolt us out of ourselves and into new surroundings with a new outlook on old issues.

All who worked at Bohr's Institute knew his definition of an expert as "a man who knows through his own bitter experience some small fraction of all the mistakes that can be made in his field"; of a workroom as "a room where no one can keep you from working"; and of a pessimist as "one who generally predicts correctly, but gets no satisfaction out of it." For a peaceable way to work out a disagreement nothing was more useful than his definition of a deep truth as "a truth whose opposite is also a deep truth." No one who knew Bohr can forget him in dialog. With five percent of jokes he mixed ninety-five percent of utmost seriousness. He intermingled delight in the lessons of the past with an immense pressure to clear up the puzzle of the present.

The puzzle and the participants: where did they come from? The issue central for the month or for the year Bohr distilled out of dialog with those who were themselves distillers of issues: present and former collaborators and special visitors from all over the world.

Bohr knew that nobody can be anybody without somebodies around. Among the somebodies—for one extended period or another—were Paul Dirac, Rudolf Peierls and E.J. Williams of Britain, Hendrik Casimir, Paul Ehrenfest and Hans Kramers of the Netherlands, Werner Heisenberg and Lise Meitner of Germany, Leon Rosenfeld of Belgium, Wolfgang Pauli of Switzerland, Vladimir Alexandrovitch Fock, George Gamow and Lev Landau of the Soviet Union, Oskar Klein of Sweden, Yoshio Nishina of Japan, and David Dennison, John Slater and Llewellyn Thomas of the United States.

Jumping into the center of the struggle. Dialog provided the climate for advance. Commitment and diligence supplied the driving force. A principle as old as Bohr's soccer dictated the action. Plunge into the middle of the scrimmage! Run with the ball!

There were two balls at the center of the physics arena in the 1910s, the atom and the nucleus. Bohr scored again and again with the nucleus, and still more with the atom. From the mid-1920s onward, he struggled—successfully—with all his force for a mastery of the message of the quantum. After 1943 he threw himself into an endeavour of vital concern to the larger community, how best to come to terms with the reality of nuclear weapons.

2. The atom

How does it come that the name of Niels Bohr is imperishably tied to the atom? Many in the 1910s toyed or worked at one or another model of the atom. Even to work at the atom, however, is very different from having Bohr's day and night diligence, Bohr's overriding commitment to solve the riddle of stability, whatever the cost. A secret of his success was the intensity with which he felt the difficulty. Why is the atom stable at all? That was his all-consuming concern, day after day, week after week.

Bohr's doctoral thesis dealt with the electron theory of matter, and in particular with the magnetic properties of a metal. Why does magnetism persist? If it arises from electric charge in motion, why does it not damp out and fade away? "Are there," Bohr asked himself, "forces in nature of a kind completely different from the usual mechanical sort, forces that might keep the electron going forever?" What other possibility was there to make sense of magnetism?

The issue of stability took for Bohr a still more urgent form with Ernest Rutherford's great discovery in late 1910. He, with his collaborators, found that the positive electric charge of an atom and almost all its mass is concentrated in a nucleus ten thousand times smaller than the atom itself. What did this finding mean for the structure of the atom: To understand more deeply what the very question meant, twenty-seven-year-old Bohr moved from Cambridge—where he had gone because of J.J. Thomson's interest in the structure of the atom—to Manchester. He joined the group of eager young discoverers gathered around the forty-one-year-old Rutherford.

The tall, ruddy New Zealander, admired by Bohr, became his mentor and lifelong friend. The spirit of the place was epitomized by Rutherford's statement that others "play games with their symbols but we, *here*, turn out the real solid facts of nature". With what delight Bohr greeted each new finding of each new week! What splendid young colleagues!

One Manchester friend was George de Hevesy, whom Bohr was later to bring to Copenhagen. In 1912 Hevesy worked out the electrochemistry of radioactive substances. Bohr explained Hevesy's systematics in terms of the concept of nuclear isotopes, nuclei with the same outer electronic structure but with different masses and different radioactivities for the central nucleus. These findings and many more, however, left unexplained a puzzle now more urgent than ever.

What keeps matter from collapsing? That great collection of negatively charged electrons that circulate in essentially free space around the positively charged nucleus to constitute a "solid body": Why doesn't it fall together and disappear in a microscopic fraction of a second?

We find many a wild idea of how nature prevents this "electric collapse" in that premier journal of physics of the 1910s, the *Philosophical Magazine*. Give up Coulomb's law for the force between charged particles? Or abandon the familiar expression for the radiation of energy by an accelerated charge? Who hesitated at such suggestions? In contrast to those who made these and other radical proposals, Bohr was what we might call a daring conservative: conservative against postulating any change in the battle-tested laws of physics, but in the application of them, daring.

Bohr did not give up the inverse-square force between electron and nucleus, as did J.J. Thomson. He did not try to claim that a charged particle will circulate in an orbit without radiation. However, he did insist that the quantum must be as essential to the Rutherford atom as it is to the Planck heat radiation. Immediate confirmation that this was the right way to think he found in the very simplest dimensional arguments about atomic sizes and the energy of binding of electrons. Having made clear to himself this wonderful point of the importance of the quantum, Bohr could go on to the next issue, the spectroscopic evidence on hydrogen, ready to appreciate this message as no one before ever had. The circular and elliptic orbits came at the end of this explanation, not at the beginning.

Bohr's very industry on the atom almost kept his findings from the world. We are mistaken if we think of the clarification of the hydrogen atom as the be-all and end-all of his labors at Ernest Rutherford's Manchester center. Week after week went by and still Bohr held off from any publication. Rutherford's expostulations grew stronger. "But," Bohr protested, "nobody will believe me unless I can explain every atom and every molecule." Rutherford was quick to reply, "Bohr, you explain hydrogen and you explain helium and everybody will believe the rest."

Rutherford, it is well known, did not trust theoretical men. "When a young man in my laboratory uses the word 'universe'," he once thundered, "I tell him it is time for him to leave." "But how does it come," he was asked on another occasion, "that you trust Bohr?" "Oh," was the response, "but he's a football player."

We recall that Bohr's original semiclassical theory did not even succeed in explaining the spectrum of helium. Nevertheless, it took less than ten years after his original publication for his general concept of the atom to sweep the field.

The failure of the theory to predict correctly the spectrum of neutral helium was redeemed in part by its triumph in identifying and explaining the slightly shifted spectrum of ionized helium and by the subsequent more precise spectroscopic confirmation of this diagnosis. However, it meant much more for the world's acceptance of the new atomic theory that it made sense and gave reasonable results for the structure of atoms all the way up and down the Periodic Table.

Rutherford, to the end of his life, followed Bohr's work in atomic physics with intense interest. For example, upon the discovery by Johannes Stark of the surprisingly large effect of electric fields on the structure of the lines in the spectrum of hydrogen, Rutherford wrote to Bohr:

> "I think it is rather up to you at the present time to write something on the Zeeman and electric effects, if it is possible to reconcile them with your theory."

Few today know the immense toil on atomic theory in the Copenhagen of the late 1910s and early 1920s. It gave us, before the advent of wave mechanics and Hartree fields, such concepts as the screening number, the self-consistent atomic field, the order of the building of the elements—and even the Pauli exclusion principle before the Pauli exclusion principle! Bohr was never content with pioneering a new domain of physics. He had the doggedness and sense of order to insist that the new idea be tested and exploited to the full to provide a completely harmonious account of a whole domain of experience.

Bohr throughout his life took immense care, and showed a unique ability, to make statements that repay intensive study: repay, because they combine maximum emphasis on what is known with maximum circumspection about what is unknown. Nowhere does this care show earlier, with greater force, than in Bohr's first paper, in 1913, on the structure of the atom:

> "The principal assumptions used are:
> (1) That the dynamic equilibrium of the systems in the stationary state can be discussed by help of the ordinary mechanics, while the passing of the systems between different stationary states cannot be treated on that basis.
> (2) That the latter process is followed by the emission of a *homogenous* radiation, for which the relation between the frequency and the amount of energy emitted is the one given by Planck's theory."

The discovery of the structure of the atom marks an immortal step in mankind's eternal search for a world of understandability. Copernicus and Darwin had taken away solidity. Bohr brought it back. Copernicus had dethroned man from the center of the universe. Darwin had taken away plan for the origin of man himself. Bohr built for man, and for the first time, a solid floor for comprehending that physical world in which we live and move and have our being. His theory, though not fully complete or perfect, allowed mankind to resume faith in a world built on regularities. Atoms understood, one could move up to chemistry; and chemistry understood, one could move up to biology and the other sciences.

Under the floor of atomic physics many an unknown still reposed, and among them two where again Bohr led the way: the structure of the nucleus and the mystery of the quantum.

3. The quantum

Many a mystery was left unsolved by Bohr's achievement in understanding the structure of the atom. When an electron jumps from a large orbit to a smaller one, what is it doing as it passes from the one to the other? How does it know to which orbit to jump? And when to make the jump? Today, thanks to no one more than Bohr himself, we have learned that these questions are largely meaningless. Where did the art come from of asking in this domain questions that will be sensible and that will give sensible answers? Out of the dialogs that Bohr carried on year after year in the 1920s at Copenhagen with concerned colleagues from near and far. We are amazed to see how far Bohr could advance into this uncharted territory without benefit of modern quantum theory. We remember the words of Albert Einstein, after he first met Bohr, "I am now reading your great works and, whenever I get stuck anywhere, I see your eager young face before me, smiling and explaining." We remember, too, Einstein's words about Bohr late in life, he "has the highest form of musicality in the sphere of science."

Ideas that Bohr could only shadow forth in earlier years he could at last formulate with compelling vision when Werner Heisenberg's 1925 matrix mechanics, Erwin Schrödinger's 1926 wave mechanics, and Max Born's 1926 probability interpretation of this formalism came to the service of science. When does the electron jump from the Bohr level of higher energy to the Bohr level of lower energy? A wrong question, he could now explain. No device that can measure the energy or frequency or wave length of the radiation given out in this transition with the required precision can co-exist with a device that will measure the time of emission of this radiation. Where is the electron located during the time of the transition? That, too, Bohr could now explain and expound in terms more compelling than those used in his great 1913 paper—and in terms more quantitative, too.

Heisenberg had joined Bohr in the work of making these considerations more quantitative. While Bohr was away in February 1927, Heisenberg wrote up his own conclusions for publication in a paper that did not satisfy Bohr when he saw it. Heisenberg tells us:

> "He pointed out to me that certain statements in this first version were still incorrectly founded, and as he always insisted on relentless clarity in every detail, these points offended him deeply."

So Heisenberg improved his paper but still made its theme indeterminism or "Unbestimmtheit", often inadequately translated as "uncertainty" or the "uncertainty principle." In contrast, Bohr continued his struggle—in endless dialogs with Heisenberg, Pauli, Oskar Klein and other close colleagues—to formulate the deeper lesson of quantum theory. Only in September 1927, at the Como International Physical Congress, was Bohr prepared to make his first statement on the more fundamental point that nature is trying to teach us.

In today's language we might put Bohr's point in these terms. Nature does not exist "out there", independent of us. We ourselves, through our choice of means of observation, are inescapably involved in what is going on. We are not observers only. We are also participators. Or as Bohr puts the central ideal, it is impossible to make "any sharp separation between the behavior of atomic objects and the interaction with the measuring instruments [employed]."

To Bohr the point at issue is not some technical detail of physics. It is human knowledge itself. It is not for nothing that he titles a 1934 collection of his essays, *Atomic Theory and the Description of Nature* and two later collections both *Atomic Physics and Human Knowledge*. All his life he struggled with the problem of knowledge. As a result of the climactic years of 1925–1927, he came to a strong position:

> "... however far the phenomena transcend the scope of classical physical explanation, the account of all evidence must be expressed in classical terms. The argument is simpl[e]...[B]y the word, 'experiment,' we refer to a situation whereby we can tell others what we have done and what we have learned ... [T]herefore, the account of the experimental arrangement and the results of the observations must be expressed in unambiguous language ... [that is, using] the terminology of [every day] classical physics."

How does Bohr's principle of complementarity fit into this view of knowledge? It is an absolutely central point. Nature—or, in more direct terms, our knowledge of nature—is so built, Bohr tells us, that

> "any given application of classical concepts precludes the simultaneous use of other classical concepts which in a different connection are equally necessary for the elucidation of the phenomena."

We know how uncomfortable these ideas made Einstein. He, who in 1905 had been the first to teach us that "God plays dice," had by 1927 become so upset by wave mechanics, indeterminism, and complementarity that he turned to the directly opposite motto, "God does not play dice."

Einstein's opposition to the new views marked the beginning of a great debate between him and Bohr. In all of the history of thought in recent centuries I know no dialog between two greater men, over a deeper issue, reaching over a longer period of time, at a higher level of colleagueship. It extended over the twenty-eight years from 1927 to Einstein's death in 1955. In the first six years Einstein brought up one idealized experiment after another to prove the logical inconsistency of quantum theory. Each Bohr turned around—often in a dramatic encounter—to establish more strongly than ever the soundness of the theory. Beginning in 1933 Einstein sought to show that quantum theory is incompatible with any reasonable idea of "reality". To this objection Bohr replied in effect that, "your concept of reality is too limited."

No idea that Einstein brought forth in this period was more interesting than the so-called Einstein–Podolsky–Rosen experiment. In it, two particles or photons fly away in different directions from a common starting point. According to quantum theory, what one can say with certainty about the one particle depends on which complementary feature of the second particle one has chosen to measure. At least half a dozen different versions of this experiment have been proposed and performed, and hundreds of papers have been written on the subject. The conclusions of quantum theory have been firmly upheld.

The debate over the EPR experiment had one great fruit. It moved Niels Bohr yet another step ahead in formulating the central idea of quantum theory, the concept of the elementary quantum phenomenon. In today's language, we can say that, "No elementary quantum phenomenon is a phenomenon until it is brought to a close by an irreversible act of amplification, such as the electron avalanche of a Geiger counter or the click of a photodetector or the blackening of a grain of photographic emulsion." Speak of the particles flying apart in the EPR experiment as endowed with this, that, or the other polarization; with this, that, or the other direction of vibration? Wrong! We have no right to attribute such a direction of polarization to either particle in all its long flight from point of production to point of detection. The direction of polarization is there, but we do not know it? No. We mistake the whole nature of things if we attribute a direction to the polarization of either particle in default of a suitable measurement.

How hard it has been to spread this central concept of the elementary quantum phenomenon! How many fruitless discussions and papers there still are today which seek to derive "communication at a speed in excess of the speed of light" or "action at a distance" or some other revolutionary doctrine from the EPR experiment!

The plain fact is that the revolutionary doctrine is quantum theory itself. That theory stands battle-tested today, more than half a century after its original formulation. No one has ever been able to find any logical inconsistency in it. No purported disagreement between experiment and the predictions of the theory has ever stood the test of time. Today quantum theory stands as the overarching principle of twentieth-century physical science, and the elementary quantum phenomenon stands forth as its most revolutionary feature.

In a taped interview at Carlsberg the late afternoon before his unexpected death, Bohr declared:

> "...they [certain philosophers] have not that instinct that it is important to learn something and that we must be prepared to learn something of very great importance... They did not see that it [the elementary quantum phenomenon] provides an objective description—and that it was the only possible objective description."

Piet Hein speaks for all of us when he says:

> "I'd like to know
> What this show
> Is all about
> Before it's out."

No deeper salient has any man ever captured in the realm of the unknown than did Niels Bohr and his great colleagues. No retreat is possible. At the tip of the salient flutters in the breeze the flag he set up, with his message of the elementary quantum phenomenon. Where more clearly than here lies the entry to that great territory beyond, still awaiting conquest, the mystery of existence itself?

Surely existence is so ramified that only mathematics will be able to bring it all in order, so preposterous that only philosophy will be able to see the grand plan of it, so quantum-connected that only physics will be able to put to it the right questions!

4. The compound nucleus, the liquid-drop model and the collective model

Loyalty to every great issue that he had ever dealt with was a hallmark of Niels Bohr and the institute he founded. Loyalty to the physics of the atom, in all its marvelous development over the years. Loyalty to the issue of the quantum, that ever-pursued Merlin, ever-changing shape during pursuit. And loyalty to the structure of the nucleus, the subject to which Bohr had already begun making important contributions during the early days with Rutherford. Developments were soon to show that his contributions to elucidating the structure of the nucleus had only begun.

Around Eastertime in 1935 Christian Møller returned to Copenhagen from a visit to Rome and the group of Enrico Fermi, Eduardo Amaldi, O. D'Agostino, Ettore Majorana, Franco Rasetti and Emilio Segré and reported their astonishing finding. Slow neutrons, passing through selected materials like silver, cadmium and boron, and interacting with the atomic nuclei of those substances, encountered effective target areas or "cross-sections" enormously larger than the cross-sectional area of the nucleus itself. How come?

Møller was only about a third of the way through reporting these results to the score of colleagues in the little conference room of the Institute when Bohr rose and intervened. How could such large probabilities for the nucleus to intercept a neutron be at all compatible with any picture of the nucleus as an open planetary system? Surely something very important lay hidden in the new findings. "Now it comes," he said, as he paced back and forth, "now it comes, now it comes." And suddenly it really did come. Then and there he sketched out the concepts of what came to be known as the *compound-nucleus model* of nuclear reactions. It has been employed decade after decade since that time in dozens of laboratories to understand hundreds of nuclear reactions.

According to the new model, the addition of energy or a new particle to a target nucleus promotes that system to a new state, a compound nucleus. That compound nucleus can get rid of this energy by one or another of numerous competing mechanisms: re-emission of the original particle with the original or reduced energy; emission of a photon and simultaneous drop to a lower energy level or to the ground state of the nucleus itself; or emission of a proton, neutron or alpha particle. Know for a given nucleus its possible energy levels and know for each energy level the probability per second that the nucleus will change this, that, or the other way: to know these is, according to the idealized compound-nucleus model, to know all of nuclear physics. Even today this simple program offers the simplest known method to forecast the yield in almost any transmutation experiment.

All those nuclear energy levels and their break-up probabilities, however: how were they to be predicted? What is still today the most economical road to an approximate estimate is provided by the *liquid-drop model* of the nucleus. There had been two or three proposals in the past to compare the normal nucleus to a quiescent liquid drop. Bohr now had a powerful motive to transform a conceptual toy to a work-a-day tool. How better than by studying the modes of vibration of the drop could one predict the excitations of the nucleus? And how more simply than via the concept of evaporation from the surface could one understand the relative emission rates of this, that, and the other particle? Nuclear physics of a newly quantitative form was on the march!

No other workable tools were available four years later but the compound nucleus and the liquid drop to understand the mechanism of fission. Those concepts plus the leadership of Niels Bohr sorted out by mid-1939 the vitally different roles of uranium-235 and uranium-238. Still more important for the future, those ideas made it possible to forecast the fissility of plutonium-239 before that element had even been seen, let alone manufactured, in what was to be mankind's first great venture into alchemistry, with all its fateful consequences.

Bohr's contributions to nuclear physics did not stop with the war and its aftermath. He had a significant influence, both jointly with his son Aage, and jointly with one and another close collaborator, in bringing into the world the *collective* or *unified model* of the nucleus and nuclear reactions, that ties together in a larger unity the *independent particle* model of Hans Jensen and Maria Mayer with the older compound-nucleus and liquid-drop models of the nucleus.

5. The Niels Bohr Institute

What is a responsible institute? The puzzles of the nucleus, after the war, won more attention than the remaining puzzles of the atom. The worldwide nuclear research budget grew at least a hundredfold. The Blegdamsvej Institute, despite a modest increase in funding and manpower, remained tiny by comparison. How then does it come about that Copenhagen, with Niels Bohr gone, is still recognized as the world center for the understanding of the nucleus as well as the atom and the quantum?

How come? Because Bohr's spirit remains. Commitment, diligence and openness are its watchwords. Jokes still light the way. No more responsible man of science has the world ever seen than Niels Bohr. No more responsible physics institute does there exist in all the world today than this one at Copenhagen, his living reincarnation.

How does the responsible institute greet the finding of a discrepancy? With welcome.

How does it respond to an unexpected finding? With joy.

How does it greet an outsider's totally new and successful way of looking at things? With full openness.

Why? Why this welcome to a problem, a paradox or a total upset? Because great advance, Bohr's example teaches us, comes only out of problem, difficulty or apparent paradox. Yes, for the responsible institute as for the responsible individual, commitment only makes its soul. Only diligence creates its genius. Only willingness to give open jury trial to every new development gives it its power to move successfully into the unknown.

6. Weapons and the open world

From nuclear weapons to the doctrine of the open world. September 1, 1939 was a day to remember. World War II began. The green-covered journal, *Physical Review,* published that day Niels Bohr's paper on "The Mechanism of Nuclear Fission," destined to influence the wartime nuclear physics programs of the Atlantic powers and Germany, the Soviet Union, and even Japan. The paper written, Bohr had returned from Princeton to Copenhagen. In the coming time of crisis he wanted more than ever to be with his family, his colleagues and his fellow countrymen. Invitation after invitation to stay overseas in safety he turned down.

No one who wants an example of what it is to be loyal to one's own country can do better than to look at Niels Bohr. In Denmark's darkest days those who had taken her liberty were trying to snuff out her soul. To overwhelming force, the people of the country and their king presented unbending moral resistance. At the center of their spiritual unity stood Denmark's men of learning under the leadership of Niels Bohr. To hold high Denmark's values Bohr and other leaders created and disseminated at their peril a great book entitled *Danish Culture*.

To belong to Denmark, truly to belong to Denmark, as H.C. Andersen put it in Bohr's favorite song and Denmark's national anthem:

In Denmark I was born
There is my home
There are my roots
From there my world goes,

always meant, in his view, a special opportunity—and special responsibility—to exert a constructive influence on the relations between the larger powers. Therefore —when under peril to his life he escaped in a small boat in late 1943 and came into touch with the progress of the atomic bomb—he turned, as no one else so effectively could, from the dangers of the weapon to its possibilities for bringing about a better world.

In a memorandum to President Franklin D. Roosevelt on the third of July, 1944, Bohr urged that America and Britain inform other allied nations about the new weapon before it was used, and consult with them on measures of control. Agreement on control, he also emphasized, would work for a greater openness between nations. The same points he made even more strongly in a long discussion with the president in August 1944 and in a subsequent memorandum. Bohr's ideas form the foundation of present-day thinking about control.

How can we know Bohr, the man, unless we conceive in imagination the hours, days and months that he devoted to developing the interlinked doctrines of control and openness; the leaders in America and Britain—including Winston Churchill—with whom he conferred; his repeated wartime crossings of the Atlantic —Europe to America on empty troop ships, America to Europe by flying boat—and the draft paper after draft paper that he prepared?

Fate frustrated Bohr's 1944 goal of man-to-man discussions among Churchill, Roosevelt and Stalin. No agreement was reached between the great powers for the control of A-bombs before the secret of their existence burst upon an unprepared world.

By the time the H-bomb began to loom upon the scene, in early 1950, Bohr's thinking had changed. He recognized that no progress toward control could take place without an atmosphere of greater confidence. Therefore, the proposed approach to the problem had to be turned about. He had believed that control of the new devices should be the first step in openness among nations. By 1950 he had realized that the approach had to be the direct opposite: first openness and then, on that foundation, control.

Few men have the vision and courage to try to induce all mankind to accept a new moral concept of such scope—the principle of the open world. But Bohr saw no other way. That was the compulsion that drove him to send to the United Nations, from Copenhagen, on the ninth of June, 1950 his famous Open World letter. It would have been more widely advertised if news of it had not been drowned out a few days later by the totally unexpected invasion of South Korea by North Korea. In his letter, Bohr reasons:

> "The very fact that knowledge is in itself the basis for civilization points directly to openness as a way to overcome the present crisis.... [F]ull mutual openness, only, can effectively promote confidence and guarantee common security...
>
> Such a stand would ... appeal to people all over the world, fighting for fundamental human rights, and would greatly strengthen the moral position of all supporters of genuine international cooperation. At the same time, those reluctant to enter on the course proposed would have been brought into a position difficult to maintain since such opposition would amount to a confession of lack of confidence in the strength of their own cause when laid open to the world."

How are we to assess the doctrine of the Open World today? All but dead? Don't newspapers forget it, historians neglect it, statesmen belittle it?

Let us tell the newspapers, those historians, those statesmen to look again and look deeper. Let us remind them that a great idea is a seed; that it lies buried underground for a time in the dark, germinating in the minds of thinking men and women; that there it gathers nourishment and strength, until in due season it bursts forth into light with all its power.

From the smallness of the seed, the onlooker has more than once underestimated the greatness of the growth.

What casual bystander who glimpsed a French thinker dreaming of a Statue of Liberty foresaw that one day two-hundred-forty million people across the ocean would look upon his Liberty, with her torch upraised to all the world, as their dearest national symbol?

Who that ever met that gentle philosopher and writer, Hu Shih, realized that his poems, set to music, would be the inspiration of soldiers on the famous Long March, turning point of the Chinese revolution?

Who that looked on Julia Ward Howe, rocking her baby's cradle in the dark hours of the night, and writing, knew that her "Battle Hymn of the Republic", would one day be counted by Abraham Lincoln as worth a division to the Union cause?

Niels Bohr's doctrine of the Open World: is it not destined to become in the fullness of time the moral undergirding for an association of Friends of Civilization, a vibrant union of liberty and social justice, a spontaneous gathering together of free peoples? More rally to the cause of the Open World each passing year.

Here and there work today those poets, writers and artists that Niels and Margrethe Bohr loved to help, to encourage and to have at their home. Out of the hands of one, shall we see some day a poem or song, on Bohr's message, of worldwide impact? From the chisel of another, a sculpture that grips the mind and moves the heart with its summons to an Open World?

Let us never forget these words in Bohr's famous letter:

> "The efforts of all supporters of international co-operation, individuals as well as nations, will be needed to create in all countries an opinion to voice, with ever-increasing clarity and strength, the demand for an open world."

7. Bohr's legacy

What lies ahead of peril and promise? We mortals do not know whence we come, who we are or where we are going. We live still in the childhood of mankind. As this beautiful blue globe of ours floats on through space, men of science search a hundred skies for what they hold of peril and promise, from the nature of a particle to the principles of communication, and from the sociology of an achieving society to the architecture of a biological molecule. The responsibility for discovery society lays on the community of science. On its findings rest the fears and hopes of all the years. Nothing has more to do with the success of the search than the warm colleagueship of the searchers, their commitment, their diligence, their openness—or, in a single word, their responsibility.

Responsible science. What makes a responsible group, a responsible university department, a responsible institute? No example ranks higher in all the world than this Copenhagen Institute with its wonderful collegiality. The way of doing science that Bohr and his institute taught the world is, beyond all Bohr's own spectacular discoveries, his greatest legacy.

That legacy is no miracle. It was built on character—the character of Niels and Margrethe Bohr and the group of colleagues Niels gathered about him.

One of his classmates and comrades through life, Ole Chievitz, said of Niels Bohr after his death: "he was a *good* man—the best man I ever knew."

The Lesson of Quantum Theory, edited by J. de Boer, E. Dal and O. Ulfbeck

Mellem Spøg og Alvor:
Personal Recollections about Niels Bohr

H.B.G. Casimir

Heeze, The Netherlands

The anecdotes I shall add to the portrait painted by John Wheeler may seem insignificant, irrelevant and even irreverent. Yet, I cherish these recollections because they help me to recreate the image of the great physicist and the generous man that was Niels Bohr.

Many speakers and writers on Bohr have emphasized the speed of his thinking, and no one will deny that he was very quick to grasp the essence of a problem and that he was often the first to see a solution, or at least the road along which a solution might be found. However, what impressed me almost more was the intensity of his untiring concentration, his looking at a problem from every side, and his unrelenting search for the best possible expression of his thoughts. Now these things are not always compatible with speed.

He might be so absorbed by a problem that he became forgetful of the world around him and he later told with great relish how, during a voyage across the Pacific, he had gone ashore—I believe it was at Hawaii—had almost missed his boat and was taken to task by an employee of the shipping line with the words: "you are the foggiest person I've ever come across." (So in our next evening performance we compared Bohr to that other great traveller, Phileas Fogg.)

His methodically looking at every side of a problem, invaluable in research, was not always helpful in daily life. I was told that when he was supervising the installation of central heating in his institute and it should be decided where to put a vertical tube, he saw so many pro's and con's of any position that the mechanic on duty finally cried out in despair: "now can't we put that damn tube at least somewhere."

Bohr's constant search for a better formulation did not lead to speed either. His special way of "dictating" a paper has often been described. Let me give just one example. I once assisted him in writing a report on candidates for a chair in theoretical physics. He wanted to praise one paper in particular and tried out a whole range of adjectives: the paper became successively beautiful, important and so on. Since the report was written in Danish the exercise certainly contributed to my knowledge of Danish laudatory terms. Det var et smukt arbejde ... et dygtigt arbejde ... et vigtigt arbejde ... et udmærket arbejde ... et fremragende arbejde ... et værdifuldt arbejde .... Finally he settled for "et lødigt arbejde." And so it went all the way. I believe it was the ninth version of the report—but it may have been the eight or the tenth—that was finally sent off.

Harald Bohr, himself a onetime star on the Danish national soccerteam said about Niels as a goalkeeper: "Joh, Niels var saamænd udmærket, men han var for langsom til at gaa ud," (Niels was in a way excellent, but he was too slow in running out) and I can imagine Niels standing there, rapidly going over the relative advantages and disadvantages of staying in his goal or running out to intercept a ball ... and coming too late. Now whether this is a correct interpretation or not, Bohr certainly liked sports and, more generally, the outdoors: skiing, yachting, and long walks in the country. When he wanted to go for a walk weather played no role. I remember an occasion when a strong wind was blowing icy rain into our face and I heard him mutter " velsignet vejr" (blessed weather). His maxim "when you see a tram you can catch it" led to many a quick sprint along Blegdamsvej. He liked felling trees and chopping wood and showed me with some pride a woodpile he had built and which looked "almost professional". He was interested in simple crafts, could watch with interest the digging of a well, for instance, and was a pretty good handyman himself. He was also interested in the physics of simple phenomena: the reflection of street lanterns in a ripply watersurface, the skipping of flat pebbles on water—playing ducks and drakes is the correct English term, I believe—or a pingpong ball dancing on a fountain. Piet Hein in a beautiful poem describes that he wants to return to earth: "...og være en gylden og tyndvægget bold/ som stiger og daler og stiger igen/ og staar paa en springvands straale." (...to be a golden and thinwalled ball, that rises and sinks and rises again, standing on a fountain jet.) I am convinced there was a connection between Bohr's understanding of simple phenomena and his outstanding gift to elucidate profound questions by careful analysis or simple examples.

Bohr had a special sense of humour. He always insisted on the element of playfulness that should enter in serious scientific studies. In many of his witty remarks there appears some indication of his ideas on complementarity. Contrary to what is often told the defense of a horseshoe over a door: "of course as a scientist one does not believe in such things, but they say it helps even if you don't believe in it" did not refer to a horseshoe of his own, but he liked telling the story. I think it originated somewhere in the United States. But perhaps we should not be sticklers for historical accuracy. Bohr himself often quoted a German saying—I believe but am not certain that its originator was von Kármán—"beim Erzählen einer wahren Geschichte soll man sich nicht zu sehr vom Zufall der Wirklichkeit beeinflussen lassen." (When telling a true story one should not be overinfluenced by the haphazard occurrences of reality.)

Bohr did not care much for detective stories. When Conan Doyle in later life became interested in occult phenomena Bohr explained to me

> "You and I, we know that we know next to nothing about impostors and how to unmask them. Neither does Conan Doyle, but he thinks he does, which puts him at a great disadvantage."

Somewhat related is his facetious theory of gambling on the stock exchange. The total sum of profits and losses must be zero. Therefore if you play completely at random you must on the average break even (apart from handling costs). Some people have real inside information; they make profits. So who are the losers? Those who *think* they know something about the stock market. Very characteristic was his favourite distinction between a simple truth and a deep truth. A simple truth is a statement the opposite of which is a fallacy. A deep truth is a statement the opposite of which is also a deep truth.

Let me now tell something about his comments on motion pictures. When I was in Copenhagen, in 1929 and 1930, Gamow, Landau and I often went to the movies together and we had a preference for lurid westerns. Sometimes Bohr came with us and his criticism was always remarkable and had often some connection with his ideas on observation. For instance after one particularly silly Tom Mix film he said:

> "I did not like that picture; it was too improbable. That the scoundrel kidnaps the pretty heroine is all right, that happens all the time. That the bridge collapses when their carriage is going over it is improbable, but not at all impossible... That the scoundrel is killed, while the heroine remains precariously suspended over the precipice is even more improbable, yet I am willing to accept it. I am even willing to accept that at that moment Tom Mix is coming by on his white horse to save her. But that at that moment there was also a man with a camera on the spot to photograph the whole scene, that is more than I can stand for."

Finally I mention his famous theory of the advantages of defensive versus offensive shooting. In a movie called "The Black Rider" (Den sorte rytter) the hero always waited until his enemy drew a gun. Then he drew his and he was always a bit quicker. Bohr claimed that this was natural: reactions are faster than decisions, and in his case that proved to be true. We bought toy pistols to try it out, the type with paper tape carrying little explosive pellets. I described this whole affair in a piece of doggerel—in German—that was my contribution to the *Journal of Jocular Physics* published in 1935 on the occasion of Bohr's fiftieth birthday. For a book of reminiscences I published a few years ago I made an English translation. Here it is:

> We went to the flicks and Niels Bohr came along,
> And we watched the Black Rider, a man bold and strong,
> In a Western picture, where guns often bark,
> But it's always the hero who first hits his mark.
> At the end of the movie Niels Bohr, deeply moved,
> Set out to explain what the plot really proved.
> "That was a good film," I can still hear him say.
> "There was really a 'pointe', it showed in what way
> In a part of the world where all villains are armed
> The innocent men are surviving unharmed.
> In truth, there's no reason for flutter or fear
> If your purpose is pure and your conscience is clear.
> When you're facing a blackguard and he draws his gun
> You quickly draw yours, shoot him down and you've won.
> The scoundrel must make a momentous decision
> And that interferes with his speed and precision,
> But for the defendant there's no such distraction,

Not a shadow of doubt can retard his reaction.
So it's easy to shoot in advance of his shot;
With his gun barely grasped he falls dead, on the spot."
We, arrogant youngsters, we ventured to doubt
This thesis of Bohr and we wished to find out
If really a deep psychological facet
Of criminal law does make virtue an asset.
So the three of us went to the centre of town
And there at a gunshop spent many a crown
On pistols and lead, and now Bohr should prove
That in fact the defendant is quickest to move.
Bohr accepted the challenge without ever a frown;
He drew when we drew ... and shot each of us down.
This tale has a moral, but we knew it before:
It's foolish to question the wisdom of Bohr.

As I said, these may seem trifling anecdotes, but the moral that one should not doubt Bohr's wisdom applies to more serious things than shooting between gunmen in westerns.

Part Three

JEST AND EARNEST

Three addresses presented at the evening gathering in Ny Carlsberg Glyptoteket October 7, 1985

edited by

Jorrit de Boer

Ludwig-Maximilians Universität, Munich

and

Ole Ulfbeck

The Niels Bohr Institute, Copenhagen

The Lesson of Quantum Theory, edited by J. de Boer, E. Dal and O. Ulfbeck
© Elsevier Science Publishers B.V., 1986

Parentesen

Asger Aaboe

Yale University
New Haven, Connecticut, USA

Ladies and Gentlemen,

I know full well that it is an act of desperation to interpose oneself between a largely Danish crowd and its feeding troughs so I shall be as brief as I can, and I hope it will be brief enough. Both the hall's columns and the "Buffet Supper" bring to mind many happy occasions in the old days at Carlsberg—they were called "Bøffelsuppe" (buffalosoup) by irreverent souls at the Institute. Which brings me to Niels Bohr's view of irreverence.

The occasion at which I heard him express it was the 25th anniversary in 1951 of *Parentesen*, the society for students of mathematics, physics, chemistry and astronomy that played a crucial role in the lives of the students and, we liked to think, of the teachers as well, particularly during the war years when outer pressures brought us all closer together. The anniversary was celebrated in the then new auditorium U at the Institute, and the entertainment was provided by us slightly older members—most of us had had our degrees for several years. It consisted in a cavalcade of events—real or fictitious—from those 25 years.

Among them were Niels Bohr's talks to *Parentesen* and we presented what pretended to be recordings of bits of two of them. They were written by Jens Lindhard, and Piet Hein was supposed to record them, for he imitated Bohr's voice very well, as in Bohr showing slides from the tour of the Far East:

> "... og her har vi det pragtfulde Kejserpalads—nej, det er Hans og en Palme" (... and here we see the splendid imperial palace—no, it is Hans and a palm tree).

However, he caught a cold, or at least his feet did, so guess who had to do it. Morten Scharff provided the sound effects in the piece on "Tippetoppen" and Aage's explanation of it.

In Lindhard's other passage Bohr was reminded of one of the deep truths:

> "—hvormed vi netop her paa Institutet saa ofte har trøstet os gennem Tiderne—at paa samme Maade, som der findes Emner af saa alvorlig Art, at man kun kan berøre dem i spøgefulde Vendinger, saadan er der ogsaa Ting saa morsomme, at man overhovedet kun kan tillade sig at omtale dem med den allerdybeste Alvor" (—in which we through the years so often have sought comfort precisely here at the Institute—that even as there are topics of such a serious nature that you can touch on them only in jocular terms, so there are also things so amusing that you may only allow yourself to mention them with the profoundest seriousness).

Bohr's response came later that evening in a charming speech congratulating *Parentesen*. He recalled that in former years foreign visitors—one suspects from Germany—had expressed their surprise at the latitude allowed the young students, but his response had been, he said: "at her paa Institutet tager vi end ikke Respektløsheden alvorlig" (here at the Institute we don't take even irreverence seriously).

During the meetings of the last several days people have shown remarkable restraint in the matter of telling Bohr stories. I told one as a prelude to urge you, if that be at all necessary, to engage in the delightful sport of exchanging Bohr anecdotes—whether they are historically true is, of course, irrelevant.

Once again, let Bohr have the last word. He said in the fifties:

> "Ja, I unge Folk anstrænger Jer for at faa mig til at se latterlig ud, men hvor meget I end prøver, kan I dog ikke faa mig gjort latterligere, end jeg ser ud i mine egne Øjne" (you young people try to make me appear ridiculous, but however hard you try, you cannot make me look more ridiculous than I do to my own eyes).

The Lesson of Quantum Theory, edited by J. de Boer, E. Dal and O. Ulfbeck
© Elsevier Science Publishers B.V., 1986

Truth and Clarity

Rudolf Peierls

Oxford, UK

Mr. Chairman, Friends,

We have heard during this symposium much about Niels Bohr's contributions to science and to other areas of serious thought. Tonight we want to remember him as a human being, with all the charm and all the amusing traits of his personality.

The room in which we have gathered is very reminiscent of Carlsberg, the home in which many of us were so warmly received by Niels and Margrete Bohr. But I like to think of the earlier, more intimate home by the Institute, in which the smaller circle was more like a family.

There we had occasion to get to know his great kindness and reluctance to hurt anyone's feelings, which, coupled with his insistence not to let any inexact or wrong statement pass, led to the famous comment: "I am not saying this in order to criticise, but this is sheer nonsense!"

But while he was intolerant of nonsense, he was interested in simple problems and simple people. He could take a genuine interest in anyone's views and talk with them without condescension. I remember an occasion when he had a serious conversation with my son, then aged four, with obvious interest.

On his attitude to the truth, we have been reminded of his saying that truth and clarity were complementary. This came out strongly in his papers, in which he tended to give all possible weight to the truth. As a result his papers were usually not easy to read. It helped if one was able to see an early draft, in which often the clarity had not yet been sacrificed to the truth. Papers always went through innumerable drafts, followed sometimes by 12 sets of proofs, and in the course of this, many changes made the paper more true but not often clearer.

He seems to have had the same attitude to other matters, to judge by the story of his visit to the site of a new extension to the Institute, when the old foreman, who knew him well, said: "Professor Bohr, do you see that wall? If you want to move it again, you must be quick, because in three hours the concrete will have set!"

I experienced some of the problems of drafting in trying to write a paper jointly with Bohr and George Placzek. There were many drafts, but it never got published. It is probably the most frequently cited unpublished paper in the literature.

As we know, he thought deeply about problems outside of physics, and he used to defend the right of scientists to take part in political and other general debates. He said:

379

"We are no wiser and no less biased than other people. But as a physicist, or a biologist, you are certain to have gone through the experience of making a confident assertion, and then being proved wrong. A philosopher, or a sociologist might never have had this wholesome lesson."

He had a fund of stories to illustrate his views. He was opposed to any form of nationalism, but he said that, if there had to be nationalism, he preferred the form in which it appeared, in the English-speaking countries, typified by the phrase: "Right or wrong, my country!" He did not agree with the sentiment, but he thought a German, or a French patriot would never admit that his country could be wrong.

Another illustration of the theme was the story of the young girl in Ecuador, who was cycling down a steep hill when her brakes failed. The cycle went faster and faster, and she almost lost heart. But then she said to herself: "I am an Ecuadorian," and this thought gave her the strength to hold on and control the bicycle until the road flattened out. Bohr commented: "If instead of Ecuadorian you say American, or German, or British, the story is not funny."

He had of course his share of absent-mindedness. In the early discussions he always had a cigar (later it became a pipe) which he tried to light while talking and not having it in his mouth. This took a lot of matches, and soon he would pat his pockets and say: "Have you got a match?" Someone would produce a box, which Bohr pocketed after using a match, and in a minute the process would repeat itself —"Have you got a match?"

I treasured for a long time as a souvenir a piece of chalk which was blackened at one end. Evidently Bohr had confused the chalk with the cigar, which he held in the same hand.

When he arrived in London during the War, after his famous flight from Stockholm, he was for a few days on his own—Aage Bohr followed later. When he had to go to meet an important person, the wise secretary of the Atomic Energy Office wrote the address and the instructions how to reach it on six pieces of paper, and said: "Professor Bohr, if you put one of these into each of your pockets, you are sure to find one when needed!"

Bohr could understand it when others were not very practical. Pauli told me about the day, after the discovery of hafnium, when the Institute had an open day and an exhibition to attract public interest. Before this started Bohr was running around putting finishing touches to the arrangements, when he saw Pauli standing rather forlorn in a corner. Bohr stopped, looked at Pauli and said: "Pauli, you are more suitable to be exhibited than to exhibit!"

Ladies and Gentlemen, I hope that my little stories may have helped you to recall the beloved personality of Niels Bohr.

The Lesson of Quantum Theory, edited by J. de Boer, E. Dal and O. Ulfbeck

Niels Bohr, the Quantum and the World

Victor F. Weisskopf

Massachusetts Institute of Technology
Cambridge, Massachusetts, USA

Niels Bohr's life as a scientist began about 1905 and lasted almost sixty years. 1905 was the year when Einstein published his paper on special relativity and on the existence of the light quantum; it was only a few years after Planck's discovery of the quantum of action. Bohr had the great luck to be present at the beginning, or perhaps mankind has had the great luck to have him at that turning point. What a time to be a physicist! He began when the structure of the atom was still unknown; he ended when atomic physics reached maturity, when the atomic nucleus was put to industrial use for the production of electric power, to medical use in cancer treatment and, unfortunately, to military and political use as the most destructive weapon man has ever conceived.

The work of Niels Bohr can be divided into four periods. In each he exerted a tremendous impact on the development of physics. The first is the decade 1912–1923, from his meeting with Ernest Rutherford until the foundation of his famous Institute of Theoretical Physics in Copenhagen. In this period Bohr introduced the concept of quantum state, created an intuitive method of dealing with atomic phenomena, and was able to explain the Periodic System of elements. In the second period, 1923–1929, he gathered around him in his new Institute some of the world's most productive physicists, who, under his leadership, developed the ideas of quantum mechanics. The third period, 1930–1940, was devoted to the application of the new quantum concepts to electromagnetic phenomena and the exploration of the structure of the atomic nucleus. Then came the Second World War and the last period of his life, in which he acted as the great leader of physics, deeply concerned with and involved in the social, political and human consequences of the new discoveries.

The second period was the time in which the quantum became fully understood. It was a heroic period without any parallel in the history of science, the most fruitful and most interesting one of modern physics. There is no paper by Bohr himself that characterizes this period as did the 1913 and 1922 papers for the first period. It was, rather, his great strength to assemble around him the most active, the most gifted, the most perceptive physicists in the world. His Institute was perhaps the first truly international institution devoted to scientific research. In lively discussions, the deepest problems of the structure of matter were brought to light. One can imagine what atmosphere, what life, what intellectual activity reigned in Copenhagen at that time. Here was Bohr's influence at its best—he found a new way of working. We see him, the greatest among his colleagues, acting, talking, living as an equal in a group of young, optimistic, jocular, enthusiastic people, approaching the deepest riddles of nature with a spirit of attack, a spirit of freedom from conventional bonds and a spirit of joy that can hardly be described. In the course of only a few years the basis was laid for a science of atomic phenomena that grew into the vast body of knowledge known to us today. During that period, Bohr coined the concept of complementarity which was to describe the relation between the observed phenomena and the atomic "reality" which they are supposed to describe. The results of different experimental arrangements turn out sometimes to be seemingly contradictory: they reveal complementary aspects of the object under investigation, aspects that defy a description in classical language, but that are predicted by quantum mechanics. All through his life Bohr was attracted by the philosophic significance of the concept of complementarity in physics and also in other manifestations of human cognition.

In the third period he turned to nuclear physics. Bohr introduced the important concept of the compound nucleus to explain nuclear reactions. The fission of uranium was discovered when Bohr was deeply involved in his studies of nuclear structure. Obviously this phenomena captured Bohr's interest, and he wrote fundamental papers on this process that had a decisive influence on the development of nuclear energy.

The work on uranium fission inevitably brought Bohr into a realm where physics and human affairs are hopelessly intertwined. He was unusually sensitive to the world in which he lived. Before many others, he was aware that science could not be separated from the rest of the world. The events of world history brought home this point earlier than expected. By the 1930s, the ivory tower of pure science had already been broken. It was the time of the Nazi regime in Germany, and streams of refugee scientists came to Copenhagen and found help and support from Bohr. Bohr's Institute was the center for everybody in science who needed help, and many a scientist found a place somewhere else—in England, in the United States—through the help of Bohr's personal actions. Then came the years of the war. Denmark was occupied by the Nazis in April 1940. Bohr was in close contact with the Danish resistance. He refused to collaborate with Nazi authorities. Soon forced to leave Denmark, he escaped to Sweden and then came via England to the United States.

In the fourth period of his life, Bohr joined a large group of scientists in Los Alamos who, at that time, were working on the exploitation of nuclear energy for war purposes. He did not shy away from this most problematic aspect of scientific activity. He faced it squarely as a necessity, but at the same time his idealism, his foresight and his hope for peace inspired many people at that place of war to think about the future and to prepare their minds for the tasks ahead. He believed that, in spite of death and destruction, there was a positive future for this world, transformed by scientific knowledge.

At that time Bohr actively engaged in a one-man campaign to persuade the leading statesmen of the West of the danger and the hope that might come from the atomic bomb. He wanted to raise nuclear technology to an international level to avoid a nuclear armaments race between powerful nations and a nuclear holocaust.

He saw Roosevelt and Churchill and other important men, and he learned quickly the difficulties and pitfalls of diplomatic life. Although he was quite able to convince a number of important statesmen, including Roosevelt, of his ideas, his meeting with Churchill turned out to be a complete failure. Bohr's great political concept did not come to any fruition.

Neither did other attempts succeed of raising nuclear technology to an international level in order to avoid a nuclear armaments race between powerful nations. Bohr ended his efforts towards international understanding on nuclear weapons with his famous letter to the United Nations, written in 1950, in which he laid down his thoughts about the necessity of an Open World. He pleaded for openness between all nations and political systems in regard to human contacts, to new ideas, and in regard to a mutual understanding of the problems faced in different parts of the world. He predicted an ever increasing nuclear arms race if such understanding cannot be realized. His predictions turned out to be tragically correct.

In the last decades of his life, Bohr spent much time in the organization of international activities in science. He participated actively in the founding of the Scandinavian Institute of Atomic Physics and the European Center of Nuclear Research in Geneva. In many ways CERN is based on the same ideas as Bohr's Institute in Copenhagen fifty years earlier, on the idea of international scientific collaboration. But now it was executed on the largest scale, in experimental and theoretical physics.

Physics became a large enterprise; large numbers of people and large machines were necessary to carry out physical research. Bohr recognized this as a logical continuation of what he and his friends had started. He was not afraid of big science if it is imbued with the same spirit as before. He saw the necessity of physics on a large scale, on an international scale. In no other human endeavor are the narrow limits of nationality or politics more obsolete and out of place than in the search for more knowledge about the universe.

He spent the last years following the results of new research, helping to get support for science from governments wherever he could, and reformulating his philosophy of complementarity. He was proud of the rebirth of European science after the ravages of war and he enjoyed his life as the grand old man of Physics.

When Niels Bohr died, an era ended—the era of the great men who created modern science. But it was Bohr himself who helped to shape the spirit and the institutions for the continuation of the scientific endeavor into the future. And it is up to us to realize his ideals in the future. He, more than anybody, knew that to succeed in this, two tasks must be fulfilled: to continue the quest for deeper insights into the riddles of nature with the same insistence and enthusiasm as he did, and to avert the threatening catastrophes engendered by today's gross abuses of military and technical applications of our great science.